MW00527182

Short Staple Yarn Manufacturing

Short Staple Yarn Manufacturing

Dan J. McCreight
Ralph W. Feil
James H. Booterbaugh
Everett E. Backe

CAROLINA ACADEMIC PRESS
Durham, North Carolina

ISBN 0-89089-853-7
LCCN 97-68959

CAROLINA ACADEMIC PRESS
700 Kent Street
Durham, NC 27701
Telephone (919) 489-7486
Fax (919) 493-5668
www.cap-press.com

Printed in the United States of America

Contents

Foreword

The Institute of Textile Technology (ITT)

The Institute was founded in 1944 by a group of textile industry leaders who recognized the need for individual firms to jointly create a private institution that would play a vital role in leading the textile manufacturing industry through the challenges that confront it. The statement of purpose of ITT was written then and still applies today:

> The Institute of Textile Technology shall establish and maintain an educational institution at the graduate level for instruction of students and for research in literary and scientific branches of learning. The Institute shall specialize in scientific instruction, research, and information service in those disciplines that apply to the materials, processes, and technologies for the manufacture of textiles, their distribution, and use.

Located in Charlottesville, Virginia, the role of ITT is to educate graduates students in the theory and practice of textile technology, to carry out research in the field, and to provide a center of information to keep the industry abreast of international developments in textiles. Student enrollment averages thirty.

To accomplish its mission, ITT employs staff and faculty totalling about sixty-five persons and was supported as of February, 1997 by thirty-two textile manufacturing companies and by ten suppliers to the industry. Over sixty-five percent of the current annual budget is provided by membership dues and sponsorship fees. Additional income is derived from extensive educational, industrial support, and information service activities.

Applied research is accomplished in the areas of raw material, quality control, yarn manufacturing, weaving, knitting, non-wovens manufacturing, yarn dyeing, fabric dyeing and finishing, energy, and the environment. Nine industry steering committees provide guidance in the direction of research in these areas. Student thesis research projects are generally chosen in response to industry needs. Twice per year in a unique workshop format that involves the manufacturing plant managers up to the chief executive officers of member companies, ITT shares the results of staff and student research and the status of other industry supported programs with the 200 or more industry leaders that attend these two-day events.

In summary, ITT is organized to effectively interface with the manufacturing, technical, and executive leaders of the textile industry as well as with the people representing the key suppliers of the industry. An understanding of the manufacturing environment and an ability to partner allow ITT staff and students to produce practical and valuable outcomes in graduate and continued education, applied research, information management, and service to the American textile industry. In 1997 the four authors of this textbook had a combined ninety-six years in association with ITT.

Roger Milliken Textile Library
Institute of Textile Technology

Acknowledgments

The authors of this textbook are grateful for the staff and graduate students of the Institute of Textile Technology who contributed to the research that supports the recommendations and conclusions presented. Special commendations and applause are offered for Sherry L. Beasley who labored alongside the authors, typing and retyping countless times our constant stream of revisions and additions. And sincere appreciation is due to the many machinery and instrumentation suppliers that allowed their pictures and drawings to be used or recreated for inclusion in this work.

Short Staple Yarn Manufacturing

Introduction

The Integrated USA Textile Industry

Textiles are among those basic products that are absolutely essential in any advancing or advanced society. The growing or production of fibers, the manufacturing of yarns from fibers, the production of fabrics from yarns or fibers, the preparation and coloration and finishing of fabrics, and the fabrication of a multitude of products from fabrics generally describe the textile industry for most of the world. In the USA, however, the integrated textile complex is regarded as having three distinct components. There is the fiber industry which includes the growing of natural fibers such as cotton or wool and the manufacturing of synthetic or man-made fibers such as acrylic, nylon, polyester, polypropylene, and rayon (regenerated from cellulose). There is the textile products segment which includes manufacturers of yarns, fabrics, and dyers and finishers of yarns and fabrics. The cut-and-sew or fabricated products industry makes up the third layer of the broad USA integrated textile complex.

The USA integrated textile complex is a large and high-technology industry which is regarded as being world class in quality, delivery, safety, environmental stewardship, and customer satisfaction. Visitors in a modern fiber-to-fabric manufacturing facility will see very sophisticated and high-speed technology in place. They will witness what is regarded as the highest productivity in the world in these highly capital intensive processes. Visitors to a state-of-the-art fabricated products manufacturing facility will generally still see labor intensive processes, but these are likewise reported to yield the highest productivity in the world.

Together there are more than 26,000 manufacturing companies in the USA integrated textile complex. Among those are 39 synthetic fiber producers supplying directly or indirectly nearly 5,000 textile products manufacturers which have as their market a potential 21,300 fabricated products companies. Consumer sales from all these companies reached about $219 billion in 1993. Total wholesale shipments were additionally valued at nearly $137 billion that year. The value of the contribution of these companies to the Gross Domestic Product was about $55 billion, second only to the aerospace industry's $65 billion.

In addition to these indicators as to the significance of the USA integrated textile complex, consider the following statistics. These companies are collectively responsible for the following:

1. Providing about 12 percent of the direct manufacturing jobs available in the USA, amounting to nearly 2,000,000 jobs in 1996;
2. Paying $4 billion annually in state and federal taxes;
3. Spending $4 billion on average annually for capital improvements to advance quality, productivity, to lower costs, and to enhance response to customers;
4. Spending $3.5 billion each year on R&D; and
5. Providing products for sales at 100,000 USA retail stores.

The yarn manufacturing segment boasts an 8.2 fold machine productivity increase over the last two decades. During the same time frame the fabric manufacturers have installed higher speed machines resulting in a 6.7 fold productivity gain. Machine productivity advances in the dyeing and finishing sector have been much less dramatic with only a 40 percent gain over the last 40 years. The total textile complex has averaged more than 3.0 percent annual gains in labor productivity since 1973, compared to a 2.5 percent gain for all USA manufacturing. Labor productivity data are from U.S. Bureau of Labor Statistics.

This large and diverse industry, which ranges from fiber production through fabric manufacturing and enhancement to the fabrication of apparel, household, and a variety of industrial products,

Table 1.1. Spinning Systems by Fiber Length Processed

Spinning System	Fiber Lengths Processed (Rounded)	
	Millimeters (mm)	Inches
Cotton	Up to 64	Up to 2.5
Woolen	51 to 102	2.0 to 4.0
Worsted	51 to 102	2.0 to 4.0
Modified Worsted:		
Stretch Break	76 to 114	3.0 to 4.5
Carpet Yarns	152 to 203	6.0 to 8.0

is a very important contributor to the USA economy and to the standard of living of this nation.

About This Textbook

This publication is devoted to presenting the processing knowledge necessary to economically manufacture quality yarns from fibers that can range up to 63.5 millimeters (2.5 inches) in length. These yarns are typically referred to as short staple yarns and are generally produced on a sequence of machines that are together called the cotton system. As shown in Table 1.1 there are also woolen, worsted, and modified worsted systems that are in part so characterized because of the range of fiber lengths that are processed into yarns. A number of fiber types can be processed on each system.

Processing of fibers into short staple yarns is the subject of this textbook because these yarns are produced in large volume for use in fabric manufacturing. There are multiple processing steps, each of which has a number of critical control factors that must be known, measured, and reacted to routinely if the highest quality yarns are to be spun. This publication describes each process as well as how to best set up and manage that process. The audience addressed is primarily those persons responsible for the manufacturing environment of such a spun yarn producing facility. The text is also appropriate for students in college level textile courses. The early sections can be used as an introduction to short staple spun yarn manufacturing for high school students. This text is also appropriate for the libraries of all educational institutions where textiles in any type course offering are taught and for the libraries of all yarn manufacturing companies and plants.

An important feature is the presentation of selected data that have resulted from in-plant applied research projects. These data are used to support the suggestions presented for effectively managing the manufacturing processes. All such data, unless otherwise noted, are statistically valid at the 95 percent confidence level. Metric units such as grams, kilograms, centimeters, and meters are usually used with the English unit equivalent often expressed parenthetically. Metric to English and English to metric units conversions are provided in Appendix II. Typical processing details such as production rates, efficiencies, stop levels, resource utilization, and ambient conditions are documented. Wherever possible, benchmarks for product quality and process performance are presented as stretch targets for improvement for the short staple spun yarn producer. For each process, machinery and instrumentation suppliers that are most active with these yarn producers in the USA are listed as a reference. Illustrations for this textbook were chosen or reproduced from those available that best represent the class of machine or action described and do not imply any preference for specific machines.

There is a special jargon in each textile industry segment, and some of the terms used are not defined in most standard dictionaries relative to use in the textile industry. Therefore, in this textbook, the first time many textile terms are used, the words are defined either parenthetically or in textual explanation. Many terms are explained in a glossary of terms in Appendix III with the definitions mostly limited to their use in the yarn manufacturing segment of this industry.

The final pages of the last chapter in this textbook present a partial list of important questions that the successful yarn spinner must ask and answer in an on-going fashion. It was the intent of the authors that the reason for each question as well as the most likely answer be revealed in this textbook. Of course, many other questions have to be addressed before superior yarn making can be accomplished. There are numerous and complex interactions and interdependencies among the process and product variables that influence spun yarn manufacturing. It is therefore impossible to do a world class job in a later process if a less than superior job is done in the earlier processes. For these reasons this textbook will best serve the yarn spinner if it is initially *studied* section by section from Chapter 1 sequentially through Chapter 14. The discussions of each successive chapter build upon what is revealed in earlier chapters, so that this orderly study approach is very necessary. The text can later be used as a reference for specific processes or concepts.

Yarn Numbering Systems

A yarn, regardless of how it is manufactured or of which fiber(s) it is composed, is specified by a number which is a relative indicator of its fineness or linear density. The term yarn count is synonymous with yarn number. Since the yarn number is a relative measure or a ratio, there are no units involved. Unfortunately, there are numerous systems for numbering yarns which can result in great confusion. Fortunately, however, all systems can be categorized into two groups:

1. Indirect or reciprocal yarn numbering systems with a higher number meaning a finer (or lighter in weight) yarn; and
2. Direct yarn numbering systems with a higher number meaning a coarser (or heavier in weight) yarn.

These two categories of yarn numbering are introduced here without going into great depth. There are many more subsystems than are discussed in this text. The intent is to present what is most common in the USA textile industry.

Indirect or Reciprocal Yarn Numbering Systems

In this system the yarn number signifies the number of units of a standard length that are necessary to balance one unit of standard weight. Table 1.2 contains the standard lengths and standard weights for several popular indirect yarn numbering systems as used in the USA.

Note in the table that there are different yarn numbering standards depending on the system on which a staple yarn is spun. Generally, in the USA if a yarn is spun on the cotton system, fiber type not relevant, then the yarn is given a cotton or English (Ne)

Table 1.2. Standard Lengths, Weights, and Common Names for Indirect Yarn Numbering Systems

Spinning System	Standard Length	Standard Weight	Numbering System Common Name
Cotton	840 yards (yd)	1 pound (lb)	Cotton or English (Ne)
Woolen	1600 yards	1 pound	Woolen Run
Woolen	300 yards	1 pound	Woolen Cut
Worsted	560 yards	1 pound	Worsted (Nw)
All	1000 meters (m)	1 kilogram (kg)	Metric (Nm)

count number. If it is spun on the worsted system, the yarn is given a worsted (Nw) number and so on. A metric yarn number (Nm) can be designated for any yarn.

By definition a cotton system yarn which is 840 yards in length and weighs one pound is classified as a number 1 yarn. Such a yarn of single strand would be designated in the USA a 1/1 Ne count or a 1's Ne count. For further example, a singles yarn containing 1680 yards in length (two 840-yard lengths) and weighing one pound would be classified a 2/1 Ne or 2's Ne count yarn. If a singles yarn contains thirty 840-yard lengths and weighs one pound, it is a 30's yarn. It is generally assumed in the USA to be a cotton system or English count for a singles yarn if the number is presented in either of these formats: #/1 or #'s. The count for a singles yarn in the worsted numbering system is written as the reciprocal, 1/#.

Table 1.3 contains examples which help to demonstrate common yarn numbering practice for staple yarns spun on several of the main spinning systems.

If strands of yarn are twisted together to form a plied yarn, the yarn number must reflect the count of the individual strands of yarn as well as the number of strands twisted together. It is most common to twist together yarns of like count so that a more balanced plied yarn results. Table 1.4 contains examples demonstrating plied yarn numbering with two indirect systems.

Direct Yarn Numbering Systems

The yarn number in a direct system indicates the number of units of standard weight that are necessary to balance one unit of

Table 1.3. Various Examples of Indirect Yarn Number Determination for Singles Yarns

Yarn Numbering System	Length of Test Yarn	Weight of Test Yarn	Calculation Using Standard for a Count of 1 (Actual x Standard)	Yarn Number
Cotton	840 yd	1 lb	840 yd/lb x 1 lb/840 yd =	1's Ne
Cotton	12600 yd	1 lb	12600 yd/lb x 1 lb/840 yd =	15's Ne
Cotton	12600 yd	0.5 lb	12600 yd/0.5 lb x 1 lb/840 yd =	30's Ne
Woolen	1600 yd	1 lb	1600 yd/1 lb x 1 lb/1600 yd =	1 Run
Woolen	3200 yd	1 lb	3200 yd/1 lb x 1 lb/1600 yd =	2 Run
Worsted	560 yd	1 lb	560 yd/1 lb x 1 lb/560 yd =	1/1 Nw
Worsted	1680 yd	1 lb	1680 yd/1 lb x 1 lb/560 yd =	1/3 Nw
Metric	1000 m	1 kg	1000 m/kg x 1 kg/1000 m =	1 Nm
Metric	20000 m	0.6 kg	20000 m/0.6 kg x 1 kg/1000 m =	33.33 Nm

Table 1.4. Examples of Indirect Yarn Number Determination for Plied Yarns

Yarn Numbering System	Count of Yarn to be Twisted	Number of Strands (Plies) Twisted Together	Yarn Number
Cotton	20/1 or 20's	2	20/2
	15/1 or 15's	3	15/3
	15/3	2	15/3/2
Worsted	1/20	2	2/20
	1/15	3	3/15
	3/15	2	2/3/15

Table 1.5. Standard Weights and Lengths for Popular Direct Yarn Numbering Systems

Yarn Numbering System	Standard Weight	Standard Length
Denier	1 gram (g)	9000 meters (m)
Tex	1 gram	1000 meters
Decitex (Grex)	1 gram	10000 meters

standard length. Table 1.5 presents the standard weights and lengths for three such direct yarn numbering systems. The denier and tex systems are used widely in the USA, and the tex system is recognized as the world standard yarn numbering system. Examples of direct system yarn numbering are provided in Table 1.6.

Table 1.6. Various Examples of Direct Yarn Number Determination for Singles Yarns

Yarn Numbering System	Weight of Test Yarn	Length of Test Yarn	Calculation Using Standard for a Count of 1 (Actual x Standard)	Yarn Number
Denier	1 g	9000 m	1 g/9000m x 9000m/g	1 Denier
Denier	100 g	9000 m	100 g/9000m x 9000m/g	100 Denier*
Denier	200 g	12000 m	200 g/12000m x 9000 m/g	150 Denier*
Tex	1 g	1000 m	1 g/1000 m x 1000m/g	1 Tex
Tex	100 g	1000 m	100 g/1000m x 1000m/g	100 Tex
Tex	100 g	9000 m	100 g/9000 m x 1000 m/g	11.11 Tex*
Tex	200 g	12000 m	200 g/12000m x 1000 m/g	16.67 Tex*

* Note: Tex is 1/9 of the Denier for yarns for similar weight-to-length ratios.

Further Yarn Numbering Information

There is further discussion of yarn numbering in Chapter 12, and Section B of Appendix II contains factors for converting yarn counts among various numbering systems.

Chapter 2

Fibers Used in Short Staple Spinning

There are essentially two groups of fibers (i.e., naturals and man-made, or manufactured). In the naturals group, cotton is the predominant fiber used in short staple spinning, followed by flax. Of course there are many other natural fibers, but they are of little or no interest to the short staple spinner because of their fineness and length properties.

In the manufactured fiber group, the short staple spinning systems used today are mainly processing polyesters, rayons, and acrylics. Some of the more interesting general details and features of each fiber mentioned are given in the sections that follow.

Natural Fibers

Cotton

The world's cotton producing countries in 1996 harvested approximately 92 million bales of cotton, each weighing, for statistical purposes, 218 kilograms (kg) or 480 pounds (lb). Of this total the United States produced nearly 18.9 million bales in its Cotton Belt which extends from Southern California across the country to the east coast as far north as Virginia. In the 1990s the United States and the People's Republic of China were the largest producers of cotton, about 40 million total bales, followed by India, Pakistan, and Uzbekistan.

Of the cottons produced in the USA, typically 60 percent are used domestically and 40 percent are exported. In 1996 there were approximately 11.1 million bales of cotton spun domestically with a very high proportion (98 to 99 percent) spun on short staple spinning systems. Cotton's popularity derives from its comfort and wearing properties which include good moisture absorption, strength and elongation, shape, luster, resiliency, and dimensional stability.

About 98 percent of the cottons grown in the USA are classified as Upland varieties which are relatively uniform in diameter, approximately 16 microns, but vary in length from 22 to 32 millimeters (mm). These cottons have moisture regain values of 7.5 to 8.5 percent, are low in luster and resiliency, and are fairly tough with strength levels that can range from 23.4 to 36.0 grams per tex and have elongation-at-break levels of between 4.0 and 9.0 percent. An interesting property of cotton is that its strength increases when moisture is added (e.g., cotton's strength can increase by 15 to 20 percent when wet).

With fiber properties similar to those above, it is no surprise that cotton is suitable for a number of different end uses. Cottons grown in the USA differ in physical properties from region to region such that the diverse crop harvested is capable of fulfilling the needs of textile producers of twine and shoe laces to fine shirtings and sheetings. The fabrics produced from cotton are comfortable to wear, have excellent laundering characteristics, display good color-fastness properties, are heat resistant, and have excellent flexibility, durability, and absorption qualities.

Flax

Perhaps the oldest known fiber used in the world, flax (or linen) is mainly grown in Europe where a temperate climate with sufficient moisture and overcast weather helps produce a crop of nearly 609 thousand metric tons annually. Of this amount the textile producers in the USA consumed 3.9 thousand metric tons of flax raw material and semi-manufactured products (e.g., linen yarn, sliver, and roving) in 1994.

Flax fibers are coarse when compared to cotton and can range in length from more than 300 mm to approximately 500 mm. These fibers, however, can be cut into short lengths for use on short staple spinning machines. Flax usually is a pale sandy color with high natural luster. It is also very strong with tenacity values of near 54.0 grams per tex and has a moisture regain value of approximately 12 percent which enhances its comfort properties. A negative characteristic for flax fiber, however, is its high stiffness coefficient which equates to low elasticity, flexibility, and elongation properties.

Flax is used to produce fabrics that can be light in weight, yet both strong and comfortable to wear as apparel. These fabrics are easily laundered and tend to remain neat in appearance during use. Some textile end products for which flax is used include draperies, upholstery, apparel, and a wide array of domestic and industrial fabrics. As an interesting note, flax is grown in the USA for seed, not fiber. The seed is used to produce linseed oil.

Manufactured (Man-Made) Fibers

Manufactured fibers generally are produced by extruding a polymer through a multi-holed spinnerette, cooling the fiber, drawing it down to the desired denier and tenacity, and cutting it into a length such that it can be used on short staple spinning systems. In 1996 USA textile plants consumed nearly 5.6 million metric tons of manufactured fibers. Of this amount 801 thousand metric tons were spun on short staple spinning systems. The most popular manufactured fiber was polyester, followed by rayon and acrylic. Many other man-mades can be and are being spun on cotton spinning systems, but production volumes are relatively small at the present time.

Polyester

Polyester staple fiber first became available in small quantities to USA textile processing plants in 1951. It immediately became acceptable to the consumer because of its ease of care and excellent crease resistance. In 1996 the USA polyester textile plant usage was 61 percent in apparel, 20 percent in industrials, and 19 percent in carpet/household products. Total domestic polyester production was 1.81 million metric tons in 1995, a value which reflects steady growth over many years. About 1.09 million metric tons were for staple, tow, and fiberfill. For multifilament and monofilament yarns, usage was 0.72 million metric tons.

Polyester staple can be made into any denier and cut length combination desired for an end-use requirement. The strength and elongation values can vary widely, depending upon the extent of the drawing down operation, to obtain different tenacities and deniers. For example, low tenacity polyester may have strength values of 3.0 to 4.0 grams per denier with elongation values in the range of from 40 to 50 percent. High tenacity polyesters have strength values that can exceed 6.0 grams per denier with elongation values in the 20 to 35 percent range. The degree of luster that the fiber possesses is controllable during the formulation of the polymer solution. The moisture regain of polyester is very low, less than 0.5 percent. This lack of moisture absorbency is one of the primary reasons that polyester apparel is not as comfortable as cotton apparel and can result in electric static in certain applications where low humidity is present. Polyester fiber has excellent elastic recovery and resiliency properties which make it ideal for fabrics where ease of care and wrinkle-free appearance are desired.

Blends of cotton, rayon, or flax (linen) with polyester fibers are ideal for some men's and women's apparel. Such blends capture the excellent properties of two quite different fibers to produce fabrics that are comfortable to wear, exhibit absorbency while reducing static charges, are strong and easy to care for, and that retain fabric shape and size for that wrinkle-free appearance.

Polyester fibers are also used in home furnishings, in industrial fabrics such as conveyor belts, fishing nets, and ropes, and in automotive applications for upholstery, headliners, and tire cord.

Rayon

Rayon is a fiber that is manufactured from regenerated cellulose, principally wood pulp and cotton linters (fibers that are too short for yarn or fabric manufacturing). These raw materials are chemically worked, and the solution is then forced through a spinnerette containing many thousands of small holes. The strands thus formed then pass through a dilute acid bath causing the pure cellulose to coagulate into usable filaments which are then cut into staple fibers suitable for short staple spinning. In 1995 the USA consumed 94 thousand metric tons of rayon/acetate on the cotton spinning system.

The properties of rayon fibers are mostly controllable during their manufacture. The spinnerette holes govern the fiber's denier, and modifications to the chemicals in the coagulating bath and variations in aging time govern the tenacity. Deniers can be of the apparel type, i.e., 1.0 to 3.0 denier, and tenacities can range from 1.5 to 2.4 grams per denier for regular rayon to 3.4 to 5.5 grams per denier for high-wet modulus rayon. Rayon has a moisture absorbency of approximately 14 percent, has low resiliency, high elongation-at-break (6.5 to 30 percent), and low elastic recovery. When wet, regular rayon loses approximately 50 percent of its strength.

Rayon is primarily used in fabrics for apparel and home furnishings (e.g., draperies and other interior design end uses). Garments of rayon are comfortable to wear because of their high moisture regain potential, and can be laundered and handled much like cotton. Rayon also readily accepts finishes to make resultant fabrics stable and to possess easy care characteristics.

Blends of rayon with cotton, polyesters, acrylics, and nylons are common and produce fabrics having the desired qualities of the individual fiber classes themselves.

Acrylics

The first successful introduction of acrylic fiber was by Monsanto Chemical in 1954. Globally, 3.1 million metric tons of acrylic fiber are produced today. It is made up of linear polymers that are formed by addition polymerization of at least 85 percent by weight of acrylonitrile. Acrylic fiber production is similar to polyester with respect to the processes of pumping a dissolved polymer through holes in a spinnerette and solidifying the filaments in either hot air or in a coagulating bath. The filaments are drawn while still hot to the appropriate tenacity and denier, and then are cut to the desired staple length. In 1996 USA textile plant consumption of acrylic and modacrylic staple, tow, and fiberfill amounted to 223 thousand metric tons. Modacrylic fiber is a manufactured fiber in which the fiber-forming substance is any long chain synthetic polymer composed of less than 85 percent, but at least 35 percent by weight of acrylonitrile units. Available statistics do not separate acrylic and modacrylic fibers.

During the manufacture of acrylics, the fiber length and denier can be controlled in a manner similar to that described in the discussion of polyester. Acrylic fibers are available in different lusters as well as colors that are produced by adding pigment or dyestuff to the polymer before extrusion.

Acrylic fibers have very high elongation values, up to 50 percent, but strength levels are generally low and fall in the 2.5 to 3.0 grams per denier range. The moisture regain of acrylics is less than 2 percent, which makes it subject to static electricity when the humidity in the air is low, and also contributes to its difficulty in dyeing. Acrylic is a bulky fiber with soft hand and warmth. Fabrics made from acrylics are durable, do not lose their shape, have excellent abrasion resistance, and are easy-care with respect to laundering and ironing.

End uses for acrylics include fleecewear, socks, sweaters, sportswear, blankets, and upholstery, and in blends with other fibers, pile fabrics, sport shirts, ski clothes, and so forth.

Chapter 3

Typical Process Steps for Short Staple Yarn Manufacturing

General Processing Functions

The actual process sequence required in the manufacture of short staple spun yarns is influenced by several factors. Among these influences are the initial raw material characteristics, the spinning system on which the yarns are to be spun, the specifications of the fabrics to be manufactured from the yarns, the fabrication methodology, and the end use planned for the resultant fabrics. Even with the machine and sequence options that are dependent on those influences, there are common basic process functions that must be accomplished. Fiber property assessment, bale warehousing, bale laydown management, separation of fiber masses into progressively smaller tufts, effective fiber mixing, impurity removal, and efficient material transport between machines are essential initial process functions. Fibers are next practically individualized and then reassembled in order to randomize fiber mixing and to begin the orientation of fibers along the machine axis and parallel to one another in the delivered intermediate strands. The weight per unit length of fiber assembly strands is progressively reduced by means of controlled fiber mass attenuation, called drafting, which also promotes the desired fiber parallelization. As the weight per unit length gets too low for interfiber cohesion to hold the strands together, twist must be added. Final drafting and twist insertion in single strands take place at the spinning process where the ultimate singles yarn characteristics are determined. At spinning, the yarn is packaged according to the capability of each spinning system type for subsequent internal use downstream or for delivery to a fabric knitting or weaving customer. Required and optional process steps are briefly introduced in this chapter.

Bale Storage and Bale Management

A cotton system spinning plant that produces 68,000 kilograms (kg) or about 149,900 pounds (lb) per week of 100 percent cotton yarn will need about 48 bales of fiber per day, allowing for expected waste in processing. If this same poundage of yarn were composed of 100 percent synthetic fibers, the need would be for as few as 32 bales per day. Blends of synthetic and cotton fibers would require a number of bales between these two extremes, depending on the ratio of synthetic and cotton fibers in the mix. This is because cotton bales generally average about 218 kg (480 lb), and bales of certain synthetic fibers weigh as much as 318 kg (700 lb).

These requirements necessitate a suitable bale inventory and bale management capability. This is more complex for cotton than for synthetics because of the variability of measured cotton fiber properties within and between bales from the same growth area, and because of the common practice of forming a bale laydown by mixing cotton bales from different growth areas. Depending on the number of fiber properties measured and considered for blending purposes and on the number of growth areas from which cotton is used, there may be as many as 7 or 8 groupings or separations of cotton bales in the warehouse. This also is a factor in determining the number of bales needed in inventory. Typical cotton yarn manufacturing plants will use only 3 to 5 warehouse separations.

For synthetic fibers the number of warehouse groupings is determined by the number of fiber types to be mixed, lot or shipment designations, and merge identification numbers. Even though fiber property variabilities within and between bales may be lower than for cotton, there is still a need to mix bales across lot numbers or shipments to ensure consistency from day to day in average laydown fiber properties and in processing performance. Typical fiber properties measured for cottons and synthetics and used to determine bale selection from the warehouse for the laydowns to be processed are discussed in more detail in Chapter 12.

Bale Laydown

The example yarn plant above processing 68,000 kg over a 7-day week will typically group 48 bales together each day and position them at the initial processing stage. This group of bales is called the laydown. This same plant, if processing 100 percent polyester fiber, will have a typical laydown of perhaps 32 bales. These laydown sizes will allow for once-per-day bale selection, unwrapping, preparation, and positioning for processing. Labor requirements for these tasks can therefore be minimized. Preparation of bales for the laydown includes removing the straps or bands and wrappings from the bales and conditioning the bales in controlled temperature and relative humidity (RH) for at least 24 hours prior to processing. For cottons this conditioning is best done between 60 percent and 80 percent RH. Synthetics can be conditioned at lower RH, perhaps at 45 percent to 50 percent.

Such conditioning of the fibers in these unstrapped and unwrapped bales allows for fiber temperature and moisture equilibrium, fiber relaxation, and some initial stock opening, improved product weight control, and for cottons, a reduced negative effect of fiber stickiness. The amount of sugary deposits of certain insects on cotton in the field varies from field to field and from growing season to growing season. These deposits are a primary cause of potential cotton stickiness problems. Conditioning of the fibers as recommended contributes to consistency of product and process and helps reduce the variability of properties in the final yarns produced.

A goal of proper bale and laydown management is to replicate the average and range of fiber properties and processing attributes from laydown to laydown, day to day, week to week, and so on. In doing so, processing machines can be set up optimally, and yarn product consistency over time will be more likely.

The Card Room

The Card Room in a typical spun yarn plant can be defined as that department that processes fibers from bale to final product strand that will be fed to the spinning machine. This department includes the Opening Room (Blowroom in other countries) where bale laydowns are placed and where fiber tuft opening, mixing, and cleaning are accomplished. The Card Room also includes the feeding of stock to and through the actual carding process where fibers are individualized, further cleaned, and partially oriented along the machine axis in the sliver strand produced. Beyond the cards are optional lap preparation and combing processes as well as the normally used drawing processes. Rotor (open-end) and air jet spinning systems are fed strands called sliver, produced at the drawing process. Ring spinning is fed from a roving package which is produced from a drawn sliver on a machine called a roving frame. All of these processes prior to spinning are generally referred to as preparatory processes for spinning and are often collectively called the Card Room. Each process is discussed in greater detail in subsequent chapters. A block diagram of a typical sequence of machines for processing 100 percent cotton in the Card Room is provided in Figure 3.1. A typical sequence for processing a 100 percent synthetic fiber laydown is shown in Figure 3.2. And a typical sequence for the intimate blending and processing of a synthetic/cotton fiber mix is presented in Figure 3.3.

Bale Feeding

After laydown selection, preparation, and positioning, bale feeding is accomplished. Initial bale feeding is done using a set of machines in parallel called hopper feeders, or more often by a machine generally referred to as a top feeder. Hopper feeders have stock supply compartments that are either filled by manual effort or by automated machines that duplicate the human bale feeding process. A cross section of a hopper feeder is shown in Figure 3.4. This diagram aids in the understanding of how such a machine is utilized to open fiber stock. Figure 3.5 represents how these machines can be grouped together to deliver opened tufts to a conveyor belt or feed table for the purpose of stock mixing and for positioning the stock for transport to the next process. A photograph of a machine used for automated hopper feeder loading is provided in Figure 3.6.

Today the most popular machine for feeding fiber from the bales is the top feeder. Figure 3.7 is a photograph of a typical top feeder in operation, and Figure 3.8 is a detailed illustration of the working elements of a top feeder. Not only does initial bale feeding take place with hopper feeders and top feeders, but initial tuft opening also occurs which facilitates the mixing and necessary cleaning of stock which must be accomplished downstream. Hopper feeders in parallel can contribute significantly to fiber mixing. Top feeders themselves accomplish only minimal fiber mixing, but prepare the tufts for effective mixing downstream if the proper mixing machines are in place.

Opening and Cleaning

After initial bale feeding, fiber tufts must be further opened and cotton fibers must be cleaned. For 100 percent synthetics often only one additional opening process is sufficient, and that process can double as the distribution point for feeding the card chutes. For 100 percent cotton additional opening processes are the norm, progressing from less intensive action to more intensive opening action. As progressive cotton tuft opening is occurring, a simultaneous progression of intensity of impurity removal usually occurs in the same machines. Hence, these machines are referred to as opener/cleaners. Progressive opening of tufts is necessary in order to minimize damage to fibers, to minimize the creation of fiber entanglements called neps, and to facilitate progressive cleaning and mixing of the fiber tufts. Figures 3.9 and 3.10 are demonstrations of typical lower intensity opener/cleaners, often called coarse cleaners because they most effectively remove the heavier and larger impurities. Figures 3.11 and 3.12 are diagrams of fine cleaners which have a more intensive action and are effective at removing the remaining heavy and dense particles as well as the less dense and smaller impurities. Figures 3.13 through 3.15 are examples of intense cleaners that generally combine several action points that progressively open and clean. Impurities of concern in cotton are plant leaf, stem, and seed particles, neps, short fibers less than 12.7 millimeters (mm) or less than ½ inch in length, and immature or not fully developed fibers. Impurities in synthetics, which are predominantly removed at carding, include fiber bundles or chips, tangled fibers, short fibers, and extra long fibers.

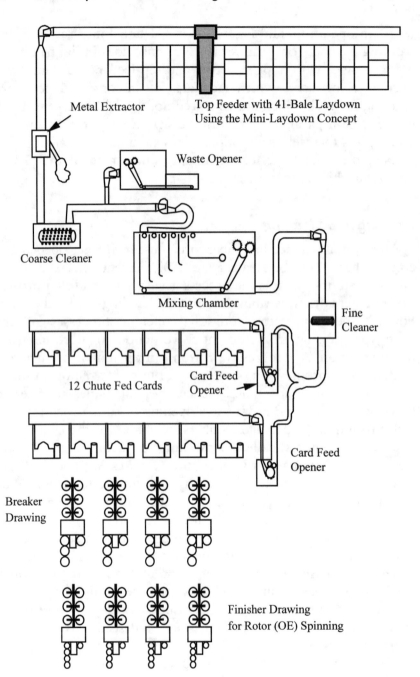

Figure 3.1. Diagram of a possible machinery layout for a Card Room processing 100 percent cotton for rotor (open-end) spinning.

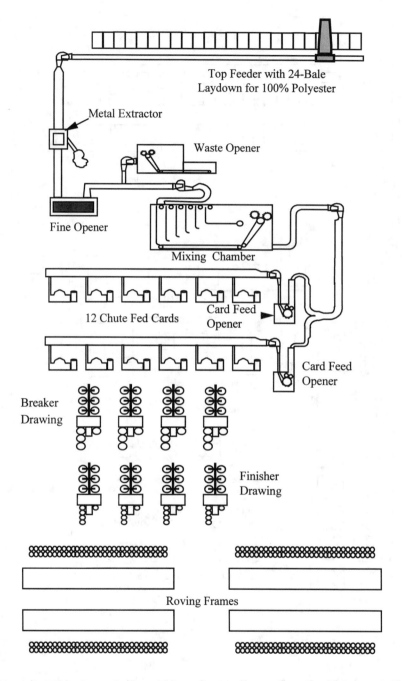

Figure 3.2. Diagram of a possible machinery layout for a Card Room producing 100 percent polyester for ring spinning.

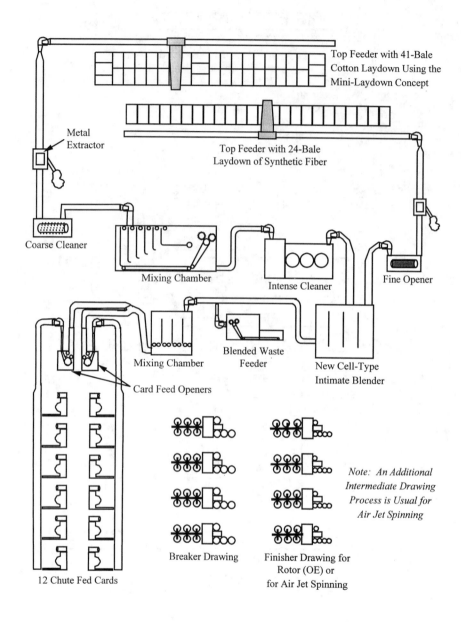

Figure 3.3. Diagram of a possible machinery layout for an intimate blend Card Room producing sliver for rotor or air jet spinning.

1. Feed table
2. Internal feed lattice
3. Light barrier
4. Baffle plate
5. Brush rolls
6. Spiked lattice
7. Cleaner roll
8. Evener roll
9. Stripper roll

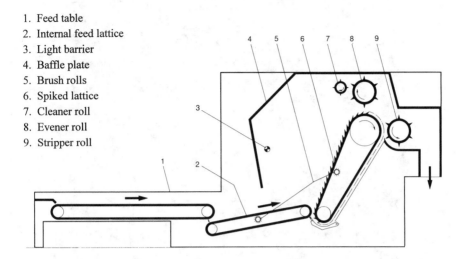

Figure 3.4. Diagram of the Super Bale Opener GBC, a hopper feeder with extended feed apron. (Courtesy Truetzschler GMBH & Co.)

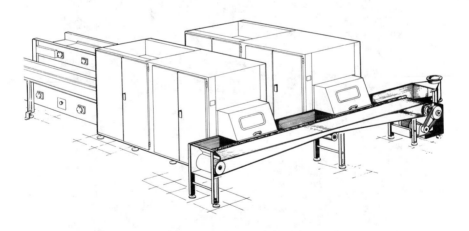

Figure 3.5. Diagram of two B10/1 Blending Feeders and a B20/1 Transport Lattice onto which the feeders deliver stock for sandwich mixing. (Courtesy Fratelli Marzoli & C. spa)

Figure 3.6. Photograph of a Bale-O-Matic in operation for feeding bale stock into hopper feeders.

Figure 3.7. Photograph of the Automatic Bale Opener (ABO) in operation. (Courtesy Crosrol Ltd.)

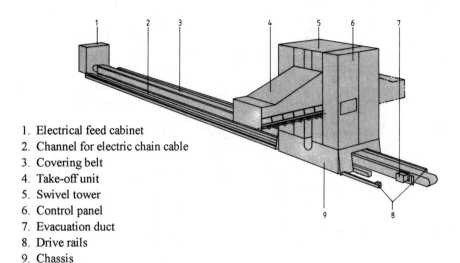

1. Electrical feed cabinet
2. Channel for electric chain cable
3. Covering belt
4. Take-off unit
5. Swivel tower
6. Control panel
7. Evacuation duct
8. Drive rails
9. Chassis

Figure 3.8. Illustration of the main elements of the Unifloc A1/2 top feeder. (Courtesy Rieter Machine Works, Ltd.)

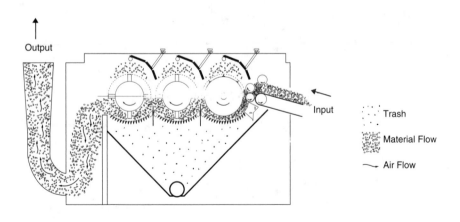

Figure 3.9. Diagram of the Three Roll Cleaner, a coarse opener/cleaner for cotton. (Courtesy Crosrol Ltd.)

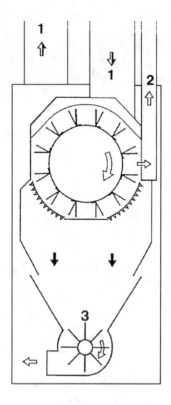

1. Material
2. Dust
3. Waste

Figure 3.10. Illustration of the UniClean B10, a coarse opener/cleaner and de-duster for cotton. (Courtesy Rieter Machine Works, Ltd.)

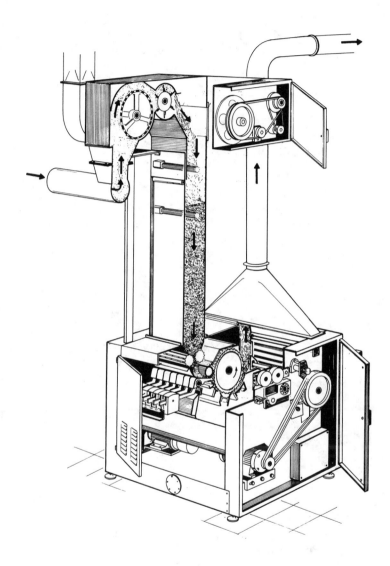

Figure 3.11. Illustration of the B35 Horizontal Opener, a fine opener/cleaner for cotton. (Courtesy Fratelli Marzoli & C. spa)

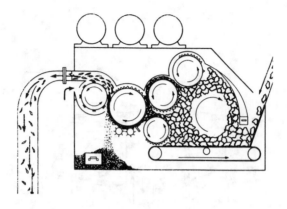

Figure 3.12. Illustration of the Rando Cleaner, a fine opener/cleaner for cotton. (Courtesy Carolina Machine Co.)

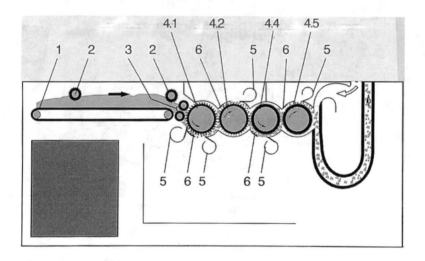

1. Feed lattice	4.4 Saw-tooth roll
2. Pressure rolls	4.5 Fine saw-tooth roll
3. Feed rolls	5. Mote knife with suction hoods
4.1 Pinned roll	6. Carding segment
4.2 Needle roll	

Figure 3.13. Representation of the Cleanomat CVT4, an intense opener/cleaner for cotton. (Courtesy Truetzschler GMBH & Co.)

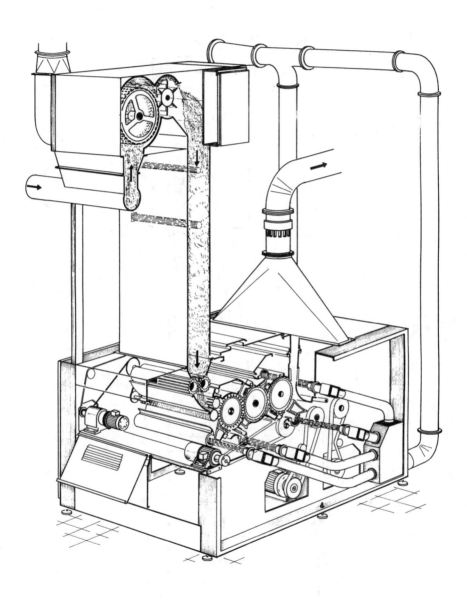

Figure 3.14. Diagram of the B37 Horizontal Opener, an intense opener/cleaner for cotton. (Courtesy Fratelli Marzoli & C. spa)

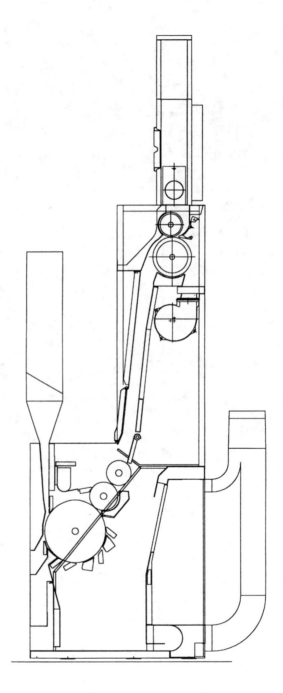

Figure 3.15. Schematic diagram of the UltraClean machine, an intense opener/cleaner for cotton. (Courtesy John D. Hollingsworth on Wheels, Inc.)

Mixing of Fibers from Bales in the Laydown

Because of the significant variability in fiber properties within and between bales, there must be opportunity for effective mixing of fibers together from every bale of the same fiber type in the laydown. This can be accomplished in several ways in the Opening Room. Hopper feeders deliver tufts that can be dropped onto the conveyor or feed table on top of the tufts from all prior hoppers in the parallel set. This is one form of sandwiching tufts from different bales. In the reserve section of the hopper feeder, stock from multiple bales is rolled and tumbled, contributing somewhat to mixing. Various size hopper feeders, also called blending reserve hoppers, are available for use. A variation of the hopper feeder, called a Laydown Cross Blender, is a very effective fiber mixer and is graphically demonstrated in Figure 3.16.

A popular machine for mixing tufts from bales of the same fiber type is known as a cell mixer. These machines mix stock by depositing tufts from a prior process, sequentially or randomly, into a set of parallel vertical chambers. Stock is simultaneously removed at the bottom or delivery end of the chambers. This provides stock output that has fiber tufts from each chamber. These machines are also called time-delay blenders or blender/reserves. Figures 3.17 through 3.20 portray typical cell mixers. Two such mixers are often used in tandem or in series, one after the other, so as to more effectively mix or fold the fiber tufts.

Blending of Different Fiber Types

The prior section addresses mixing of tufts from bales of the same fiber type, applicable to fibers processed as a 100 percent mix of one fiber type. This can be prior to intimate blending with another fiber type, or as a 100 percent mix at least through the carding process. The same mixing machines described above can be used to further homogenize dissimilar fibers after an initial weight-based metering out and proportionate blending of these dissimilar fiber types. Such a blending of fibers by weight (directly or indirectly weighed) in the Opening Room is referred to as intimate blending. There are currently three machine systems in use for the weight-based blending of different fiber types. These are weigh-pan hopper feeders, belt-weighing machines, and chamber type, pressure sensor equipped machines. Examples of each system are provided in Figures 3.21 through 3.24.

Carding

Opened, mixed, and cleaned fiber tufts are transported by air ducts from the final stage of the Opening Room processes to a chute that prepares a fiber mat for feeding the card. Figures 3.25 through 3.27 are representations of the card chute and card with labeling of the working elements of these machines. The card has often been referred to as the "heart" of the spun yarn manufacturing process. It has also been stated that "well carded is half spun". Both of these statements suggest, and rightly so, that the carding process is critical in the manufacture of spun yarns. It is so critical in part because it is at the card that fibers are actually individualized for straightening and cleaning, and it is at the card that initial fiber strand properties and variabilities are established. These properties and their variabilities significantly influence the processing of the fibers downstream and the ultimate quality of the yarns produced.

The function of the chute is to form a continuous and even mat of small fiber tufts to be fed into the card. A feed plate and feed roll present the fiber mat to the card. The card separates small individual fiber tufts into individual fibers to partially orient the individual fibers in a longitudinal direction, to remove contamination that may have been nested within the fiber tufts, to open or remove nep-like structures, to mix the fibers intimately, and to produce a continuous strand of defect-free sliver, with correct weight, usually precisely coiled into a sliver can.

Lap Preparation and Combing

An option for the processing of 100 percent cotton is lap preparation and combing. Simply, combing is primarily done to obtain the benefits associated with the removal of short fibers (less than 12.7 mm) from the stock. Also, during the lap preparation and combing processes, fibers are oriented, hooked fiber ends emanating from carding are straightened out, and significant further fiber mixing is accomplished. Laps are produced by combining and drafting multiple card slivers, perhaps as few as 16 slivers to as many as 48, to form a fiber sheet either 267 mm (10.5

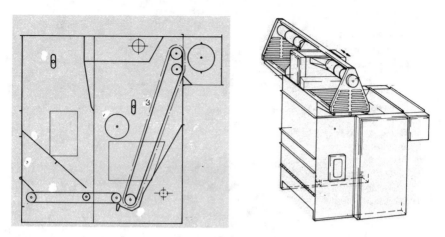

Figure 3.16. Illustrations providing cross-sectional and three-dimensional views of the Laydown Cross Blender LCB. (Courtesy John D. Hollingsworth on Wheels, Inc.)

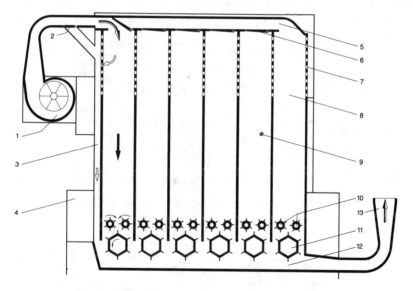

1. Fan TV 425	8. Blending chamber
2. Air by-pass	9. Light barrier
3. Exhaust air duct	10. Delivery rolls
4. Program control/pressure switch	11. Opening roll
5. Feed duct	12. Blending channel
6. Closing flap	13. Material suction tunnel
7. Perforated plate	14. Exhaust air outlet

Figure 3.17. Diagram of the working elements of the Multi-Mixer MPM. (Courtesy Truetzschler GMBH & Co.)

1. Feed pipe
2. Vertical filling trunks
3. Conveyor belt
4. Conveying roller
5. Upright lattice
6. Blending chamber
7. Stripper roller
8. Take-off roller
9. Filling trunk
10. Opener/cleaner roller
11. Waste collector
12. Fiber delivery
13. Exhaust air piping
14. Exhaust air outlet

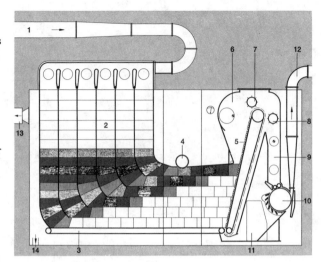

Figure 3.18. Cross-sectional diagram of the Unimix B7/3, a combined mixing and cleaning machine. (Courtesy Rieter Machine Works, Ltd.)

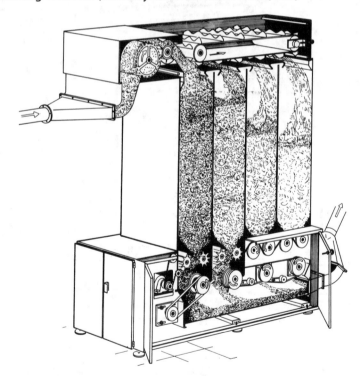

Figure 3.19. Illustration of fiber tuft mixing in the B142 Automixer. (Courtesy Fratelli Marzoli & C. spa)

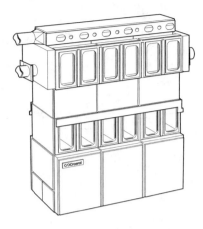

Figure 3.20. Illustration of the Six-Chamber Blender 6CB. (Courtesy Crosrol Ltd.)

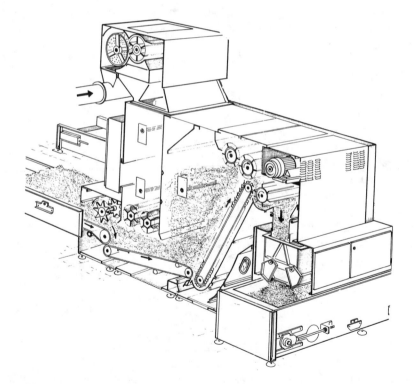

Figure 3.21. Drawing of the working elements of the B13 Blending Feeder. (Courtesy Fratelli Marzoli & C. spa)

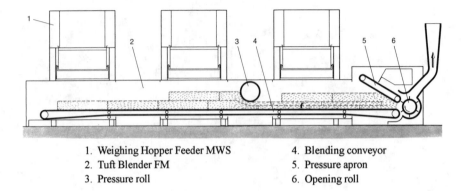

1. Weighing Hopper Feeder MWS   4. Blending conveyor
2. Tuft Blender FM   5. Pressure apron
3. Pressure roll   6. Opening roll

Figure 3.22. Diagram of a series of weigh-pan hopper feeders for the intimate blending of dissimilar fiber types. (Courtesy Truetzschler GMBH & Co.)

A. Feed section
B. Metering section
C. Opening section

1. Feed chute
2. Glass pane
3. Sieve drum
4. Dummy drum
5. Decompression zone
6. Pair of drafting rollers
7. Measuring probe
8. Measuring unit
9. Conveyor belt
10. Pair of feed rollers
11. Opening roller
12. Blending channel

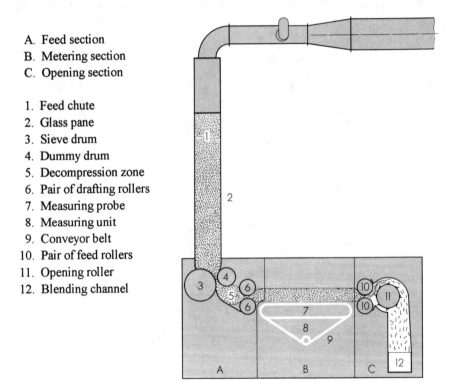

Figure 3.23. Simplified sectional drawing of a belt-weighing machine for intimate blending known as the Fiber Metering Device B0/1. (Courtesy Rieter Machine Works, Ltd.)

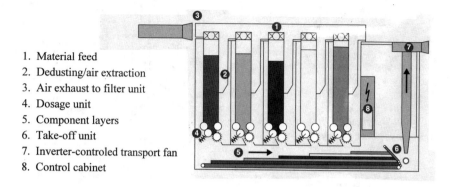

1. Material feed
2. Dedusting/air extraction
3. Air exhaust to filter unit
4. Dosage unit
5. Component layers
6. Take-off unit
7. Inverter-controled transport fan
8. Control cabinet

Figure 3.24. Illustration of the UniBlend A80, a chamber intimate blending machine. (Courtesy Rieter Machine Works, Ltd.)

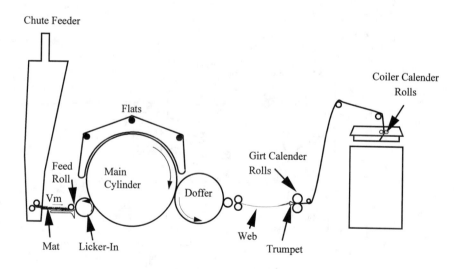

Figure 3.25. Diagram of the main parts of the typical short staple card.

1. Distribution duct
2. Dust extraction duct
3. Upper air outlet
4. Material reserve trunk
5. Fan
6. Feed roll
7. Opening roll
8. Control unit
9. Adjustment of nominal pressure
10. Adjustment of base speed
11. Pressure transducer
12. Feed trunk
13. Lower air outlet
14. Take-off rolls
15. Card feed roll
16. Speedometer

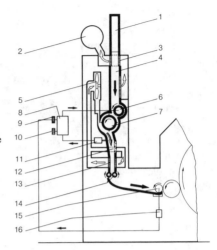

Figure 3.26. Detailed drawing of the Exactafeed FBK 533 Tuft Feeder for chute feeding a card. (Courtesy Truetzschler GMBH & Co.)

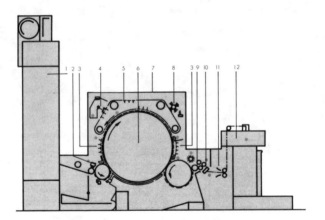

1. Aerofeed-U chute
2. Unidirectional feed
3. Stationary carding plates
4. Main flat cleaning
5. Reverse-running flats
6. Main cylinder
7. Complete machine casing
8. Flats after-cleaning zone
9. Web detaching/gathering
10. Web detaching/gathering cont.
11. Integrated leveling
12. Can coiler

Figure 3.27. Illustration of the working elements of the C4 card. (Courtesy Rieter Machine Works, Ltd.)

inches) or 305 mm (12 inches) wide. The sheets weigh between 58 and 74 grams/meter (g/m) or from about 825 to 1050 grains/yard (gr/yd), depending on width, and are rolled into laps weighing between 14 and 28 kg (30.86 to 61.73 lb). Figures 3.28 through 3.31 are illustrations of lap preparation and combing machines.

Cotton slivers delivered by the card still contain small trash particles, neps, and a number of short fibers. In addition, the individual fibers are not straight and parallel, and most fibers have hooks on at least the trailing end. The purposes of combing are to separate long fibers from shorter ones, to straighten and increase the parallelization of the longer fibers, and to remove a waste called noil which consists of trash particles, neps, short fibers, and some long fibers. Noil extraction varies from 5 to 25 percent of the weight of laps fed, depending on the end use of the yarn. Removal of 5 percent noil is called scratch combing, 6 to 9 percent noil is called half combing, 9 to 15 percent noil is normal combing, and removal of 15 to 25 percent noil is referred to as fine combing. Normal and fine combing of cotton stock can significantly improve resultant yarn uniformity, strength, and appearance and can allow finer (lighter) yarns to be spun.

Drawing

At this time the drawing process is still essential for the manufacture of nearly all short staple spun yarns. There is a vigorous effort to contribute to reduced costs by eliminating process steps, and drawing has been a target in this effort. However, because straight and parallel fibers are desired in the strands fed to spinning, drawing continues to be an essential process. Even cotton slivers that were prepared as a lap and have been combed still require at least one passage through a draw frame, not because of unoriented or irregularly arranged fibers, but because of inherent piecing faults in combed slivers.

The main processing section of the draw frame is essentially a series of pairs of nipping rolls, usually with a steel bottom roll and a rubber or similar compound covered top roll in each pair. See Figures 3.32 through Figure 3.34 for illustrations of draw frame working elements and drafting roll configurations. Fibers are straightened by drafting the sliver between these pairs of rolls. Drafting is accomplished as a successive pair of nipping rolls turns with faster surface speed than the prior pair. The top roll is either spring or pneumatically loaded for constant pressure against the bottom roll in each pair. This gives a measure of control of the movement of fibers between each pair of rolls. Attenuating the sliver in this manner reduces the weight per unit length of the slivers fed, and the fibers are forced into a more parallel position.

Multiple slivers are fed into the roller drafting zones, and this combining of several slivers, called doubling, reduces the long-term mass unevenness of the slivers. This doubling also offsets the weight per unit length reduction caused by drafting. An additional advantage of the doubling and roller drafting accomplished at the draw frame is the further mixing of fibers from multiple deliveries of a prior process, thus improving the homogeneity of the fiber mix.

Drawing of slivers in preparation for the roving or spinning processes is usually done using from one to three successive passes through draw frames. If three passes are used, the first pass is called breaker drawing, the second is intermediate drawing, and the final pass is always known as finisher drawing. There are a number of factors that determine whether zero, one, two, or three processes of drawing are needed. These and other critical drawing issues are discussed in detail in Chapter 6.

Roving

Roving is the name given to both the process and the product of that process. The roving product is the supply package for the ring spinning frame. The roving frame is a roller drafting machine which further attenuates the sliver and reduces the weight per unit length of the strands substantially, since only one sliver is fed per roving delivery. Therefore, no doubling takes place at roving. For the first time in the preparatory processing of the fiber assemblies, the material weight per unit length will be low enough to require twist to be inserted. The roving frame also accomplishes this by means of a revolving flyer. For this reason the roving frame is also sometimes called the fly or flyer frame. It has also been referred to as a slubber. An end-view schematic representation of a roving frame is shown in Figure 3.35.

The fiber mass in the draft zone of the roving frame has now been reduced to a point where aprons can be used to improve control of fiber movement in these draft zones. Use of single or double apron systems improves the consistency and quality of the roving and also allows for higher process speeds. Aprons are made of rubber or other flexible and durable materials and are demonstrated in several possible roving frame drafting zone configurations in Figures 3.36 and 3.37.

Spinning Room

The department of the short staple yarn manufacturing plant that actually converts a roving or sliver into yarn is usually called the Spinning Room. Today there are three types of spinning systems mainly used for short staple spun yarn operations in the world. These are ring spinning which is still dominant even though the basic design was developed in 1830, rotor (open-end) spinning which was first used in production in the mid 1960s, and air jet spinning which was commercially employed in the USA around 1980. Wrap spinning is a less-used system which is popular in the manufacture of sewing thread. And a still developing system is known as friction spinning. The three major spinning systems are introduced here, and are documented in great detail in Chapters 8, 9, and 10. The minor systems are briefly addressed in Chapter 11.

Ring Spinning

Figures 3.38 through 3.40 are illustrations of ring spinning frames. The critical process control areas of the ring spinning frame include the creel, the drafting mechanism, the twist insertion components, the yarn package windup elements, and the ambient conditions surrounding these spinning frame zones.

Yarn formation at ring spinning is accomplished as follows. Roving is drafted to a considerably lower weight per unit length by the spinning frame roller and apron drafting elements. The material delivered from the front roll of the drafting zone requires significant turns of twist per inch of material in order to consolidate the fibers into a yarn and to provide strength to the yarn

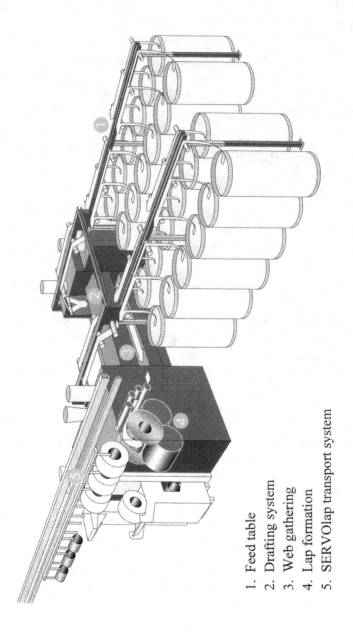

1. Feed table
2. Drafting system
3. Web gathering
4. Lap formation
5. SERVOlap transport system

Figure 3.28. Illustration of the Unilap system for forming comber laps from once-drawn card sliver. (Courtesy Rieter Machine Works, Ltd.)

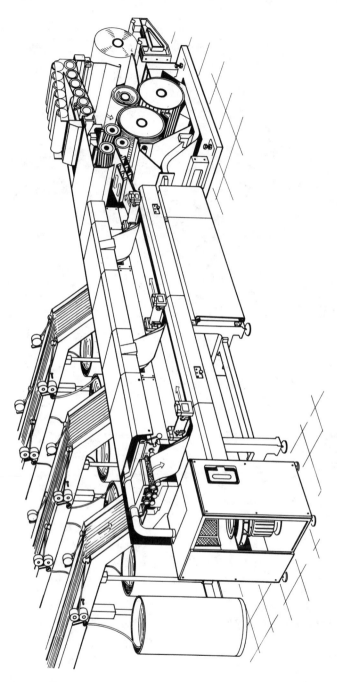

Figure 3.29. Illustration of the draft and lap winding zones of the Sliver Drawing Winder SR34. (Courtesy Fratelli Marzoli & C. spa)

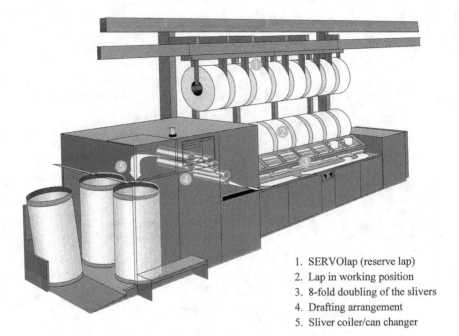

1. SERVOlap (reserve lap)
2. Lap in working position
3. 8-fold doubling of the slivers
4. Drafting arrangement
5. Sliver coiler/can changer

Figure 3.30. Illustration of a combing machine with working and reserve laps. (Courtesy Rieter Machine Works, Ltd.)

 1. Delivery table
 2. Control cabinet
 3. Delivery rollers
 4. Sliver trumpet
 5. Delivery rollers
 6. Detaching rollers
 7. Clearer roller
 8. Top nipper
 9. Bottom nipper
10. Nipper shaft
11. Circular comb
12. Brush for circular comb
13. Central suction channel
14. Cam shaft
15. Feed rollers
16. Reserve lap tray

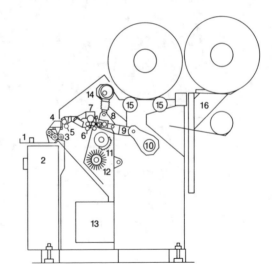

Figure 3.31. Simplified cross section of the E7/5 comber. (Courtesy Rieter Machine Works, Ltd.)

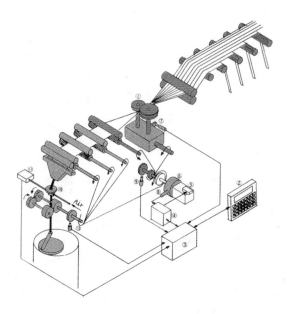

1. Tongue and groove
2. Micro-terminal
3. Control unit
4. Power unit
5. Servo motor
6. Differential gear
7. Distance sensor
8. Tacho-generator (constant)
9. Variable tacho sensor
10. FP trumpet
11. FP-MT preamplifier
12. Calender tacho sensor

Figure 3.32. Detailed illustration of the working elements of a Cherry Hara High-Speed Draw Frame D-600 with autoleveler. (Courtesy Hara Shokki Seisakusho, Ltd.)

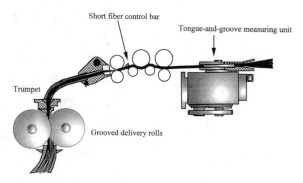

Short fiber control bar

Tongue-and-groove measuring unit

Trumpet

Grooved delivery rolls

Figure 3.33. Sliver processing elements of the SH 800 draw frame with autoleveler. (Courtesy Vouk Macchine Tessili spa)

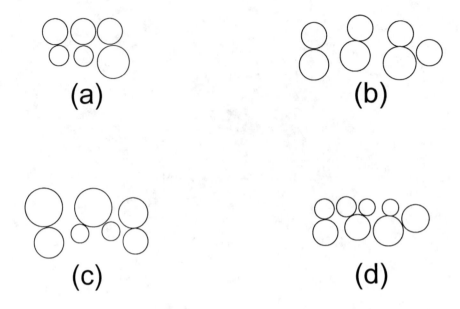

Figure 3.34. Drafting roll configurations for (a) Rieter RSB 951, (b) Truetzschler HS 900, (c) Vouk SH, and (d) Zinser 720 draw frames, sliver processed from left to right.

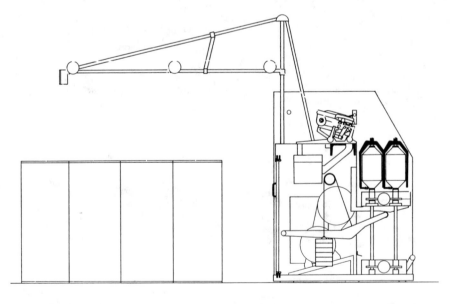

Figure 3.35. End-view diagram of a roving frame. (Courtesy Rieter Machine Works, Ltd.)

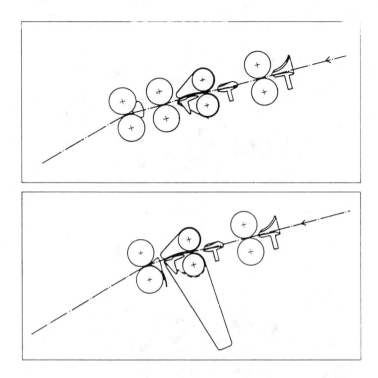

Figure 3.36. Illustrations of 4-roll-2-apron and 3-roll-2-apron drafting systems for a roving frame. (Courtesy Zinser GMBH)

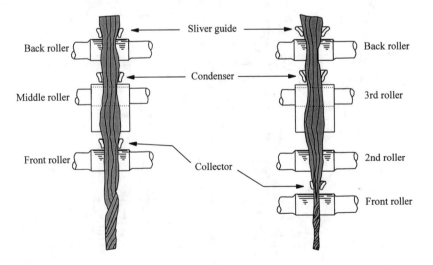

Figure 3.37. Illustration of a 3-roll and 4-roll drafting arrangement for a roving frame. (Courtesy Toyota Automatic Loom Works, Ltd.)

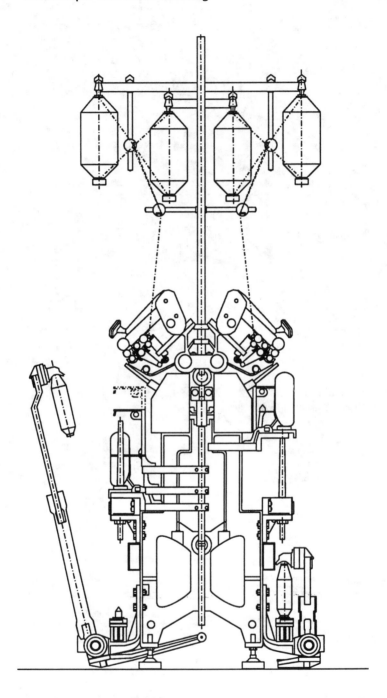

Figure 3.38. Sectional view of a ring spinning frame with automatic full yarn bobbin doffing. (Courtesy Zinser GMBH)

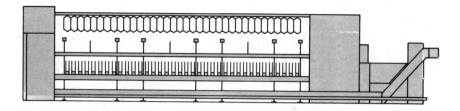

Figure 3.39. Side-view illustration of a ring spinning frame that will be linked to a winding machine so that doffed spinning bobbins can be automatically transported to the winder. (Courtesy Rieter Machine Works, Ltd.)

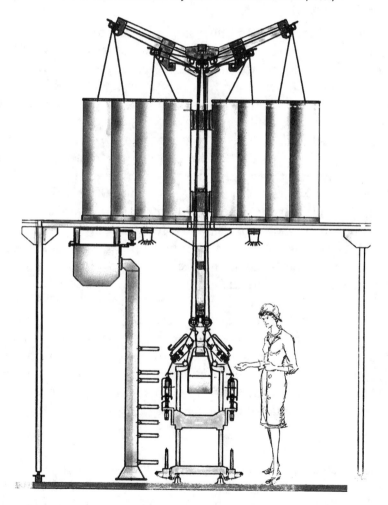

Figure 3.40. Illustration of ring spinning from sliver cans positioned above with the RingCan system. (Courtesy Spindelfabrik Suessen)

strand formed. Twist is inserted by the action of a metal traveler which moves around a stationary ring. The spindle is centered in the ring and supports the relatively small tube on which the yarn is wound. Yarn is threaded through the traveler and onto the tube, and the actual rotation of the yarn package by the spindle causes the yarn to wind up and to pull the traveler around the ring. A view of the twist insertion and windup components of ring spinning is provided in Figure 3.41.

Rotor Spinning

Figure 3.42 is a detailed illustration of a rotor or open-end spinning machine. Figure 3.43 is a drawing which represents the components of yarn formation in rotor spinning. Finisher sliver is fed to an opening roll (combing roll) or opening beater which is covered by either metallic wire or pins. The opening roll separates the sliver into individual fibers and also contributes to impurity removal through the action of centrifugal forces on any heavier-than-fiber trash particles that are present. Fibers are transported with air assistance through the fiber transfer tube into a rapidly revolving rotor which can turn up to 150,000 revolutions per minute (rpm). Fibers are impinged into the rotor groove on top of each other to create the needed mass for the desired yarn. The tail of a feeder yarn is introduced into the rotor. This tail is set in motion by the revolving rotor, and the revolution of this yarn inserts twist into the strand extending out of the rotor. As this feeder yarn is drawn out of the rotor, fibers are wiped from the rotor groove and are twisted into the yarn end, creating new yarn. Note that the exposed yarn end is rotated for twist insertion, compared to the need for the entire yarn package to be rotated as in ring spinning. This difference allows a much larger package to be built at rotor spinning. Productivity of a rotor spinning position is up to 10 or more times that of a ring spinning position.

Air Jet Spinning

The mechanism of air jet spinning with the most popular installed machine can be understood with the aid of Figures 3.44 and 3.45. Finisher sliver is fed to a high speed roller and apron drafting system. Fibers delivered from the front roll of this drafting system are acted upon in an air vortex which is created in an air nozzle that is positioned near the front roll. This air vortex imparts twist to the leading free surface fibers while their trailing ends remain fixed by the front roll. A second air nozzle creates a vortex which imparts false twist to the yarn bundle with air moving in the opposite direction of that of the first nozzle. Because of higher air pressures in the second jet, the false twist migrates toward the front roll. As the yarn exits the second jet the false twist is removed, and the twist in the fibers in the core of the yarn is reduced to zero. At that point the surface fibers, which were caused to break away from the drafted fiber bundle by the first nozzle, wrap around the core fibers as these fibers return to zero twist. Wrapper fibers serve to consolidate the fiber assembly into a continuous yarn strand. The newest air jet machine has a pair of apron balloon rollers in place of the second air nozzle, as shown in Figure 3.46, but the yarn formation principle remains the same. Air jet spinning produces up to 15 times faster than ring spinning with the promise of 20 and greater times more productivity with the newer roller jet models being developed.

Winding

Yarn wound on a ring spinning bobbin must be transferred to a larger package for efficient downstream utilization. This is normally accomplished at the winding process and during this transfer, both thick and thin place defects can be sensed and cleared from the yarn strands. Figure 3.47 is a simplified drawing of the winding process.

Control of Ambient Conditions

The processing of textile fibers and yarns requires reliable control and consistency of temperature and relative humidity (RH) in the process areas. It is especially imperative that the right relative humidity be determined for each area. The temperature is also important for processing, but is often set so that operators and support personnel working in the area will be comfortable. There are certain processes where precision of control is more critical than for other processes. Spinning is one such process where even the control allowed by the instruments in use today may not be tight

Table 3.1 Typical Temperature and Relative Humidities in Short Staple Spun Yarn Plants

Process	Temperature, %F			Relative Humidity, %		
	Cotton	Polyester	Polyester/ Cotton Blends	Cotton	Polyester	Polyester/ Cotton Blends
Opening	82	80	80	56	55	55
Carding	82	80	80	52	48	48
Predrawing	82	—	—	51	—	—
Combing	82	—	—	56	—	—
Finisher Drawing	82	80	80	50	45	45
Roving	82	80	80	50	45	45
Ring Spinning	83	83	83	44	40	41
Rotor Spinning	83	83	83	58	56	58
Air Jet Spinning	—	78	78	—	38	38
Winding	80	80	80	52	50	50

enough. Normal control is within ± 2 degrees Fahrenheit (± 1 degree Celsius) temperature and ± 2 percent RH. Table 3.1 contains typical ambient conditions maintained in spun yarn manufacturing facilities. These are not necessarily the targets for a specific site, but are the norms observed.

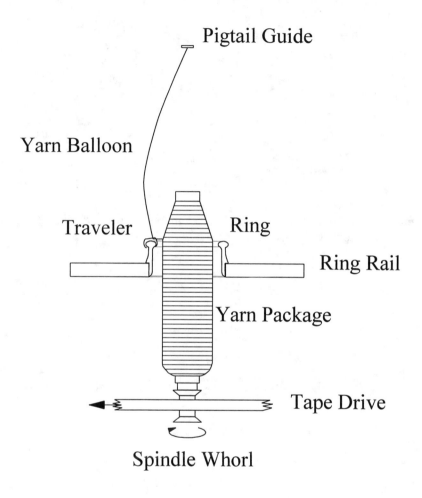

Figure 3.41. Illustration of ring spinning twist insertion and yarn windup components.

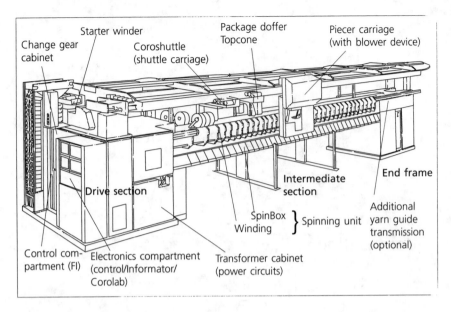

Figure 3.42. Detailed illustration of the elements of the Autocoro 288 rotor spinning system. (Courtesy W. Schlafhorst AG & Co.)

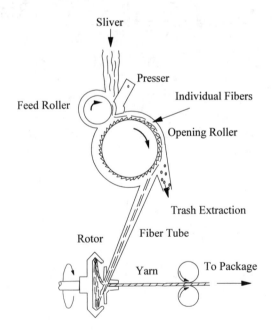

Figure 3.43. Diagram of yarn formation and twist insertion mechanisms of rotor spinning.

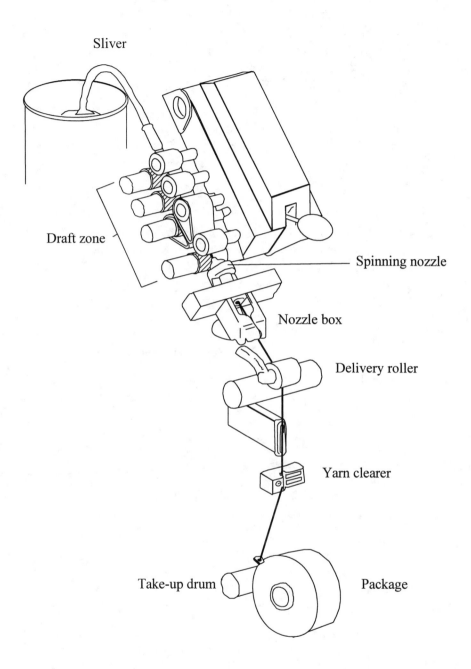

Figure 3.44. Illustration of the sliver-to-yarn process of the MJS 802 air jet spinning system. (Courtesy Murata Machinery, Ltd.)

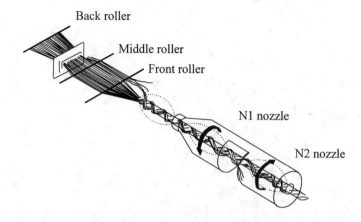

Figure 3.45. Illustration of the double-nozzle yarn assembly mechanism of MJS air jet spinning machines. (Courtesy Murata Machinery, Ltd.)

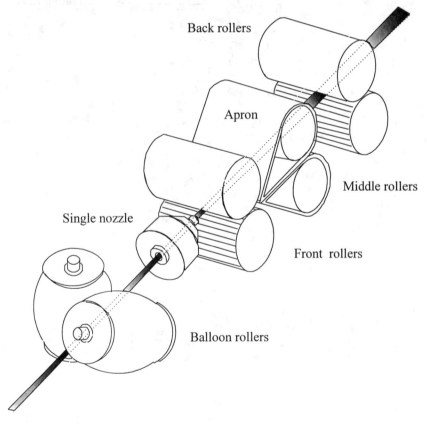

Figure 3.46. The single nozzle and balloon roller concept of the 804 RJS spinning system. (Courtesy Murata Machinery, Ltd.)

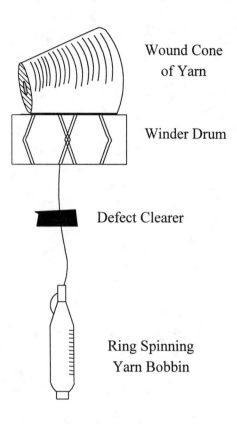

Wound Cone
of Yarn

Winder Drum

Defect Clearer

Ring Spinning
Yarn Bobbin

Figure 3.47. Simplified illustration of the winding process.

The Opening Room

General Functions and Goals of the Opening Room

The functions and goals in the Opening Room of a short staple spun yarn manufacturing plant are as follows:

1. Bale laydown preparation, placement, and replenishment;
2. Feeding of bale stock into, through, and between successive machines in the opening line;
3. Progressive opening of fiber tufts;
4. Mixing or homogenizing of fibers so that all the bales of a specific fiber type are represented in a cross section of the stock in the card mat;
5. Blending by weight the premixed fibers of different types, if intimate blending is desired;
6. Cleaning impurities from the fiber tufts, especially required for cottons;
7. Proper preparation, feeding, and blending in reworkable fibers from downstream processes and/or fibers reclaimed from cleaning and carding machine wastes;
8. Accomplishing the required feeding, opening, mixing, and cleaning with minimal damage to the fibers;
9. Maintaining appropriate temperature and relative humidity in fiber material storage, staging, and processing areas;
10. Keeping all machinery, controls, and auxiliary equipment in well-maintained, like-new condition, and set to specifications; and
11. Continuous effort to improve processes and products of the Opening Room.

These functions and goals are elaborated upon in this chapter in sufficient detail to be used as a handy guide by the persons responsible for the success of this manufacturing department. The ultimate goal is to produce the intermediate fiber assemblies that can be further processed into yarns of superior quality with all processes performing at maximum potential efficiencies. This means doing everything the best way consistently, beginning with the foundational processes in the Opening Room. Mistakes at any yarn manufacturing step are not easily compensated for in downstream processes. Mistakes are instead usually magnified and compounded by subsequent processes. This makes the benchmarking of all measurable process and product performance criteria an absolute necessity. Targets and the known best achievable performances are included where applicable.

Bale Laydown

Fibers are delivered to the yarn manufacturing facility in compressed bales. Bales of cotton are wrapped in either burlap, cotton, polyethylene, or polypropylene bagging and a series of metal straps secure the wrap and the compressed state of the fibrous mass. It is also possible to buy cotton with the wrapping on the outside over the straps. Synthetic fibers are purchasable in cardboard or synthetic wrappings, and strapping of synthetic material is generally used.

Bale Conditioning

All fibers will process in the Opening Room machinery most consistently and with least chance of damage if the bales are unstrapped and unwrapped and allowed to condition for at least 24 hours prior to introduction into the process machinery. This requires a conditioned staging area where temperature and relative humidity are controlled. It also requires floor space to be allocated for this purpose. When a top feeder is used for initial bale feeding, there is usually space set aside for the current laydown and for the next laydown. If the number of bales in the laydown allows for at least 24 hours of production time, then staging the next laydown for conditioning is easily accomplished. Figure 4.1 is an illustration of this approach when using a top feeder.

Figure 4.1. Photograph of a top feeder working on a wedge-shaped cotton bale laydown with bales visible in the background that will be added one at a time to replace an exhausted bale. (Courtesy Truetzschler GMBH & Co.)

Bale Side Cleaning and Heavy Particle Removal

From time to time certain cotton bales arrive at the unwrapping stage with soiled bale sides or ends. This can come from improper bale handling where the bale is allowed to slide across the ground or floor in transport. Various kinds of debris can be imbedded in the fibers on the bale outer surface. Dirt, grease, rubber, and wood splinters are among these contaminants often observed. After wrapping removal and before strap removal, if possible, these contaminants should be cleaned from the bale. Failure to do so creates the potential for these contaminants to become part of one or more yarn strands, which can be graded as defects in the fabrics in which these yarns are used. These contaminants can also increase unplanned process stoppages.

Other heavier contaminants can get into the bales and/or early processed stock, especially if reworkable fibers are fed back in near or at the initial bale feeding step. These contaminants are referred to as heavy particles and may include metal, plastic, wood, and other debris. Units called heavy particle separators can be installed in the process line. These units generally take advantage of inertia of these heavier particles in order to remove them from the opened stock in transport. Magnets are also used but will naturally remove only metal particles. Magnets must be cleaned of removed particles on a reasonable frequency. The particles will otherwise catch fiber tufts and form tags in the stock transport ducts. Such fiber tags roll and twist in the airstream and get longer and more voluminous over time and can create unwanted air turbulence and even stock chokes in the transport ducts. Figures 4.2 through 4.6 are examples of various methods for heavy particle removal. The greatest dangers in failure to provide for removal of

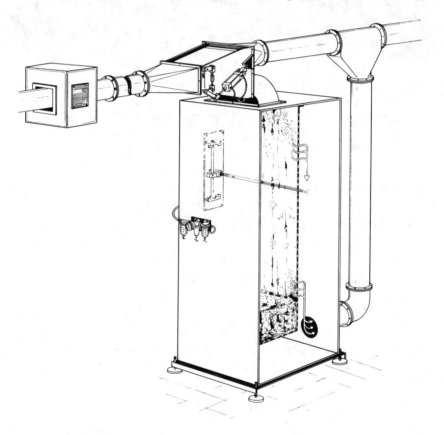

Figure 4.2. Illustration of the B160 Metal Detector for the opening line. (Courtesy Fratelli Marzoli & C. spa)

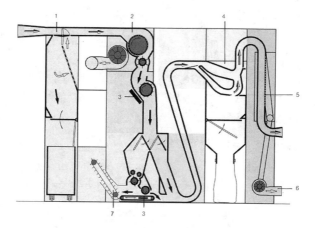

1. Fire protection flap
2. Condenser LVSA
3. Metal director
4. Heavy particle separator
5. Air separator
6. Dust-laden air
7. Ferrous metal particle separator

Figure 4.3. Illustration of the Securomat SC, a station for fire detection and for separation of heavy particles out of a cotton stream in the opening room. (Courtesy Truetzschler GMBH & Co.)

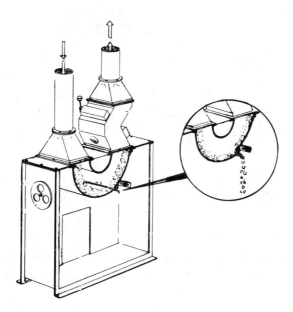

Figure 4.4. Schematic representation of the Separator of Heavy Particles B170 with arrows indicating the direction of cotton stock flow. (Courtesy Fratelli Marzoli & C. spa)

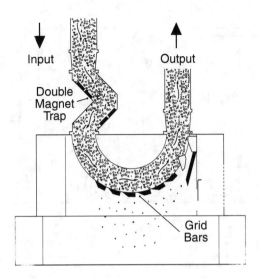

Figure 4.5. Illustration of the principle of operation of the Heavy Particle Separator HPS. (Courtesy Crosrol Ltd.)

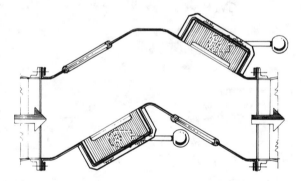

Figure 4.6. Diagram of a typical set of magnets for metal removal in fiber stock transport ducts. (Courtesy Fratelli Marzoli & C. spa)

these contaminants include potential spark creation and fires in the process line and damage to the wire teeth (clothing) or other metal surfaces of the process machines. Damaged machine components equate to quality losses, machine downtime, and unnecessary maintenance expense.

Laydown Size

Not just any number of bales in the laydown will suffice. Too few bales will make it impossible to achieve a consistent mix aver-

Table 4.1 Minimum Number of Bales Needed in a Laydown Based on Highest Individual Fiber Property Overall Variability

Overall %CV of Most Variable Fiber Property	Number of Bales Needed
10	16
12	23
14	32
16	41
18	52
20	64
22	78
24	92

age of all the measurable fiber properties being managed. Too many bales will do the same unless the downstream mixing machine capabilities are correct for the number of bales to be mixed. Hence, there are both statistical and practical considerations for determining how many bales should comprise the laydown.

The statistical considerations take into account the highest overall variability of a measured fiber property. The overall variability number used should be the highest possible fiber property variation within and between bales in a laydown-size sampling from the warehouse population from which the laydown is to be selected. Such overall variability of a single fiber property can range up to over 20 percent. Typical variability is between 10 and 14 percent. Table 4.1 is an aid to understanding the influence of fiber property overall coefficient of variability on the minimum number of bales needed for the laydown. Number of bales were calculated using the equation[4.1] for sample size.

$$n = \frac{t^2 v^2}{e^2} \quad (4.1)$$

Where in this application:

n = Number of bales needed to consistently average out the fiber property;

t = Desired precision of the estimate;

v = Overall coefficient of variability of the fiber property expressed as a percentage; and

e = Risk to be tolerated of rejecting the true answer in this calculation, also expressed as a percentage.

For the construction of this table, a "t" value of 2.0 (indicating 95 percent precision) and an "e" value of 5 percent were used. This really means that 95 percent of the time the number in this table will actually be within 5 percent of the number of bales needed, generally acceptable in the industry for this type of calculation. Higher precision required and lower error tolerated will cause the number of bales needed to increase.

In typical laydowns of cotton, if the variability of a fiber property measurement ranges between 10 and 20 percent, the proper number of bales in a laydown will range from 16 to 64, respectively. Performing this calculation is not a standard routine practice for the textile yarn manufacturing facility. Therefore, a starting point or default of 40 has been recommended to provide for the large majority of sites an appropriate number of cotton bales in a laydown for successful property averaging. Forty bales of cotton will generally also provide a modest factor of safety for the yarn spinner. For similar reasons 24 bales have been recommended as the starting point or default for laydowns of synthetic fibers.

Practical considerations for determining the number of bales in a laydown include total laydown processing time desired, floor space availability, bale feeding machinery capacities, and the mixing power of downstream machines. In any case the statistically calculated minimum number of bales takes precedence, and equipment should be installed to effectively mix whatever number of bales are used in the laydown.

Processing time desired can dictate the number of bales in the laydown according to the calculation in this equation:[4.2]

$$\text{Number of bales in laydown to last n hours} = \frac{\text{Kg processed in n hours}}{\text{Average net Kg / bale}} \quad (4.2)$$

Where in this application:

n = A specified number of hours.

It is the practice of many yarn manufacturing operations to select the number of bales that will allow once-per-day laydown preparation, conditioning, and placement. Other facilities have reasons for shorter or longer than 24-hour laydown process time.

Depending on which type bale feeding machinery is used, there will be some maximum number of bales that can practically be serviced. A hopper feeder has a specific volume capacity and can accommodate only a certain maximum number of bales to be represented in the actual opening and mixing compartment. Stock charges from each bale placed in the hopper feeder ideally range from 50 to 100 mm (2 to 4 inches) in thickness, so only so many can be handled by the hopper feeder.

For many older facilities, floor space availability is a serious constraint. The number of bales in the laydown should be chosen with consideration of floor space needed for both the laydown in process and the next laydown being conditioned. Remember that it is wise, whatever the practical considerations, to never put fewer bales in the laydown than the statistical calculation suggests. Putting a few more than is calculated will provide a safety factor which is also wise, provided the mixing power of downstream machines is adequate.

The mixing power of the machines downstream from bale feeding must provide at a minimum that every bale is represented in a cross section of stock in process, sampled after the last mixing step. The laydown should not contain more bales than can effectively be mixed downstream with an adequate safety margin for mixing capacity applied. This means that there must be an objective method for assessing machine capability for mixing the stock. Such a method has been established and is elaborated upon later in this chapter in the section titled "Blending and Mixing of a Single Fiber Type."

Cotton Laydown Bale Positioning

When it was the norm for cotton bales to be fed initially into a hopper feeder, bale positioning was important because of the natural selectivity of these machines for feeding preferentially the heavier tufts and the tufts of finer fibers. For this reason many laydowns were planned so that bales of like density of compression, as an example, would be placed behind one hopper feeder. Similarly, bales grouped behind a hopper would have like fiber fineness as another example.

Bale positioning with top feeders is an opportunity because feeding is done from so few bales at a time. Feeding is generally from one to three bales at a time, although top feeders and laydown configurations are available so that 4 to 8 bales can be fed from at one time. It is logically not wise to have a top feeder to feed from multiple consecutive bales of similar fiber property averages if there is a wide range of those averages across the laydown. The opportunity for bale positioning at a top feeder goes beyond such an obvious example. There is a concept known as "mini-laydowns" which represents the best approach to cotton bale positioning for top feeding. The objective is to feed sequentially from bales in a small group so that, on average, the fiber mass from each bale group is identical to that of every other bale group, with as little variation from group to group as is reasonably possible, and with each group being near the overall average of the laydown.

With the aid of a computer and appropriate software, up to 5 critical properties can be managed to meet this objective for an example 40-bale laydown which is divided into five 8-bale groups. Group-to-group fiber property averages can be maintained within 1.5 to 2.0 percent of the laydown average. A tremendous benefit of such a mini-laydown approach to bale positioning is the return of a significant blending contribution to the initial bale feeding step which was diminished greatly when hopper feeders were replaced with top feeders. This bale positioning technique will actually reduce the effective size of the laydown and may demand less mixing to be performed by downstream machines for the same level of homogenizing of fiber tufts. Further research needs to be done in order to quantify for publication this effect and the many other expected benefits of using the mini-laydown concept. Reduced product and process variabilities should be the primary benefit of this procedure.

Cotton Bale Preparation Automation for the Laydown

Beginning in 1991 several machinery builders began exhibiting technology for automating single bale preparation and transport to the cotton laydown. These systems were refined by the 1995 exhibitions and were being installed in a few USA spun yarn manufacturing plants. The general features of these systems include the following:

1. Automation of bale strap cutting and removal;
2. The possibility of bale conditioning in a queue of reserve bales;

3. Automation of transport of bale from strap removal station to the reserve or active laydown;
4. Automation of replacing exhausted bales in the active laydown;
5. A top feeder that creates a wedge-shaped laydown to facilitate automated bale handling; and
6. Handling of up to 20 bales per hour per system.

An illustration of a wedge-shaped laydown is provided in Figure 4.7 and a photograph is shown in Figure 4.8. The idea of using such a wedge shaped laydown so that one bale at a time can be re-

Figure 4.7. Illustration of the wedge-shaped bale laydown.

Figure 4.8. Photograph of a top feeder for feeding from a wedge-shaped laydown.

placed is not new. This has been done in manual bale feeding operations. It is a relatively new concept, however, for the top feeder, and the possibility now exists for replacing two, three, or four bales at a time. The latter capability should assist in easier management of the mini-laydown approach to bale positioning that is discussed in the immediately preceding section.

There are pros and cons of such an approach to cotton bale feeding. The positives are as follows:

1. This approach all but eliminates the natural shock to the system associated with changing out an entire laydown. Changing out one to four bales at a time ensures near constant laydown average properties as long as the newly introduced bales have characteristics that match the exhausted bales.
2. The need to adjust the milling head bite depth throughout the laydown is eliminated. Differing densities of stock at the top, middle, and bottom of bales will be averaged out and all densities will be fed from each pass of the top feeder head.
3. Handling of one bale at a time increases the chance of success of bale preparation automation.
4. Fewer electronic sensors are needed for the top feeder, making it potentially less expensive.

The cons of the wedge-shaped laydown include these observations:

1. The concept is not appropriate for plants that have a number of frequent mix changes. The approach is most ideal for plants that run one cotton laydown description for very long periods of time, measured in weeks and months.
2. The concept of the mini-laydown is more difficult, but not impossible to manage when replacing a number of bales that is less than the number of bales composing the mini-laydown group.

Critical Factors in Virgin Stock Feeding

No single function of fiber handling and processing is an end to itself. There is considerable interaction and effect of one function on the manner in which other functions are accomplished. In the Opening Room there is significant interdependency among stock feeding, fiber tuft opening, mixing, and cleaning for effective performance of each of these functions. Most of those interdependencies are made clear throughout this chapter.

Virgin stock refers to fibers fed from the newly opened bales and to these same fibers at any point in the downstream processing line. Once processed and removed from the main stock flow, fibers and associated matter are referred to as reworkable, non-reworkable, or reclaimable depending on the processing source of the fibers which were separated from the main stream. In this section the focus is the feeding of virgin fiber tufts. There are several critical factors in the feeding and transport of fibers to and through the Opening Room machines. These include tuft size requirements for effective processing at specific machine types, production rates, and the run time of process equipment.

Importance of Tuft Size

It is an objective of the Opening Room processes that fiber tufts are progressively opened into smaller and lighter tufts. That, of course, is accomplished by the opening function of the machines, but tuft size has an effect on stock transport, on production rates of process machines, and on the opening, mixing, and cleaning performance of these machines.

Lighter tufts are easier to transport through air ducts and require less air velocity in transport than do heavier tufts. Certain opening and cleaning machines are designed for heavier clumps of fibers, while others are designed to perform best on lighter tufts of fibers. That is why certain machines are used preferentially either early or later in the process line. Figures 4.9 and 4.10 demonstrate machines designed to be used primarily as the first or second stage of cotton opening and cleaning. Figures 4.11 and 4.12 represent machines that are most often utilized as later opening/cleaning steps because they are designed for that purpose and work best with lighter tufts. Intense opener/cleaners, introduced in Chapter 3, will generally be used as the second or only cleaner in the line. The openness of tufts delivered from a top feeder can be too much for effective processing in machines designed for larger fiber clumps. At the same time, this level of openness makes machines

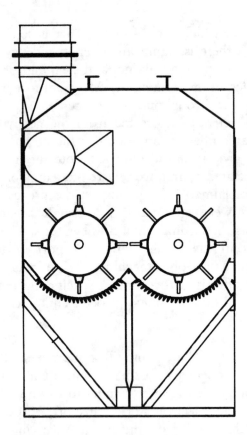

Figure 4.9. Cross section of a Model WR2 Roller Cleaner that is designed for use early in the cotton opening and cleaning line. (Courtesy John D. Hollingsworth on Wheels, Inc.)

that work best on lighter tufts more appropriate for use earlier in the line. This intimates how tuft size and machine type must be considered together. Failure to do so can lead to less than achievable process performance and to potential damage to the fibers.

Lighter tufts generally equate to lower 100 percent kg per hour productivity through a machine than do heavier tufts. This means that tuft size, machine feed rates, and other machine settings are interdependent. It takes more energy to remove impurities from and to further open a heavy clump of fibers than it does to clean or further open a much lighter tuft of fibers. In fact it is easier to remove impurities from lighter tufts simply because there are fewer fibers to interfere with impurity escape from the fiber tuft. Finally, large

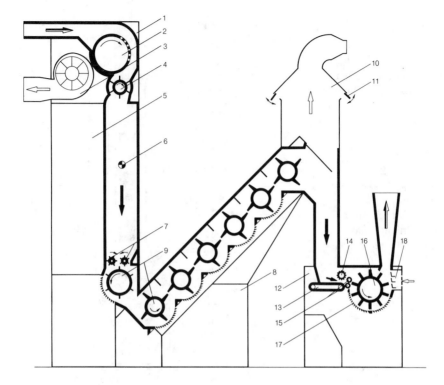

1. Condenser LVS	7. Feed rolls	13. Inner feed lattice
2. Dust cage	8. Step Cleaner SRS 6	14. Pressure roll
3. Fan	9. Cleaner roll and grids	15. Feed rolls
4. Finned roll	10. Microdust suction	16. Porcupine beater
5. Feeding Unit BE	11. Fresh air flaps	17. Two-part grid
6. Light barrier	12. Cleaner RN	18. Air inlet

Figure 4.10. Diagram of a Step Cleaner and Cleaner RN system, best used early in a cotton opening and cleaning line that is fed by hopper feeders. (Courtesy Truetzschler GMBH & Co.)

and heavy clumps of fibers can not be as homogeneously mixed as can the smaller and lighter tufts. When any processing parameter is changed, care must be taken to measure the impact on the accomplishment of each of the desired Opening Room functions.

Importance of Stock Production Rate

Stock feed rate is roughly equivalent to the production rate of an opener/cleaner type machine. Feed rate may or may not ap-

1. Material supply
2. Feed head with fan
3. Exhaust air piping
4. Material delivery
5. Laminar chute with material
6. Plain drum
7. Dust cage
8. Feed rollers
9. Knife grid
10. Opening and cleaning beater
11. Suction duct
12. Waste chamber
13. Driving motor
14. Waste removal pipe

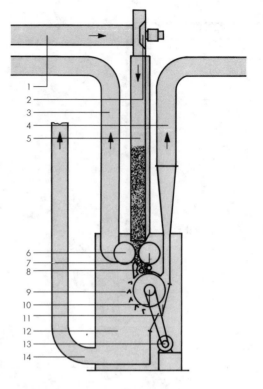

Figure 4.11. Diagram of the ERM Model B5/5, a fine opener/cleaner that is best used midway in a cotton opening and cleaning line. (Courtesy Rieter Machine Works, Ltd.)

proximate the production rate of a machine that holds reserve stock. Delivery rate in kg per hour as stock is being delivered from any Opening Room machine is always roughly equivalent to the instantaneous production rate of that machine. For an opener/cleaner, the amount of waste removal has negligible impact on the production rate calculation.

All machines, whether openers, opener/cleaners, reserves, or mixers have a production rate recommendation for best machine performance. This is not usually the maximum specified by the machine builder. That maximum is mechanically possible, but does not necessarily coincide with most effective opening, cleaning, or preservation of fiber quality. A general rule of thumb that is applicable to many opener/cleaners is not to exceed 365 kg (800 lb) per hour of instantaneous stock throughput rate per meter of

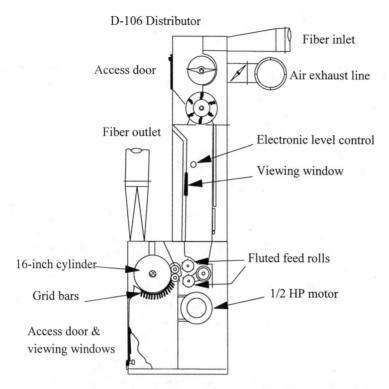

Figure 4.12. Illustration of the 310 Vertical Fine Opener that is normally used late in a cotton opening and cleaning line. (Courtesy Fiber Controls Corp.)

machine width. Hopper feeders perform their functions best at or below a delivery rate of 120 kg per hour per meter of hopper width (80 lb per hour per foot of hopper width). Top feeders, reserves, and mixers can generally operate nearer their mechanical limits than can opener/cleaners. It is important to distinguish between instantaneous production rate and net production per unit of time. The following discussion on run time helps with this distinction.

Importance of Process Machinery Run Time

Another key measure related to stock feeding is process machinery run time, defined as the percentage of time available that a machine actually processes stock. This equation[4.3] is used to calculate run time:

$$\text{Machine Run Time, \%} = \frac{\text{Total time machine processes stock}}{\text{Total time of test}} \times 100 \quad (4.3)$$

Opener/cleaner run time should be as high as possible, and continuous run time would be the ideal. For stock reserves and mixers in which opening or cleaning is not done, run time is not as critical, but higher is usually better.

There are proven advantages to continuous run time. With stock processing through a machine 100 percent of the time, there are the following advantages:

1. There is relatively constant output product bulk density or tuft openness.
2. The lowest instantaneous stock throughput rate (kg/hour) is achieved.
3. There is less wear on starters and motors.
4. Lower energy amounts are required.
5. There is less variability in the output product quality and in downstream process performance.

Consider by way of the data in Table 4.2 the impact of run time on production rate of an opener/cleaner that is set up to actually deliver 300 kg (660 lb) of stock in one hour. These data reveal that, for a certain net production needed, reduced run time can cause the required rate of stock throughput to increase dramatically, perhaps well beyond the effective range of the machine. Increased throughput rates coincide with reduced fiber tuft opening and with reduced impurity removal effectiveness.

Machine run time can be monitored electronically if such a modern system is installed, or with the use of two stop watches. One watch must time the test duration, preferably 30 minutes or more, and the second watch must be started and stopped to coincide with the initiation of stock processing and cessation of stock processing, respectively. Some machines have feed rolls that start and stop, governing stock flow through the machine. In these cases feed roll starting and stopping can be the basis for timing with the second watch. Other machines have no feed rolls and stock flow is determined by the run time and output of the prior machine. In these cases the second watch is used to time actual stock flow through the subject machine, or the run time of the

Table 4.2 Effect of Opener/Cleaner Percent Run Time on Instantaneous Production Rate

Run Time, %	Instantaneous Production Rate Required of Machine When Processing 300 Kg (660 Lb) of Stock in One Hour	
	Kg/Hour	Lb/Hour
100	300	660
90	333	733
80	375	825
70	429	943
60	500	1100
50	600	1320
40	750	1650
30	1000	2200

prior machine is established which will be the same for the subject machine.

The advice of this discussion is to never accept lower run time than can be attained. The advantages of higher run time are too significant to ignore.

Fans and Condensers for Stock Transport

Stock transport between Opening Room processes is usually done via air ducts. The air stream necessary to carry the fiber tufts can be created by stock transport fans or by fans in perforated screen condensers. Fans blow the stock, and condensers pull the stock through the ducts. Air should be supplied from the room where the process machines operate. This air is later filtered, reconditioned, and put back into the process area in order to help control temperature and relative humidity (RH). Transport components are pictured in Figures 4.13 through 4.16. There is a tendency to operate the fans and condensers at too high a rate, creating excess air velocity and the accompanying ill effects. These ill effects include the potential for excessive fiber tuft rolling and spiralling in the ducts resulting in increased stock neppiness, the potential for the recompaction of tufts which reduces the tuft openness achieved earlier, and the obvious added energy costs.

Air velocity for stock transport needs to be the minimum necessary without causing chokes in the transport ducts. This recommendation is more likely to be achieved with consistently and well-opened fiber tufts and with high machine run times, which

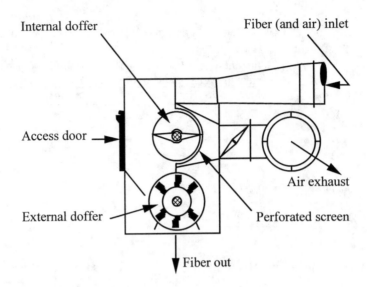

Figure 4.13. Diagram of the D-106 Pneumatic Distributor for stock transport and dust extraction. (Courtesy Fiber Controls Corp.)

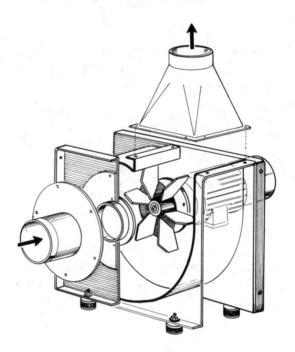

Figure 4.14. Illustration of the B151 Motorfan for stock transport with material entering and exiting as indicated by the arrows. (Courtesy Fratelli Marzoli & C. spa)

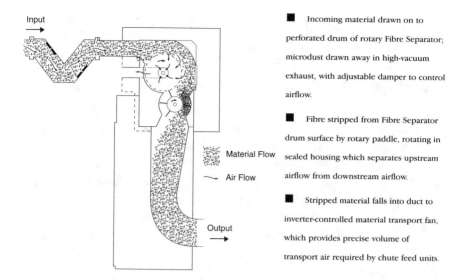

Input →

■ Incoming material drawn on to perforated drum of rotary Fibre Separator; microdust drawn away in high-vacuum exhaust, with adjustable damper to control airflow.

■ Fibre stripped from Fibre Separator drum surface by rotary paddle, rotating in sealed housing which separates upstream airflow from downstream airflow.

■ Stripped material falls into duct to inverter-controlled material transport fan, which provides precise volume of transport air required by chute feed units.

Material Flow

Air Flow

Output →

Figure 4.15. Illustration and explanation of operation of the Dust Remover/Airstream Controller DR/ASC. (Courtesy Crosrol Ltd.)

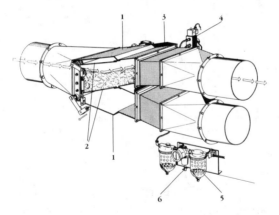

1. Window for air intake from the room.
2. Gate to allow or to deviate material passage.
3. Pneumatic cylinder to drive the gate.
4. Electrodistributor for the compressed air to drive the cylinder upon intervention of the photocell.
5. Cleaning and lubricating unit for compressed air.
6. Pressure reducer and manometer.

Figure 4.16. Diagram of the B92/6 Directional Pneumatic Deviator, also called a y-valve distributor, for directing material flow alternately between two transport ducts. (Courtesy Fratelli Marzoli & C. spa)

together mean less mass to transport at any given instant in order to meet a required net production amount.

Opening of Fiber Tufts

Opening of fiber tufts or flocks is best done progressively and gradually. A main reason to accomplish tuft opening (or tuft size and weight reduction and shape alteration) progressively is to avoid potential over-aggressive mechanical manipulation of the fibers. Such aggressive action can damage fiber length characteristics and generate additional neppiness in the stock. The specific aim is to thoroughly and uniformly open fiber tufts to facilitate impurity removal and to facilitate mixing with minimum fiber damage and nep generation. There are 7 basic mechanical opening actions used for this purpose in the Opening Room: grabbing, plucking, stripping, beating or striking, raking, and carding. These actions can be used individually but are most often used in combination in an Opening Room machine and are appropriately defined and graphically demonstrated in the sections that follow.

Grabbing as an Opening Action

Grabbing of layers of stock from the top of a bale is an opening action used by bale feeders designed to simulate and automate the manual feeding of bale stock into hopper feeders. A diagram of the mechanism for grabbing as an opening action is provided in Figure 4.17.

Plucking as an Opening Action

Plucking can be defined as the gentle removal of smaller fiber tufts from a large and compressed mass of fibers like a bale. Top feeders use rolls called star, milling, or plucker rolls to lift tufts from the top of the bales in the laydown. Figure 4.18 represents the machine components of a top feeder that are used to pluck as an opening action.

Stripping as an Opening Action

Hopper feeders use a stripping action as the primary means for opening fiber tufts. Tufts are in a sense self-regulating in the presentation to the vertical pinned apron of a hopper feeder. Larger and denser clumps are presented preferentially to this apron. At the top

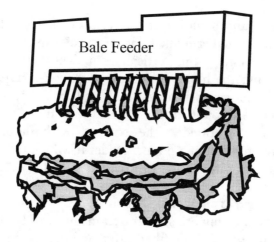

Figure 4.17. Illustration of grabbing as an opening action.

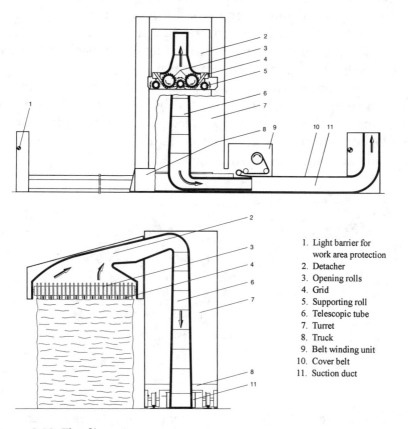

1. Light barrier for work area protection
2. Detacher
3. Opening rolls
4. Grid
5. Supporting roll
6. Telescopic tube
7. Turret
8. Truck
9. Belt winding unit
10. Cover belt
11. Suction duct

Figure 4.18. The fiber plucking mechanism of the Blendomat BDT 019 Programmable Bale Opener. (Courtesy Truetzschler GMBH & Co.)

of the vertical apron movement a reciprocating comb, spiked or pinned roller, or smaller pinned apron act to strip larger portions of the tufts off the pins of this vertical apron. These stripped fibers are kicked back into the hopper reserve and mixing compartment and will repeat the process when again selected by the pins of the vertical apron. Small amounts of the original tufts remain in the pins of the apron and are doffed at the exit end of the hopper feeder. A graphic representation of a hopper feeder in which this stripping action to open fiber tufts occurs is provided in Figure 4.19.

Beating or Striking as an Opening Action

Beating and striking are similar actions but can be differentiated relative to the opening of fiber tufts or flocks. Beating best describes the separation of a fiber tuft that occurs between roller spikes or pins moving in opposite directions at the opening zone between two rollers. Fiber tufts are in a free-fall or loose state as this beating action takes place. An example of a machine that uses this type beating action is shown in Figure 4.20.

Striking appropriately describes the action of the spikes, blades, pins, or wire teeth of a rapidly revolving roller against a fringe of fibers held by feed rollers or by a feed roller and feed plate. An example of a machine with components that utilize striking as an opening action is presented in Figure 4.21.

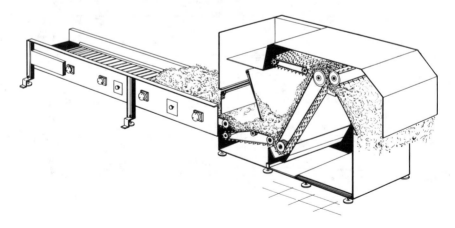

Figure 4.19. Illustration of a hopper feeder with stripping as an opening action taking place at the top of the vertical pinned apron. (Courtesy Fratelli Marzoli & C. spa)

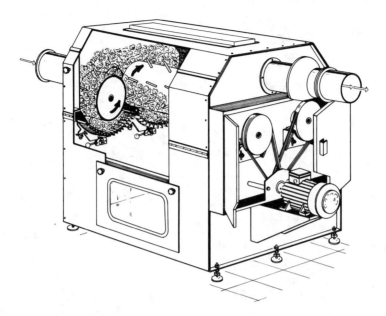

Figure 4.20. Illustration of the B31/1 Two-Beater Opener which utilizes a beating action to open fiber tufts between two revolving pinned rollers. (Courtesy Fratelli Marzoli & C. spa)

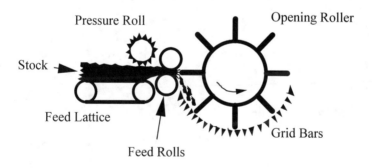

Figure 4.21. Illustration of striking as an opening action.

Raking as an Opening Action

By comparison to other fiber tuft opening actions, raking provides much less actual tuft separation. This action occurs as fiber tufts are pulled across stationary grids placed under the opener rollers. A common grid bar and opener roller configuration is shown in Figure 4.22.

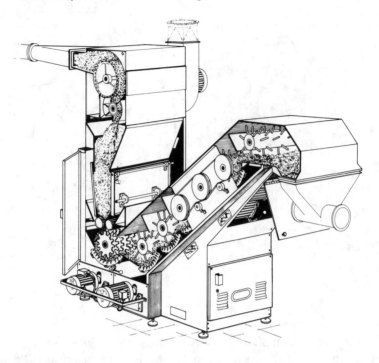

Figure 4.22. Illustration of the B51/1 Six-Step Cleaner with several opening actions including the raking of stock tufts across grid bars. (Courtesy Fratelli Marzoli & C. spa)

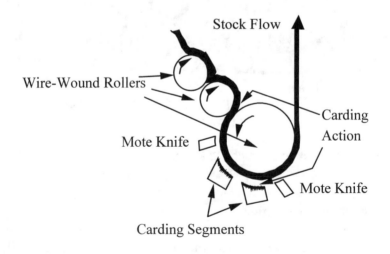

Figure 4.23. Illustration of machine components that card as an opening action between the wire-wound rollers and between the carding segments and cylinder.

Carding as an Opening Action

Some newer machines use wire-wound rollers that move fibers against stationary wire-covered plates, creating the classic textile processing carding action. Older machines opened stock between two revolving wire-covered cylinders as well. Figure 4.23 demonstrates the wire-wound rollers, carding plates, and opening zones of typical machines that utilize carding as a tuft opening action.

Opening Action Intensity Versus Machinery Decisions

Opening action intensity is an important consideration when choosing opener/cleaners and when deciding how to sequence these machines in the Opening Room. Less intense opening is generally done early in the process line or first in a machine with multiple opening zones. Lower intensity of opening action should nearly always precede opening action(s) of higher intensity. This creates the careful progressive opening and gradual tuft size reduction desired. The natural intensity associated with a specific mechanical opening action can also be magnified with the use of more aggressive opener rollers. Table 4.3 is an attempt to rank opening intensity by opening action. The same is done in Table 4.4 for opening intensity by popular opener roller type. These rankings are not to be construed as final but do offer general guidelines for decision making. Example drawings of various opener roller types are shown in Figure 4.24.

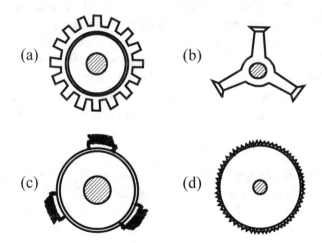

Figure 4.24. Illustrations of the (a) porcupine, (b) three-blade, (c) Kirschner three-lag, and (d) wire-wound opener/cleaner beaters.

Table 4.3. Subjective Rankings of Fiber Tuft Opening Intensity by Mechanical Opening Action

Opening Intensity	Mechanical Action
Most Intense	Carding
	Striking
↑	Beating
	Plucking
	Stripping
	Raking
Least Intense	Grabbing

Table 4.4. Subjective Rankings of Fiber Tuft Opening Intensity by Opener Roller Type

Opening Intensity	Opener Roller Type
Most Intense	Fine Saw-Tooth Wire
	Medium Saw-Tooth Wire
	Coarse Saw-Tooth Wire
↑	Needles
	Pins
	Partially Pinned
	Blades
	Short or Thin Fingers
Least Intense	10-cm to 15-cm Fingers

Openness Measurement

Studies at the Institute of Textile Technology have resulted in the development of a short staple fiber tuft openness tester. Specific volume of a fibrous sample is determined. The apparatus is a 4000-milliliter Pyrex beaker having an inside diameter of 152 mm and a plexiglass disc with added weight in the handle so that the disc totals 200 grams in weight. Air escape holes are drilled into the plexiglass disc, and the disc is machined to slightly less than the inside diameter of the beaker. The apparatus is illustrated in Figure 4.25. Typical process contributions to openness are provided in Table 4.5.

The test procedure is as follows:

1. Obtain random sample from processing point, being careful not to disturb relative openness of the specimen.
2. Fill cylindrical beaker as uniformly as possible with test stock.
3. Place 200-gram disc in cylinder and allow load to settle to its final position (approximately 5 seconds). NOTE: If disc

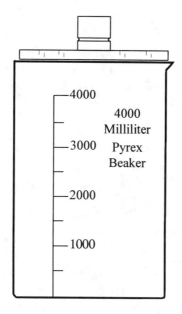

4000

4000
Milliliter

3000   Pyrex
Beaker

2000

1000

Figure 4.25. Drawing of the ITT Openness Tester apparatus.

settles at a sharp angle, carefully reposition the stock so the disc will settle evenly.

4. Record volume to which sample is compressed. See Figure 4.26.

5. Remove disc and test sample, and weigh and record sample weight.

6. Eight to ten replications are generally necessary per test location for 95 percent statistical confidence in the resulting data.

Table 4.5. Expected Process Contributions to Improved ITT Openness Index

Process Machine	Expected Change in Openness Index
Hopper Feeder	+40 to +50
Top Feeder	+80 to +85
Coarse Opener/Cleaner	+5 to +10
Fine Opener/Cleaner	+15 to +25
Intense Opener/Cleaner	+20 to +30
Transport Fan	±0
Condenser	±0
Cell Mixer	±0

| 1. Remove lid of apparatus and fill with fiber tufts taken from the sampling position. Fill beaker to top. | 2. Replace lid on full beaker and lower slowly by hand until the 200-gram lid is supported by the stock. | 3. Read off the volume in the beaker using the scale printed on the outside. If the lid rests in an unlevel position, read at center point of lid or reposition the stock so lid will stay level. |

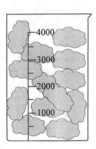 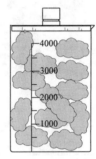 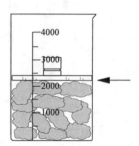

Figure 4.26. Illustrations and initial steps for determining fiber tuft specimen volume which is converted to specific volume and then to the ITT Openness Index.

7. Calculate for each test sample the ITT Openness Index as follows in this equation:[4.4]

$$\text{Openness Index} = \frac{\text{volume of stock (in ml)}}{\text{weight of stock (in grams)}} \times \text{specific gravity of fiber} \quad (4.4)$$

8. Average the results of the eight to ten replications to obtain the Openness Index for stock from each respective sampling point.

Specific Gravity of:

Cotton	=	1.54
Rayon	=	1.51
Polyester	=	1.38
Acrylic	=	1.17
Polypropylene	=	0.92

NOTE: For an intimately blended stock, use the formula[4.5] below to calculate approximate Specific Gravity of the blend:

Blend specific gravity = (specific gravity of 1st component × decimal % in blend) + (specific gravity of 2nd component × decimal % in blend)   (4.5)

Process Influences on Machine Opening Effectiveness

The opening effectiveness of a fiber processing machine is influenced by several factors. Increasing throughput kg per hour tends to reduce opening effectiveness and vice versa. Therefore, increased run time generally improves openness of the delivered stock. More intensive mechanical action and more intensive opener roller types result in greater machine contribution to stock openness. Excessive moisture in the stock will tend to reduce opening effectiveness as will excessive air velocity used for transporting the stock through the air ducts. Improper use of air through machines can also influence fiber tuft opening. Finally, the openness of stock fed to a machine will influence the degree of openness improvement through that machine.

Blending and Mixing of a Single Fiber Type

Practical Definitions of Blending and Mixing

Before further discussion, definitions of how the terms "blending" and "mixing" are and should be used will be helpful. Blending and mixing are often used interchangeably, but really do not mean the same thing. Blending of a single fiber type has as its basis the exact measurement of important fiber properties and scientific proportioning and combining of those fibers. Proper blending of fibers in this case contributes to predictable and reproducible yarn physical properties. Effective blending of a single fiber type is then the condition of finding each fiber property in a cross section of yarn in proportion to its presence in the mix, and with variations of these proportions only as a consequence of random selection.

Blending also refers to the combining of desired proportions of dissimilar fibers on the basis of weight. Before this weight-based combining of different fiber types is done, each individual fiber type must be blended according to the definition above. If dissimilar fibers are blended by weight in the Opening Room, it is called intimate blending. When accomplished at a later drawing process, this weight-based combining of different fiber types is called draw frame blending. It is common to call either of these ultimate fiber assemblies the "blend" or "blended stock" regardless of whether intimate blending or draw frame blending is used.

Mixing is more mechanical than scientific, although the mixing power of a machine can now be expressed with a number. To complicate matters this number is referred to as a Blend Factor. Mixing is a means to get to good blending. The contribution of a machine to the homogenization of the tufts, so that fibers from each bale are uniformly dispersed throughout the mix, is best labeled as mixing. A synonym of mixing as used in some other parts of the world is the term "folding".

Reasons for Blending of a Single Fiber Type

Effective measurement-based blending is paramount in cotton processing because of the wide variations in the cotton fiber. These variations result from differences in varieties, soils, rainfall, irrigation practices, fertilizers, chemical treatments, temperatures, insect damage, growth seasons, and harvesting and ginning methods. Variations in synthetics may not be as pronounced as those in cotton, yet still exist and can definitely affect process and product performance. These variations include differences in fiber length distribution, fiber denier, crimp, finish level, and even bale density.

Fluctuations in the fiber properties of a mix being processed can lead to erratic processing performance, variable yarn quality, the sudden appearance or continuous high levels of yarn defects, and faulty dyeings. Scientific blending should be considered the most important single operation in cotton yarn manufacturing. It also receives top priority in the processing of synthetics. Whatever the raw material, blending is one of the essential means by which the yarn manufacturer creates and reproduces a functional end product.

Mixing Power or Blend Factor of Machines

At one time there was popular advice that the more the number of bales in the laydown, the better the chances for uniform yarn properties. That is absolutely not good advice unless the mixing capability of machines in the line allows for additional bales in the laydown. To assess the match between number of bales in the laydown and the mixing machinery, a numerical expression is needed for approximating the mixing power of a single machine and for a sequence of machines. Such a number exists for specific mixing machines, and a simple mathematical treatment of these individual machine numbers can yield a number that approximates the mixing power of the complete installed line of machines.

Called "Blend Factor", the numerical expression of the mixing power of an Opening Room machine was first revealed to the staff of the Institute of Textile Technology in 1990 in an article written by Ferdinand Leifeld of Truetzschler GMBH & Company. Table 4.6 contains Blend Factor numbers for typical fiber mixing machines, as utilized by ITT yarn manufacturing researchers.

The total mixing capability of a line of sequential machines is expressed as the product obtained by multiplying together the Blend Factor numbers for each machine in the line. Opener/cleaners or other machines that do not mix stock are assigned a Blend Factor of one. Observe the example in Figure 4.27. The significance of the calculated number "36" is that this example line has a mixing capability that would allow each of 36 bales in a lay-

Table 4.6. Blend Factor Numbers for Various Fiber Mixing Machines

Machine Type or Name	Assigned Blend Factor #
Hopper Feeder:	
61-cm or 2-ft width, volume 0.68 m³	1.33
91-cm or 3-ft width, volume 1.02 m³	2
91-cm or 3-ft width, volume 1.53 m³	3
91-cm or 3-ft width, volume 1.70 m³	4
122-cm or 4-ft width, volume 2.04 m³	5
Top Feeder:	
Average number bales fed from instantaneously equals Blend Factor	1–8
Cell Mixers, Compartments Fed Sequentially:	
4 compartments or chambers	4
6 compartments	6
8 compartments	8
10 compartments, etc.	10
Hollingsworth Laydown Cross Blender has a Blend Factor equal to the number of bales fed per pass across the LCB	Usually 18+
Rieter Unimix	
6 Chambers x 3 for Apron Mix Section	18
Fiber Controls Big BINN:	
91-cm or 3-ft width	3
122-cm or 4-ft width	4
152-cm or 5-ft width	5

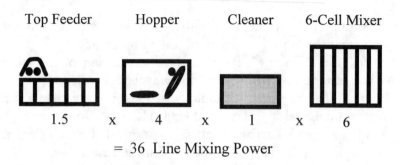

Top Feeder	Hopper	Cleaner	6-Cell Mixer

1.5   x   4   x   1   x   6

= 36 Line Mixing Power

Figure 4.27. Calculation of line mixing power (Blend Factor) for machines in tandem.

down to be represented once in a cross section of material taken after the final mixing point. Therefore, such a sequence of machinery would not be suitable for more than 36 bales in the laydown. The line would be most appropriate for 18 bales, which would theoretically allow each bale to be represented twice in the cross section of material taken after the final mixing point. This provides for a margin for error that is not a customary practice in textile operations, but is advisable because it ensures complete confidence that adequate mixing will take place in support of the necessary scientific blending plan. The need for such a margin for error is negated if the mini-laydown concept is used. This concept is discussed earlier in this chapter in the section titled "Cotton Laydown Bale Positioning".

Observe another example opening line in Figure 4.28. Note that the total mixing capability is calculated by:

1. Determining the numerical representation of mixing capability for each sequence of machines; and
2. Adding together the numbers for mixing capability of these parallel sets of machines.

The point of this example is that Blend Factors are multiplicative for machines in sequence and are additive for machines in parallel at the point where the stock is first combined into one stream. This must be kept in mind when determining system mixing capability because there are numerous opening line machine configurations used in the textile industry. Diagrams of typical mixing machines may be reviewed in Chapter 3. Hopper Feeders

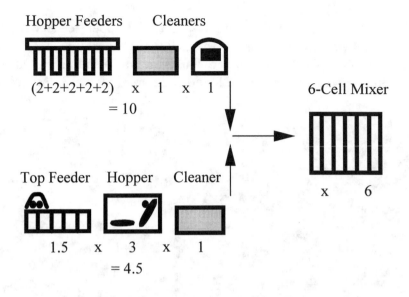

Hopper Feeders   Cleaners

(2+2+2+2+2)   x   1   x   1   6-Cell Mixer
        = 10

Top Feeder   Hopper   Cleaner

    1.5   x   3   x   1   x   6
            = 4.5

(10 + 4.5) x 6 = 87 Line Mixing Power

Figure 4.28. Calculation of line mixing power (Blend Factor) for a combination of machines, some in series and some in tandem.

are shown in Figures 3.4 and 3.5. Figures 3.16 through 3.20 demonstrate other popular time-delay mixers.

Intimate Blending of Different Fiber Types

The direct or indirect weighing out of dissimilar fiber types to create a blend is referred to as intimate blending when this is accomplished in the Opening Room. Intimate blending of two different fibers, polyester and cotton as an example, requires additional machinery for the opening lines. The polyester is fed from its bale laydown, with adequate mixing of fibers from the different bales, and is opened sufficiently to nearly match the measurable openness index of the cleaned cotton with which it will be intimately blended. The cotton will at the same time have been fed from its bale laydown, progressively opened, cleaned, and mixed together in separate machines from those processing the polyester. The well-mixed cotton and polyester fibers will be brought together at the weigh blending stage. There are three primary machine systems available for this weight-based blending of dissimilar fibers.

The most predominate system is a set of hopper feeders, each equipped with weigh pans on the delivery side. Figure 4.29 is a photograph of one such hopper. This intimate blending system is also illustrated in Figures 3.21 and 3.22 in Chapter 3. Adjacent

Figure 4.29. Photograph of a weigh-pan hopper feeder. (Courtesy Crosrol Ltd.)

hoppers are typically fed with a different fiber. The number of hoppers for each fiber type and the weight of the fiber charge deposited in each pan combine to determine the blend ratio of the two fibers in the mix. The sandwich mixing method is again used on the feed table to begin the process of homogenizing the weighed out tufts of dissimilar fibers.

A second weight-based blending system utilizes one tuft-weighing machine for each fiber type. Figure 4.30 is an example of this type intimate blending system. Also, see Figure 3.23. Weighed out tufts from each machine are combined in the air stream which transports the delivered stock to the next process.

Another innovative intimate blending system was introduced in the USA in 1995. A schematic drawing of this system is provided in Figure 3.24 in Chapter 3. Using vertical chambers that are capable of delivering from 3 to 300 kg (6.6 to 660 lb) per hour each, it is possible to produce very low percentage blends of dissimilar fibers. For example, with 4 chambers it is possible to produce a one percent blend, popular in some specialty yarns. The thickness of the fiber mat delivered from each chamber is measured by a pressure sensor, and delivery speeds can be adjusted in milliseconds. The measured pressure will have been calibrated to the cor-

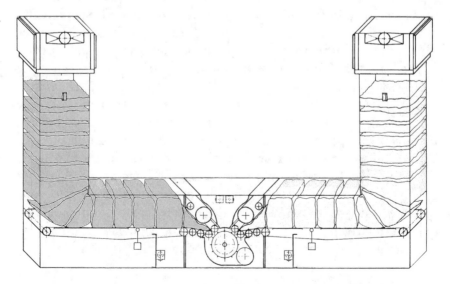

Figure 4.30. Illustration of the intimate blending of two different fibers with the MasterBlend system. (Courtesy John D. Hollingsworth on Wheels, Inc.)

responding weight per unit volume of tufts delivered. This machine can utilize up to 8 chambers and can process more than one blend at a time while feeding up to 4 lines of carding machines.

Subsequent to the weight-based blending of dissimilar fibers, homogenizing or mixing is begun using cell mixers or hoppers or both. Really effective homogeneous mixing of similar or dissimilar fibers occurs first at the downstream carding process.

Different fiber types can also be blended by weight in sliver form at the drawing processes. This is naturally called draw frame blending. Intimate blending in the Opening Room is generally preferred where enhanced dye shade consistency of resulting yarns and fabrics is a must. This is because the intimately blended fibers are generally more homogeneously mixed than are the draw frame blended fibers. Even so, there are many products and end uses where a draw frame blend is quite sufficient and would be the technique used, primarily for reasons of flexibility and lower total cost.

Cotton Cleaning

Early Research Contributions

The Shirley Institute in England pioneered research on the opening and cleaning of cotton in the early to mid 1930s. This research revealed the need for more effective machines for both opening and cleaning of cotton. It was this research that first linked the degree of opening to the resultant degree of cleaning. It was further learned that accomplishing the necessary opening and cleaning with a few machines was preferred over the use of longer lines of machines. Other important findings of this research included the conclusions that repetition of the same opener/cleaner machine in the line is comparatively ineffective and that significant cleaning of cotton must occur prior to the carding machine. These early investigations stimulated ongoing research in the opening and cleaning processes and have influenced the design of successive generations of more effective opener/cleaners. Machines used predominantly or that are relatively new on the market in the 1990s are the focus of the machinery references that are included in this section.

The Importance of Cotton Cleaning

Much research has been done and many reports have been written documenting the value of cleaning cotton to very low levels of dust and visible impurities. These visible impurities are called trash by most cotton yarn spinners. Research at ITV Denkendorf and at the Institute of Textile Technology has produced the data shown in Figures 4.31 through 4.33 which validate the need for effective cotton cleaning. As remaining trash levels in the material increase, downstream process stops and resultant prod-

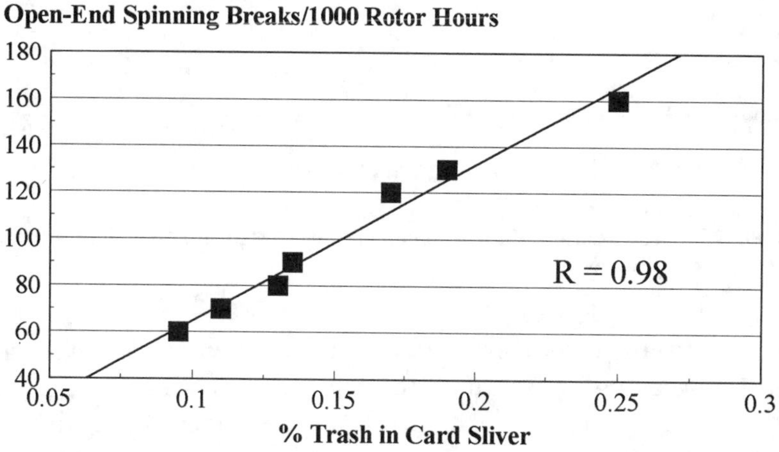

Open-End Spinning Breaks/1000 Rotor Hours

$R = 0.98$

% Trash in Card Sliver

Figure 4.31. The relationship between percent trash in card sliver and the number of spinning breaks per 1000 rotor hours. *Note:* From "Trash Content of Card Slivers Related to the Frequency of Broken Ends in Rotor Spinning." by Artzt, Azarschab, & Maidel, Textil-praxis International, 45(11), p. 1146.

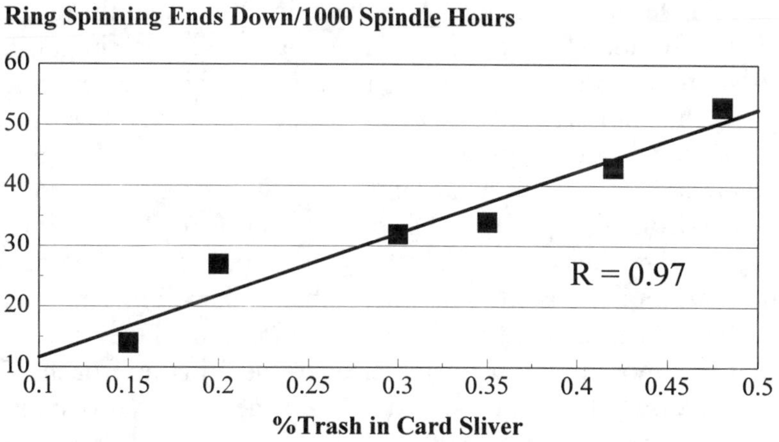

Ring Spinning Ends Down/1000 Spindle Hours

$R = 0.97$

%Trash in Card Sliver

Figure 4.32. The relationship between percent trash in card sliver and the number of ends down at ring spinning.

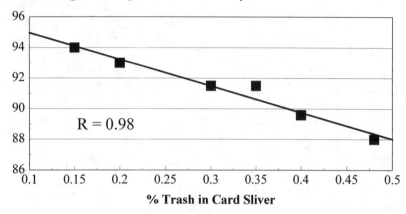

Figure 4.33. The relationship between percent trash in card sliver and yarn performance at weaving.

uct quality are negatively affected. Ends down at spinning and stops at weaving or knitting are major cost drivers. The expense and effort to properly clean cotton must be balanced against the performance of resultant yarns at these and other high cost operations.

Mechanisms of Trash Removal

There are basically four forces used to separate or dislodge trash particles from the cotton stream being processed through the Opening Room. These are impact, gravitational, centrifugal, and aerodynamic forces which are applied in combination in cotton cleaning machines. Opening actions and intensity influence the cleaning effectiveness of a machine. Insufficient stock opening means less cleaning will be accomplished. The objective is to gently but thoroughly open the fiber tufts so that foreign particles can be exposed and dislodged. When the more intense carding action is used to open the stock, fiber-to-fiber separation can be approached which allows for aggressive cleaning of the cotton. Several of the most modern opener/cleaners use this approach. The forces necessary to clean the cotton are generated by the interaction of machine components including beaters, grid bars, mote knives, and by air currents that are produced either internal or external to the machine.

Use of impact forces in combination with gravity to facilitate trash removal is common in coarse cleaners which are placed early in the line and function best with larger tufts of fibers. Please refer to Figures 3.9 and 3.10 in Chapter 3 and to Figure 4.20. The revolving beater spikes move the larger tufts across the grid bars, impacting the tufts repeatedly and quickly against successive grids. Exposed trash particles, particularly those that are larger and heavier, are dislodged by this action and fall through the openings between the grid bars. A modest amount of centrifugal force may also be at work here, but gravity is the main force that moves the separated trash away from the cotton stream in cleaners using these type components.

Centrifugal and aerodynamic forces act on well-opened fiber streams and the exposed trash particles. Please refer to Figures 3.13, 3.14, and 3.15 in Chapter 3 and to Figure 4.23. The carding roller and mote knife arrangement interact to generate these forces for extracting both large and small trash particles. As the fibers, along with attached or trapped trash particles, are accelerated and moved very rapidly by the revolving carding roller, a significant air current is produced. The tip of the mote knife, which extends the full length of the carding roller, causes the air speed to increase considerably through the space between the knife and the roller. This serves to keep most of the fibers on the wire teeth of the roller as the foreign matter is propelled by centrifugal force to the underside of the mote knife. It is possible for the air current under the mote knife to pass across the knife and rejoin the faster moving air above the knife. Unattended to, this will allow trash and dust particles that were dislodged to recombine with the main cotton stream. Use of a collecting shroud and vacuuming away the waste beneath the mote knife is the most popular solution applied in modern cleaners using these components.

How Much Waste to Remove

All opener/cleaners extract fibers, also called lint, along with the foreign matter. The composite of trash, dust, fiber fragments, and fibers removed is called waste. Depending on machine design, on the amount of waste removed, and on other factors, the fiber component of opener/cleaner waste can range from well below 40 percent to more than 70 percent by weight. The average today is about 50 percent fiber in these wastes. More fiber in the waste makes the waste more costly. This adds an economic consideration to the technical factors that help determine how to set up opener/cleaners and how much waste to remove. The economic concern is best addressed by understanding the costs that will be added in downstream processes with failure to effectively remove all foreign matter from the cotton. When such a total system review of potential cost is undertaken, a decision to thoroughly and completely clean the cotton will nearly always be the profitable decision. The target Opening Room contribution to thorough cotton cleaning would be the removal of 50 to 70 percent of the impurities by weight in the cotton bales. Only 30 to 40 percent cleaning effectiveness is achieved in the Opening Rooms of most USA cotton yarn plants with older generation machines. More modern machines can usually be easily and economically justified when total cost from bale through the ultimate product is considered.

There is an easy way to estimate the amount of waste that must be removed in the Opening Room in order to achieve a prescribed level of cleaning of the cotton. This equation[4.6] can be used for such an estimate:

$$\% \text{ waste needed} = \frac{\text{average \% trash in bales} \times \text{\% trash to be removed}}{\text{average \% trash in waste}} \quad (4.6)$$

If there is 2.50 percent visible trash in the cotton bales and it is desired to remove 50 percent of this trash prior to carding and if the wastes contain 50 percent trash on the average, then the calculation is as follows:

$$\% \text{ waste needed} = \frac{2.5\% \times 50.0\%}{50.0\%} = 2.5\%$$

If more intense Opening Room cleaning were desired, say 70 percent cleaning effectiveness, then the estimate of the percent waste to be removed would be as follows:

$$\% \text{ waste needed} = \frac{2.5\% \times 70.0\%}{50\%} = 3.5\%$$

This same approach can be used to estimate the amount of waste that must be removed at the later carding process. A reason for such an estimate is to provide a starting point goal for machine setup and for discussion with the often hard-to-convince cost department that significant waste must be removed in order to achieve effective cotton cleaning.

Percent Waste Removal Calculation

The amount of waste removed at each cotton cleaning stage must be known for proper Opening Room evaluation. This information is used in conjunction with lint and trash content and fiber length distribution data for waste and for the stock in process for proper assessment of waste extraction effectiveness.

Percent waste is calculated with this equation:[4.7]

$$\% \text{ waste} = \frac{\text{waste removed, kg}}{\text{total stock processed, kg}} \times 100 \quad (4.7)$$

Analyzing Waste Constituencies

For best assessment of waste extraction effectiveness, the constituencies of the waste must also be determined. Additionally, the length distribution of the fibers or lint in the waste must be known. Viewing the content of cotton processing waste from a purely general and theoretical basis, the machine waste should be mostly hard impurities (leaf, stem, sand, etc.) and partially short undesirable fibers and neps. With the opener/cleaners of current design, this ideal waste is only possible at extremely small percentage removal. In truth an ideal waste content which coincides with effective trash removal is not yet practical. The optimum waste constituencies for each opener/cleaner will be a response to material, cost, quality, processing, and end-product considerations.

Components of cotton opener/cleaner wastes are most often determined by the referee procedure known as the Shirley Analysis Non-Lint Test Method. This test and others are briefly summarized in Chapter 12 of this textbook. Detailed description of the Shirley Analyzer test is available in the standards publication of the American Society for Testing and Materials (ASTM) as standard number D2812. Results of the Shirley Analyzer test are percent lint, percent trash, and percent cage loss of the test sample. Percent trash is also referred to as visible trash, visible loss, or visible foreign matter. Cage loss is often called invisible loss, and is composed of mostly dust and ultra-short fibers similar to peach fuzz which escape through the perforated cylindrical screen of the Shirley Analyzer. The sum of trash and cage loss is known as the non-lint content of a test specimen.

Cleaned lint from a Shirley Analysis can be tested for fiber length distribution, realizing that there is modest fiber breakage during the Shirley Analyzer testing process. The Suter-Webb Array technique (ASTM D1440) is still the referee method. Other popular instruments used for characterizing fiber length distribution are the Zellweger Uster Almeter (ASTM D5332) or their AFIS instrument with L & D (length and diameter) module. The AFIS, which stands for Advanced Fiber Information System, has no ASTM designation as yet, but is a very widely used instrument for testing of multiple fiber characteristics.

Short fiber content is the percent of fibers in a specimen that are less than 12.7 mm (1/2 inch) in length. This number by itself does not mean much. Even though the short fiber content of a waste sample is a relatively high percentage, this does not always indicate efficient short fiber removal from the stock in process. The short fiber content of entering and exiting stocks must also be determined before the quality of waste can be adequately assessed. For example, if the cotton fed to a cleaner contains 12 percent short fiber and the delivered cotton contains 20 percent short fiber, the short fiber content in the waste provides little information regarding extraction effectiveness. Such a set of data would indicate that the machine is breaking fibers, and this would first be rectified before attempts at the optimization of machine waste content.

The amount of longer spinnable fibers in the waste is useful information. With most opener/cleaners, the higher the waste percentage removed, the higher the spinnable fiber content in the waste. Documenting the amount of longer fibers in the waste also allows for more accurate calculations of waste value. It may make economic sense to reclaim the spinnable fibers from opener/cleaner wastes and to reuse them in the same or other fiber mix. Many cotton yarn production facilities are increasing net fiber yield by reclaiming and reprocessing fibers from wastes. With this practice, the amount of waste to be removed is no longer as constrained by the system economic considerations which may be difficult to calculate.

Cotton Cleaning Efficiency Calculation

The effectiveness of a machine or series of machines at removing visible trash from cotton stock is expressed mathematically as cleaning efficiency. This can be determined by the following equation:[4.8]

$$\% \text{ cleaning efficiency} = \frac{\% \text{ trash in stock fed} - \% \text{ trash in stock delivered}}{\% \text{ trash in stock fed}} \times 100 \quad (4.8)$$

Another equation[4.9] can also be used to calculate percent cleaning efficiency:

$$\% \text{ cleaning efficiency} = \frac{\% \text{ waste removed} \times \% \text{ trash in waste}}{\% \text{ trash in stock fed}} \quad (4.9)$$

Both equations require the numerator to be a representation of the amount of trash removed by the machine or sequence of machines being evaluated. A difficulty is in knowing the numbers in these equations with accuracy. With the Shirley Analyzer, averaging the resultant test data from 3 cotton waste samples allows a confidence interval of about ± 0.5 percentage points around that average at a 95 percent probability level. Increasing the number of samples to 9 and averaging the test results improves the accuracy of the average, but only to ±0.3 percentage points. This means that very much testing must be done in order to have confidence in the resulting data. The reasons for this are the natural wide variations in the trash content among cotton tufts, inherent variations in testing procedures and instruments, and inconsistencies in sampling and sample preparation.

Sampling to Estimate Cleaning Efficiency

Because of the chances for inaccuracy of trash content data, care must be taken in the use of these data, regardless of test instrument type used. At best, an estimation of cleaning efficiency can be derived using trash content data from tests on samples of stock entering and exiting a machine or sequence of machines. Test instruments other than the Shirley Analyzer are more popular today for non-lint testing of in-process cotton, although the Shirley Analysis is the recognized referee method. This is because the tests on the more modern instruments are done on specimens of a few grams, whereas the Shirley Analysis requires 100 grams per specimen. These newer and more rapid testing instruments include the AFIS-T (trash module), MTM (Microdust and Trash Monitor, becoming obsolete at this time), and the MDTA (Microdust and Trash Analyzer), all offered by Zellweger Uster. The following data explain this procedure:

1. Stock fed to a machine tests at 1.20 percent visible trash content.
2. Stock exiting that machine tests at 0.90 percent trash content.
3. Cleaning efficiency % $= \dfrac{1.20\% - 0.90\%}{1.20\%} \times 100 = 25\%$.

This method provides a reasonable estimate of cleaning efficiency *only* when numerous samples of stock are tested for a somewhat reliable average of the percent trash component. A current best recommendation would be to take 10 samples entering and 10 samples exiting the process over a half-hour period with before and after samples taken simultaneously if possible. Less sampling will make the resulting calculations a less dependable estimate. It is the tendency in practice for one-shot sampling before and after processes for non-lint analysis. This sometimes results in negative cleaning efficiency calculations which is not a practical possibility if trash is in fact present in the waste. The need for adequate sample sizes can not be overstressed when the resulting testing data are to be used for assessing machine or process line performance.

Sampling for a More Precise Estimate of Cleaning Efficiency

A more precise method for determining cleaning efficiency is called the trash reduction approach. The starting point is to perform enough non-lint tests on the bale (or other starting point stock) so that the average trash content is known with reasonable certainty. The equation[4.1] shown in the earlier discussion on laydown size should be used to determine the sample size required for estimating the average so that 95 percent of the time it is known within 10 percent of the true average. This will likely be accomplished if every bale is sampled or if 25 samples of mixed stock are taken at a single point over a 30-minute period. Each sample will be divided into several specimens for the non-lint testing. This is a lot of work and is one reason that sampling is not generally done properly for this type analysis.

Once the starting point stock trash content is known as described above, the next testing step is to perform the Shirley Analysis on the machine wastes. The Shirley Analyzer is still the best instrument for waste analysis at this writing. Test results on machine waste are much more reliable than are the same test data on stock in process. This is because a 100-gram sample of waste represents many kg of stock processed through the machine. The actual number of kg represented depends on the percent waste removed at the specific machine. For example, a 100-gram sample of waste from a machine removing only 0.10 percent represents 100 kg (220 lb) of stock processed. For a machine removing one percent waste, the 100-gram waste sample represents 10 kg (22 lb) of stock processed. Usually two 100-gram samples constitute a standard test set for any Shirley Analysis, thereby doubling the kg of stock represented by the test results. It is more statistically sound to analyze waste because less sampling must be performed for there to be confidence in the resulting data.

Another key measurement in this more precise approach for determining cleaning efficiency is the actual percent waste removed by the machine or machines being analyzed. This is best determined as the waste to be tested is being generated by the machine(s). A detailed example of how to perform this entire experiment in a modern cotton-containing yarn manufacturing plant is presented in a later section of this chapter that is titled "Example Plan for Effectively Analyzing a Modern Cotton Opening Line". At this point, the data calculations are reviewed as follows:

1. Percent trash in bale laydown tests at 1.5 percent on average from 40 bales sampled.
2. Opener/cleaner (O/C) No. 1 removes 0.6 percent waste and this waste is composed of 50.0 percent trash.
3. Opener/cleaner No. 2 removes 2.0 percent waste and this waste is composed of 30.0 percent trash.
4. Percent Cleaning Efficiency (CE) of opener/cleaner No. 1 is

$$\% \text{ CE} = \frac{(0.6\% \times 50.0\%)}{1.5\%} = 20.0\%$$

5. Percent trash fed to opener/cleaner No. 2 is in this case the same as percent trash in stock exiting opener/cleaner No. 1;

$$\% \text{ trash in stock fed to No. 2 O / C} = 1.5\% - \frac{(0.6\% \times 50.0\%)}{100} = 1.2\%.$$

6. Percent CE of opener/cleaner No. 2 is

$$\% \text{ CE} = \frac{(2.0\% \times 30.0\%)}{1.2\%} = 50.0\%.$$

7. Percent trash in stock exiting opener/cleaner No. 2 is

$$\% \text{ trash remaining} = 1.2 - \frac{(2.0\% \times 30.0\%)}{100} = 0.6\%.$$

8. Percent CE of the 2-machine line is

$$\% \text{ CE of line} = \frac{(1.5\% - 0.6\%)}{1.5\%} \times 100 = 60\%.$$

Notice that the cleaning efficiency of the line is 60.0 percent which is not the sum of the individual machine cleaning efficiencies. Cleaning efficiencies are not additive for calculating the effectiveness of a sequence of machines. Beginning stock trash content minus ending stock trash content, expressed as a percentage of beginning trash content represents the cleaning efficiency of a line of machines.

The above approach is called the trash reduction method because the trash fed to subsequent machines is based on subtracting out the trash that is removed at the immediately preceding cleaning process. And the data are more reliable because of the reasons explained above regarding the amount of stock processed represented by a waste sample. With this approach for determining cleaning efficiencies, only one point at the start requires the numerous samples to be taken and tested for non-lint content. This is the method that should be used from time to time to assess the trash removal effectiveness of the cotton opener/cleaners in an Opening Room. Important times for such a thorough study are after cotton crop changes or laydown average property shifts, and after significant changes in machines such as replacement, resequencing, major overhauls, resetting, or production rate adjustments.

Process Influences on Machine Cleaning Effectiveness

The effectiveness of an opener/cleaner in removing trash from cotton stock is influenced by many factors which include the following:

1. Amount of trash in stock fed,
2. Openness of stock fed,
3. Moisture content of stock fed,
4. Instantaneous stock throughput rate,
5. Percent run time of the process,
6. Percent waste removed,
7. Waste constituency,
8. Transport air flow,
9. Machine maintenance,
10. Machine setup, and
11. Machine design.

More trash in the stock fed is generally associated with higher machine cleaning efficiencies. A more opened stock usually results in more trash extracted. Cotton trash that is not too wet is easier to remove. Decreased instantaneous stock production rate through a machine tends to increase both waste percentage and trash removal effectiveness. Higher percent run time means lower instantaneous throughput rates and improved stock opening, both of which contribute to better trash extraction. More waste removed generally means more trash extracted, all else being the same for a given machine. Of course, the greater the percent trash in the waste for a given percent waste removed, the greater the cleaning efficiency. Air flow, where applicable, that is engineered so as not to reintroduce extracted particles into the main stock flow allows for better cleaning effectiveness. Finally, a machine with a quality maintenance condition and set up optimally will clean cotton to its design capabilities if all other factors discussed are also optimized.

Dust Removal in the Opening Room

The removal of dust particles from cotton is also desired. This is accomplished to some insignificant degree as waste is extracted from the various opener/cleaners. Dust removal is more significant with machines and transport devices designed with this intention. Perforated screen condensers are effective at dust removal as are so-called dedusters that utilize exhaust air to removed dust that is released from the cotton stock by the action of tufts being opened and impacted by machine components. All the necessary dust removal is not accomplished in the Opening Room, especially for cotton to be spun on the rotor system. Additional dust removal is offered by the carding machine, the optional comber, and by the air circulation and vacuum systems of the draw frames.

Assessing and Minimizing Fiber Damage in Opening Room Processes

An objective of the Opening Room should be the most gentle handling of fibers possible while accomplishing the other desired objectives of feeding, opening, mixing, cleaning, and stock transport. Improper handling of stock by Opening Room machinery can lead to excessive fiber damage as measured by neppiness increase and by the increase in short fiber content in the stock processed. It is the very nature of nearly every opener/cleaner to increase stock neps and short fiber content to some degree. This is because the fibers, especially cotton, are individually quite fragile and with any degree of handling and mechanical manipulation are susceptible to damage. As discussed above with reference to sampling of stock in process before and after machines for non-lint testing, an appropriate number of samples must also be taken for nep and fiber length testing. Again, 10 samples taken before and 10 taken after a machine over a 30-minute time period is the recommended practice. Take a sample before and after the machine simultaneously if possible. Repeat this 10 times over the half-hour sampling period. The most popular tester today for nep counting in fibrous stock is the AFIS-N (nep module), sold by Zellweger Uster. Nep generation from the cotton bale stock through the preparation of the mat feeding the card is normally between a 50 and 100 percent increase over the nep level in the bales. This increase is substantially less for synthetics.

Nep increase percentages are calculated with the following equation:[4.10]

$$\% \text{ nep generation} = \frac{\text{neps / gram in stock delivered} - \text{neps / gram in stock fed}}{\text{neps / gram in stock fed}} \times 100 \quad (4.10)$$

Cotton short fiber content will also increase with Opening Room processing. A 12 percent short fiber content (SFC) in the bale will typically become 15 percent in the card mat. Again, with synthetics this SFC increase is substantially less. The AFIS has fiber length modules for both cotton and synthetic fibers. Also, the Almeter (AL-101) is often used for characterizing fiber length distribution. Read more about these and other tests in Chapter 12 of this textbook.

Fiber damage potential in the Opening Room is influenced by the following:

1. Fiber fineness or denier,
2. Presence of immature cotton fibers,
3. Openness of stock,
4. Compatibility of tuft size and machine action,
5. Stock instantaneous throughput rate,
6. Intensity of opening action,
7. Moisture content of stock,
8. Transport air velocity, and
9. Condition of fiber contact surfaces.

Finer fibers are generally more prone to nep creation and to fiber breakage than are coarser fibers. With cottons there is always some percentage of immature or not fully developed fibers in each bale. This percentage tends to be higher with finer fibers. Immature cotton fibers have deficient cell wall development and much lower than normal cellulose content. These undeveloped walls collapse around the center or lumen of the fiber and are easily entangled with other fibers to form neps. Immature fibers also will break more readily than will mature cotton fibers. Poorly opened stock fed to some machines will exacerbate nep generation while very small tufts fed to some machines will tend to roll and become stringy with increased neppiness. Therefore, stock openness and machine opening action and intensity must be compatible.

Increased instantaneous throughput rates may cause fiber damage, especially in machines that use feed rolls to meter fibers into a beating zone. The more intense the opening action, the more likely there will be increased nep and short fiber creation. Cotton fibers that are too dry in processing can become brittle and will be more easily broken. Transport air velocity that is more than is needed can introduce additional turbulence in material transport ducts. This turbulence can roll and twist fiber tufts and cause an increase in stock neppiness. Finally, any burred or damaged surface that the fiber tufts contact can lead to fiber damage as well.

Example Plan for Effectively Analyzing a Modern Cotton Opening Line

Generating reliable data from which to characterize the performance of a modern cotton Opening Room can be difficult. Factors which make it difficult include the following:

1. Several people are required.
2. Several man-hours are required.
3. Stock is not always at the same level in reserve chambers.
4. Stock in process is enclosed in machines and in air transport ducts that are connected for continuous stock flow.
5. Waste compartments are generally vacuumed clean automatically every few minutes.
6. Card production must be the basis for stock throughput calculations.
7. It is a somewhat dirty job.

A practical approach for correctly measuring certain machine operating characteristics and for collecting in-process and waste samples is described in the steps that follow:

1. Ensure that the current bale laydown will last for at least several hours.
2. Attempt to do this study with all or nearly all cards running that are supported by the opening line being studied.
3. Note the exact level of stock in any reserves or mixers. The ideal would be for the levels to be the same at the beginning and at the end of the study period. Otherwise, production rates, waste percentages, and machine run times can be miscalculated, especially for those machines prior to such a reserve.
4. Disconnect or block off air suction of machine waste compartments. While going to this effort to study the Opening Room, it is necessary that the cards be studied as well. The amount of waste removed at carding must be known in

order to add this back to sliver production so that the amount of stock fed to carding can be calculated.

5. Clean out thoroughly all waste compartments of the machines to be studied. This will in most cases require machine stoppage for safe access to the waste compartments.

6. Production through the Opening Room machines will be determined with card sliver production as the starting point for calculations. For best accuracy begin the test period with a weighed but empty can under each card coiler in the test lines. At the conclusion of the test period, these cans will be doffed and weighed again. Sliver pounds produced during the test will be the difference between the doffed can weight and the beginning empty can or tare weight. A reasonable estimate of the amount of sliver produced can be made using the average production rate of the cards and the total carding machinery run time during the test period. This means that one person will be needed to monitor and record all card downtime during the test period. Cards to be monitored are all those supplied by the opening line being studied. This monitoring must be done even with the more accurate weighing approach described above so that carding production demands at 100 percent efficiency can be calculated.

7. Begin the study period, at least one hour in length, with all cards running into empty cans (if applicable as described) and with an associate monitoring card downtime for the hour.

8. During the hour of study one associate should sample each bale in the laydown and place each sample in a separate and appropriately labeled container, perhaps a quart size plastic freezer bag. If the first machine after the top feeder is a stock reserve or mixer in which no waste is removed, then the sampling of each individual bale can be substituted by the sampling of stock exiting this initial reserve or mixer. Take samples once per minute until at least 25 samples are collected. Again, keep each sample separate for the required laboratory tests. These include non-lint, fiber length, and nep count tests at a minimum.

9. Samples before and after opener/cleaners should also be taken during the test period. Collect a sample of stock en-

tering the machine at the same time an associate is collecting a sample of stock exiting the machine. Repeat this every 2 to 3 minutes until at least 10 samples per location have been collected. Keep each sample separate and label appropriately with the location and time of sampling. These samples will be tested for fiber length distribution and nep content primarily, but can also be used for non-lint testing as confirmation of data to be determined from trash reduction analysis as explained earlier in this chapter.

10. During the hour other associates should determine the percent run time of each machine in the line being studied, especially the opener/cleaners. These run time checks should last at least 30 minutes for each machine. Several tests can of course be done simultaneously with enough people working on the study team.

11. When the study hour is completed, immediately doff all cards (if applicable) and stop the monitoring of card downtime. At the same instant shut down the opening line so that the waste compartments can be cleaned. Again, note the level of stock in any reserve or mixer. If not the same as at the beginning of the test, an adjustment will have to be made. For the purposes of further discussion, assume the level of stock in these reserves or mixers is the same at the beginning and at the end of the study period.

12. Collect all wastes, bag, and label appropriately.

13. Restart the machinery.

14. Determine the net weight of waste from each machine by source including card strips and undercard waste.

15. Determine the amount of card sliver produced during the test period on each card.

16. Add to the card sliver weight produced the weight of waste from each respective card to yield a total weight of stock processed by each card.

17. Determine the percent undercard and flat strip waste removed by each card. Remember that percent waste is always based on the weight of stock fed to a machine.

18. Total the weights of stock fed to all cards supplied by a single card feeding machine.

19. Repeat the addition of the weight of waste removed to the calculated weight of stock delivered for each prior machine in reverse order back to the top feeder. Use these data to calculate percent waste removed by each machine.
20. Using run time data along with the calculated total amount of material processed through each Opening Room machine, calculate the instantaneous production rate of each machine.
21. Adjust the run time for each machine to what it would need to be to support all cards when they are running and producing sliver.
22. Submit all of the many stock-in-process and waste samples to a laboratory for standard testing. All stock-in-process samples should at least be tested for fiber length distribution and nep content. The starting point bale or stock-in-process samples and the waste samples should at least be tested for non-lint content and for fiber length distribution.

It can be easily understood from the above steps that properly studying the Opening Room is a tedious task. This is one reason that these processes are often not thoroughly studied and instead abbreviated studies are attempted that can result in erroneous data. Table 4.7 contains example data and calculations of percent waste removed and production rates from such a thorough study of an Opening Room and supported line of cards. These calculated performance indicators will be used in combination with laboratory test results to complete the analysis of the example Opening Room. How this is done is explained in the appropriate prior sections of this chapter.

Reworkable Fiber Processing

Many short staple spun yarn plants reprocess fibrous wastes that are generated at carding, drawing, lapping, combing, roving, and spinning. When the yarn type and quality demands allow, these fibrous wastes can be fed back into the mix in the Opening Room. In addition, trashy wastes from opener/cleaners, cards, and combers can be processed through a fiber reclamation machine, and the cleaned fibers then can be fed back into the virgin fiber mix in the Opening Room.

Table 4.7. Example Data and Calculations from a 1-hour Cotton Opening Line Study

Process Line Characteristic During 1-Hour Study	Result
1. Machines in the line	1 Top Feeder
	1 Coarse Opener/Cleaner
	1 Mixer
	1 Intense Opener/Cleaner
	1 Fine Opener/Card Feeder
	8 Cards @40 kg/hr @ 100%
2. Card downtime	24 minutes
3. Carding efficiency	$\dfrac{480 \text{ minutes} - 24 \text{ minutes}}{8 \text{ machines} \times 60 \text{ minutes}} \times 100 = 95\%$
4. Card sliver produced	40 kg x 8 cards x 0.95 = 304 kg
5. Total carding flat strip waste	3 kg
6. Total undercard waste	5 kg
7. Total amount fed to cards	304 kg+3 kg+5 kg=312 kg
8. Average flat strip % removed	3 kg/312 kg x 100 = 0.96%
9. Average undercard % removed	5 kg/312 kg x 100 = 1.60%
10. Average total card waste %	8 kg/312 kg x 100 = 2.56%
11. Total Fine Opener waste	0.6 kg
12. Total amount fed to Fine Opener	312 kg +0.6 kg = 312.6 kg
13. Fine Opener Waste %	0.6 kg/312.6 kg x 100 = 0.19%
14. Total Intense Cleaner Waste	10.4 kg
15. Total amount fed to Intense Cleaner	312.6 kg + 10.4 kg = 323 kg
16. Intense Cleaner waste %	10.4 kg/323 kg x 100 = 3.22%
17. Total amount fed to Mixer	323 kg + 0 waste = 323 kg
18. Total Coarse Cleaner waste	1.7 kg
19. Total amount fed Coarse Cleaner	323 kg + 1.7 kg = 324.7 kg
20. Coarse Cleaner waste %	1.7 kg/324.7 kg x 100 = 0.52%
21. Top Feeder run time %	75%
22. Top Feeder instantaneous production rate	324.7 kg/0.75 = 432.9 kg/hr
23. Coarse Cleaner run time	Same as Top Feeder since no reserve between these machines = 75%
24. Coarse Cleaner instantaneous production rate	323 kg/0.75 = 430.7 kg/hr
25. Mixer run time %	96%
26. Mixer instantaneous production rate	323 kg/0.96 = 336.5 kg/hr
27. Intense Cleaner run time %	97%
28. Intense Cleaner instantaneous production rate	312.6 kg/0.97 = 322.3 kg/hr
29. Fine Opener run time %	98%
30. Fine Opener instantaneous production rate	312 kg/0.98 = 318.4 kg/hr

Reworkable Virgin Fibers

At the fiber preparation processes, some virgin stock in process can be separated from the main stream because of transport chokes, ends down, and creeling fallout. This separated stock is referred to as reworkable waste or reworkable fiber. At roving and ring spinning, a fibrous waste is generated which is called pneumafil waste. The pneumafil waste from ring spinning contains a high percentage of short fibers, so care has to be taken when planning whether to reprocess or to sell this particular fibrous waste.

For those reworkable fibers that are to be reprocessed, there are these ideal requirements to be met:

1. Premix the reworkable fibers from various sources in the same proportions as produced.
2. Pre-open the reworkable fiber to match the openness of the virgin material to which this waste will be introduced.
3. Feed in this reworkable fiber as early in the Opening Line as is possible.
4. Feed in this reworkable fiber at a constant percentage of the virgin mix.
5. Feed in this reworkable fiber continuously.

Extreme care should be exercised in the handling of reworkable fiber. These fibers, with the exception of ring spinning pneumafil waste, usually have a better fiber length distribution than does the virgin bale stock. However, these fibers will ordinarily have reduced strength from prior processing which can lead to fiber breakage. Such breakage will increase stock short fiber content which is a major detriment to processing and quality performance. If reworkable fiber and the basic stock are not mixed properly, the fibers will be opened to varying degrees and the resultant mix will have varying fiber-to-fiber cohesion levels. The latter will make it impossible to optimize fiber drafting behavior at roller drafting processes. Intermittent feeding of reworkable fiber creates even greater cohesion ranges and can also result in uneven distribution of short fibers in the resultant mix. Intermittent feeding of reworkable fiber is perhaps the worst error that can be made in the handling of this material and should not be permitted under any circumstances.

In theory the proportionate premixing of the reworkable fiber from various sources can help to minimize many of the potential problems associated with the feeding in of this material. A key to success in any textile process is consistency. Premixing reworkable fiber as described will bring a measure of consistency to this material. The reworkable fiber then must be mixed effectively and intimately with the virgin fibers. This is more likely if the reworkable fiber is opened to near the same Openness Index as is the virgin stock and if the reworkable fiber is fed in prior to the mixing machines in the line.

The handling of reworkable fiber is further complicated if this material is a blend of different fiber types. Blended reworkable fibers are routinely fed back into the basic mix, but precautions must be taken so that the effect on the resultant blend of the basic mix is known and plans are made for this effect.

There are a number of ways that reworkable fibers are prepared and fed back into the virgin mix. Each method has pluses and minuses as it relates to the 5 requirements listed above for properly handling this material. Some plants premix and bale this material and place a bale of reworkable fiber in the laydown to be fed by the top feeder. This method ensures that the premixing can be done if properly managed, and allows for a rather constant percentage to be fed into the mix. A drawback of putting one bale of reworkable fiber in the laydown is that it is really impossible to effectively blend in one outlier bale in this fashion. This approach also violates the recommendation that the laydown be divided into groups of bales that each represent the average of fiber properties of the entire laydown. Putting additional bales of reworkable fiber in the laydown will increase the percentage fed beyond the percentage generated unless very narrow bales at full height can be produced. The latter is not practical and is not likely to be done in the yarn manufacturing plant of today. When a top feeder is used, it is best to premix and bale the reworkable fiber, feed it into a waste opener/feeder, and air doff the opened material into an air transport duct that introduces the reworkable fiber to the basic stock flow after the top feeder and prior to the first mixer in the line. A disadvantage of even this approach is that the waste feeder can run empty, and allow intermittent feeding unless an alarm or stop motion is activated when this feeder runs empty. When hopper feeders are used for initial bale feeding, a waste opener/feeder is often placed in the series of machines. The best practice is to position the waste feeder in the middle of the set of hoppers, rather than at one end, the latter being most often the case. There is a better chance of effective sandwiching in of the reworkable fiber on the conveyor belt if the waste feeder is positioned centrally. Even with this approach it is still preferable to premix and bale the reworkable fiber before this material is fed into the waste opener/feeder. Again, the entire line should shut down or a loud alarm should signal if this waste feeder runs out of stock.

Reworkable Cotton Fibers Reclaimed from Trashy Wastes

Waste removed by cotton opener/cleaners and cotton cards is composed of about 50 percent fibers and 50 percent non-lint. The fibers can be substantially reclaimed for subsequent reintroduction into the basic stock mix. The rules for handling these fibers and feeding them back into the mix are exactly the same as for handling reworkable virgin fibers. A difference is in the nature of the reclaimed stock which is composed of more short fibers than are most of the virgin reworkable fiber types.

This process of reclaiming fibers from trashy wastes for reintroduction into the virgin stock is used in a small minority of USA cotton spun yarn operations. Those facilities that do reclaim fibers have experienced a satisfactory return on the investment in the machinery to accomplish this process. At this time in the USA, the Temafa Clean-Star is the most popular system for reclaiming fibers from trashy cotton wastes. A schematic of the cross section of this machine is provided in Figure 4.34. Newer on the market are the UltraClean from Hollingsworth and the Wastomat WST from Truetzschler. The UltraClean is an intensive cotton cleaner that can also be used for fiber reclamation. The Wastomat is a modification of the Cleanomat intensive cleaner for cotton and is shown in Figure 4.35.

Opening Room Benchmarks

Several of the best achievable or target Opening Room process and product performance measures have been referred to throughout this chapter. In this section these are reviewed and added to in Table 4.8. A benchmark should be the practical stretch target of improvement efforts and when a benchmark is surpassed, a new

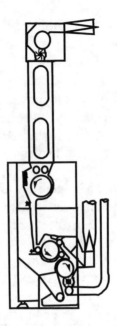

Figure 4.34. Schematic drawing of the Clean-Star system for reclaiming fibers from "dirty" cotton wastes. (Courtesy Temafa GMBH)

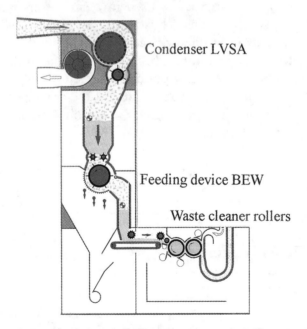

Condenser LVSA

Feeding device BEW

Waste cleaner rollers

Figure 4.35. Schematic representation of the Wastomat WST 2 for waste fiber reprocessing. (Courtesy Truetzschler GMBH & Co.)

Table 4.8. ITT Product and Process Benchmarks for World Class Performance in the Short Staple Yarn Plant Opening Room

Product Benchmarks:
- Minimum but consistent average bale short fiber content
- Bales conditioned in appropriate controlled environment for a minimum of 24 hours
- All bales represented in stock cross section from last mixer in the line
- Top feeder product at 135 or more ITT Openness Index
- Reworkable fiber opened to match virgin stock at entry point
- Card chute line distribution product at 150 or more ITT Openness Index
- Maximum 20% increase in average short fiber content from bale to card mat
- Maximum 50% increase in average nep content from bale to card mat
- 50% to 70% coarse trash removed from cotton prior to carding

Process Benchmarks:
- Cotton bales 100% High Volume Instrument tested including Near Infrared Maturity testing
- Statistically calculated number of bales in the laydown: Defaults are 40 for cotton, 24 for synthetics
- Controlled ambient conditions
- Machinery mechanical mixing power to match laydown:
  - 2x number of bales in laydown if not using the mini-laydown technique
  - equal to the number of bales at a minimum if using the mini-laydown technique
- Reworkable fibers premixed in proportion to generation amounts
- Reworkable fibers fed consistently and fed in prior to mixers in the line
- Progressive stock opening, avoiding recompression
- Cotton cleaners capable of removing seed coat particles
- Zero unscheduled machine interruptions
- Continuous run time for opener/cleaners
- Minimum air velocity for stock transport
- Maximum of 50 feet (15 meters) for transport duct length per fan or condenser
- All machines and accessories maintained in like-new condition

one is created. Continuous and incremental improvement should be the goal and the practice of textile manufacturers. And when the limit of a technology has been reached, newer and more advanced technology should be adopted.

Opening Room Machinery Manufacturers

There is still quite a variety of very old to very new machines operating in USA short staple spun yarn plants. The newer machines are generally technologically superior to the older machines, although there are definite exceptions. Most plants being

Table 4.9. Manufacturers of Opening Room Equipment Most Commonl Used in the USA

Equipment Manufacturer	Home Country
Crosrol Ltd.	England
Fiber Controls Corporation	USA
John D. Hollingsworth on Wheels, Inc.	USA
Fratelli Marzoli & C. spa	Italy
Preparation Machinery Services, Inc.	USA
Rieter Machine Works, Ltd.	Switzerland
Truetzschler GMBH & Co.	Germany

modernized today would be equipped with the newer Opening Room machines that are sold, installed, and serviced by the manufacturers listed in Table 4.9.

The Carding Process

The carding process for short staple fibers to be spun into yarn is the subject for this section. Short staple usually refers to fibers up to 64 mm (2 ½ inches) in length. Thus, short staple includes all cotton fibers. The following descriptions exclude the modern short staple cards used in the production of non-woven fabrics. Also, long staple cards as used for wool and worsted processing are not included.

The carding process for short staple fibers can best be described by considering two separate machine functions, namely, the feed chute and the card.

The Feeding of Stock to the Card

Introduction

In the early 1960s the method of feeding fibrous stock to the carding machine (typically called the "card") underwent a change from picker lap feeding to chute feeding. Picker laps, produced on a machine called a picker (scutcher in many countries of the world), consisted of a calendered blanket of fibrous stock. The blanket was about 9.5 mm or ⅜ inch thick, 96.5 to 99.0 centimeters (cm) or 38 to 39 inches wide, and approximately 55 meters (m) or 60 yards (yd) long. This blanket was wound into a roll with a lap pin through its axis, the pin having been used for handling and mounting on the card feed end. The unit weight of the blanket was usually about 465 grams per meter or 15 ounces per yard, causing the lap to weigh approximately 25.4 kg or 56 lb. No further discussion is appropriate concerning the picker as it is no longer used in developed countries.

The transition to the chute feeding system occurred rapidly because of several benefits of this new feed system as given below:

1. Reduced waste associated with laying laps onto the card;
2. The elimination of the strenuous manual task of handling the heavy laps; and
3. The elimination of card damage caused by chokes at the lickerin roll of the card when the new lap blanket overlaid the old lap tail, especially on cards running at high production rates.

One negative aspect of the chute feeding system is the fact that weight control by weighing was lost when the picker was replaced. All of the picker laps were weighed, thus giving a complete control of long-term product weight. The control of weight with the chute feed is examined more fully later in this chapter.

Stock Transport to the Chute

The typical stock flow arrangement in the yarn manufacturing plant is to have a machine called a fine opener or card feeder as the final machine unit that processes the stock before it is transported by means of a stock transport fan to the duct system that delivers the stock to the chutes. Shown in Figure 5.1 is a diagram of a typical stock transport fan. The purpose of the fine opener is to fully open the fibrous material so it can flow consistently to and through

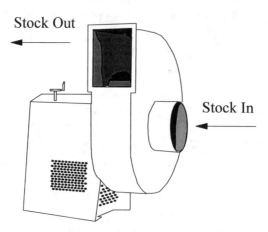

Stock Out

Stock In

Figure 5.1. Diagram of a typical stock transport fan.

the chutes. During this opening process some degree of final cleaning can be realized. This final cleaning is especially important if the stock is 100 percent cotton, or a blend with cotton. Figure 5.2 is a diagram of a fine opener. The supplier of this particular unit is Truetzschler GMBH & Company located in Monchengladbach, Germany. This fine opener, as with most machines of this type, has

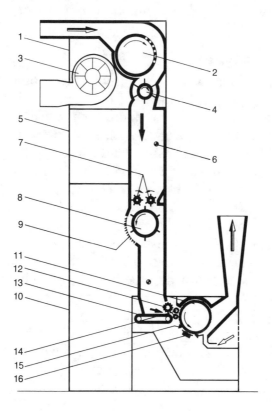

- 1. Condenser LVS
- 2. Dust cage
- 3. Fan
- 4. Finned roll
- 5. Feeding Unit BE
- 6. Light barrier
- 7. Feed rolls
- 8. Cleaning rolls

- 9. Grid
- 10. Cleaner RSK
- 11. Saw-tooth roll
- 12. Pressure roll
- 13. Inner feed lattice
- 14. Feed rolls
- 15. Mote knife
- 16. Carding segment

Figure 5.2. Diagram of a fine opener. (Courtesy Truetzschler GMBH & Co.)

a pair of feed rollers which can be regulated in speed to feed the correct amount of stock as required by the cards being supplied by the unit. One fine opener with transport fan will typically feed stock to a line of 6 to 10 cards, depending primarily on the production demand of the cards. Thus, if there were 8 cards in a line and the plant needed 32 cards for its total production, there would be 4 separate lines with a fine opener and transport fan for each line.

System of Transport Ducts to Chutes

There have been two basic transport duct systems used commercially. One, used early, was called the "recirculation system". Several machinery companies produced such systems, including Rieter Machine Works, Ltd. (Switzerland), Platt International, Ltd. (England), and Continental Moss-Gordin Gin Co. (USA). Rieter's *Aerofeed* transport recirculation system is diagramed in Figure 5.3. The fine opener (1) feeds stock to a stock transport fan (2) which supplies stock to the circular pneumatic duct (3) in the flow direction shown by the arrows. Fundamentally, the fine opener (flock feeder) together with the stock transport fan supplies some excess stock to the duct system, and the excess stock, which does not fall into the vertical chutes, is returned to the fine opener to be recirculated.

The second type of duct transport to chutes is called the dead-end system. Figure 5.4 is a diagram showing a plan view of 7 cards whose chutes are being fed with the dead-end system of transport ducting. Shown in this diagram are two separate fine openers and transport fans, each feeding the chute distribution duct from opposite ends. Because there is no recirculation of stock, the distribution duct can be blocked between two particular cards, thus allowing two groups of cards in the line to process simultaneously two different fibrous stocks. Blocking the duct between two cards is accomplished by use of a simple slide plate. The off-on action, or variable speed, of the feed roll in the fine opener is usually controlled by a sensitive pressure transducer in an upstream location in the transport duct that furnishes stock to the chutes.

The dead-end system of stock transport has now been accepted as the preferred system for a number of factors as listed below:

1. Fewer neps are generated in the stock.
2. There is less fiber breakage.

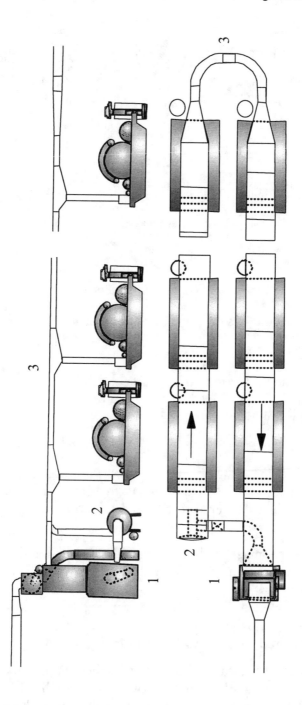

Figure 5.3. Diagram of the Rieter Aerofeed recirculation chute feed system. (Courtesy Rieter Machine Works, Ltd.)

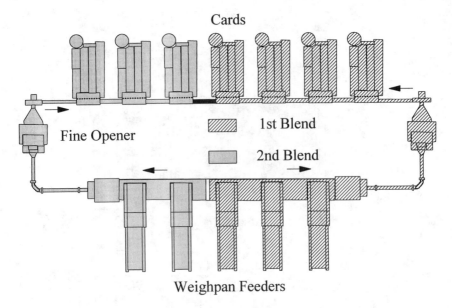

Figure 5.4. Diagram of a dead-end chute feed system. (Courtesy Truetzschler GMBH & Co.)

3. The dead-end system has greater flexibility.
4. Controls for over-feeding are more reliable.
5. Machinery installation is simpler.

The Feed Chute

The purpose of the card chute is to form a mat or fleece of fibers that when delivered to the feed table of the card is characterized as follows:

1. Has even mass in both longitudinal and lateral directions that will remain consistent during the total production period;
2. Is well opened and remains consistent in openness and weight per linear yard; and
3. Is delivered to the card feed roll in a uniform manner.

A number of the early chutes were rather simple box-like compartments, open at the top to allow stock to fall in. There were pairs of delivery rolls at the bottom to discharge the stock onto the card feed table when the feed roll of the card was operating. Stock was caused to flow down through the chute by gravity, pressure in the stock transport duct, or an oscillating compression perforated plate, or by a combination of these mechanisms. Experience gained in the textile plants in operating these simple chutes showed that product weight control was very poor and erratic. Slight changes in ambient conditions, frictional properties of the stock, openness of the stock, and changes of pressure in the distribution duct all caused the longitudinal weight of the stock on the feed table to vary considerably. This troublesome operation of the early chutes encouraged the machinery builders to develop more sophisticated designs that solved some of these problems. The diagram in Figure 5.5 illustrates the simple design used extensively in the early era of card feed chutes.

One important change for improvement was the use of a double compartment in the chute that separated the mat forming chamber in the lower position of the chute from the pressure changes that

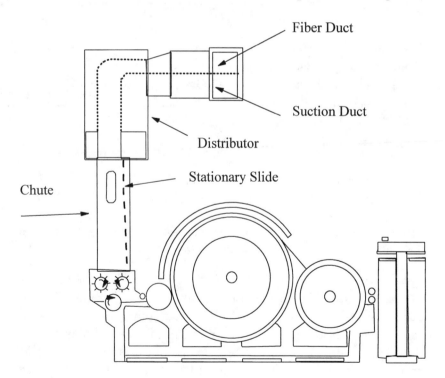

Figure 5.5. Diagram of a typical early card chute feed design.

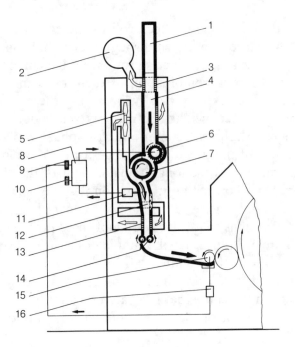

Figure 5.6. Diagram and element nomenclature of the FBK 533 card chute or tuft feeder. (Courtesy Truetzschler GMBH & Co.)

- 1. Distribution duct
- 2. Dust extraction duct
- 3. Upper air outlet
- 4. Material reserve trunk
- 5. Fan
- 6. Feed roll
- 7. Opening roll
- 8. Control unit
- 9. Nominal pressure adjustment
- 10. Base speed adjustment
- 11. Pressure transducer
- 12. Feed trunk
- 13. Lower air outlet
- 14. Take-off rolls
- 15. Card feed roll
- 16. Speedometer

occur in the transport duct that is associated with the upper portion of the chute. This more sophisticated design is represented in Figure 5.6 which is a diagram of the Truetzschler Model 533 Exactafeed FBK chute. The purpose of the fan in the chute is to keep a consistent air pressure on the fibrous stock in the lower mat-forming compartment. The white-filled arrows in the diagram trace the air flow caused by the fan. Grids in the front and back walls near the bottom of the compartment allow the excess air to flow up to the intake of the circulating fan.

The feed roll in the upper chute chamber is caused to actuate by a pressure sensor located in the lower compartment. When the feed roll is not turning, no stock gets delivered to the mat-forming compartment. In this model chute, the feed roll is driven by a D.C. motor whose speed is incrementally controlled by the pressure sensor. In this manner the mass of stock in the lower compartment can be maintained quite consistently. The air escape grids, mentioned earlier, cover the entire width of the lower chute, an arrangement which causes the stock mass to be positioned uniformly from side to side. If the stock is well opened as it enters the transport duct, this chute design will deliver a very satisfactory mat to the card feed table. There are several other chute designs which are well accepted in the industry. Some of the popular machinery builders that supply chutes are John D. Hollingsworth (USA), Fiber Controls (USA), Crosrol (England), Truetzschler (Germany), Rieter (Switzerland), and Marzoli (Italy).

Additional advantages of the double compartment chutes are as follows:

1. Lower volume of stock in lower compartment,
2. Less waste when starting up,
3. More sensitive controls, and
4. More even mat (fleece) produced.

Other factors that can cause variations in mat mass evenness include changes in fiber properties brought about by changing ambient conditions. Even changes in atmospheric pressure can cause changes in the mat weight. Low fiber openness should be avoided. A minimum fiber openness target of the mat is 140 Openness Index as measured by the Institute of Textile Technology openness test procedure, described in Chapter 4. The openness of the mat must remain consistent. Tagging or choking in the transport ducts or in the chute itself can be very disruptive to consistent mat openness and density.

The delivery rolls at the bottom of the chute are driven by a chain linkage with the card feed roll. The stock in the chute is compressed by air pressure or an oscillating perforated wall plate. The delivery rolls also apply some additional pressure to the stock

Table 5.1. Target Mat Weights under Card Feed Roll

Fiber Type	Grams Per Meter	Ounces Per Linear Yard
Acrylic	372–434	12–14
Polyester	403–465	13–15
65/35 Polyester/Cotton	434–496	14–16
50/50 Polyester/Cotton	465–527	15–17
Rayon	496–558	16–18
Cotton	496–558	16–18

as it is being forced out of the chute. Therefore, there is an expansion of the mat as it is released onto the card feed table. The turning of the feed roll of the card tends to pull the mat over the table and under itself as the mat is fed to the card lickerin. Because of these actions, it is critical that the surface speeds of the chute delivery rolls and the card feed roll be set at a ratio that will prevent pressure from being built up on the mat, and prevent the mat from being pulled (drafted) enough to cause short-term mass variation. Short-term in this case refers to inch-to-inch mass increments along the length of the mat.

An important consideration is the weight per unit length of the mat. Weight variation is usually higher on light mats, which can cause the draft in the card to be too low for optimum carding. Heavy mats can cause card drafts to be too high for good carding. In Table 5.1 are listed targets for mat weights "under the feed roll" of the card. This distinction is made because the mat on the feed table is somewhat heavier than when it has been accelerated in its movement and compressed under the feed roll. This change in mat weight can be determined by comparing the velocity of the mat on the feed table to the velocity of the surface of the feed roll.

The Card and Its Standard and Optional Components

The card is the first machine in the yarn manufacturing process that is designed to produce a continuous strand of fibers. A quotation from long ago states that "the card is the heart of the spinning mill," and all those knowledgeable in the yarn making business believe strongly in this adage. The purpose of the card is to separate fibers individually, arrange them in a somewhat longitudinal direction, remove contamination, reassemble the fibers into a continuous strand of uniform weight and quality, and coil the strand (sliver) into a suitable transport container.

The basic elements of the carding machine have not been changed substantially since the first patents were granted approximately 250 years ago. However, there have been many improvements made during these years resulting in increased production rates at higher efficiencies and improved product quality.

Probably the most dramatic change was the introduction of rigid metallic wire clothing (covering) as used on the cylinder and doffer elements in place of the flexible wire clothing normally used prior to the introduction of the metallic clothing. This change took place in the early 1940s. The flexible wire was referred to as fillet clothing. As it functioned on the cylinder surface, the fillet clothing retained some long and many short fibers. The wire interspaces eventually became clogged to the extent that the carding action (combing) deteriorated. At relatively short intervals, therefore, the cylinder clothing had to be stripped. The substitution of low-profile metallic wire for cylinders, with no deep spaces to collect and retain fibers and waste, eliminated the need for stripping, as was the case with the longer fillet wire.

Other changes and improvements that have been developed in the later years include the following:

1. Roller doffing at the doffer cylinder instead of the earlier employed oscillating comb doffer (this change became necessary as production rates were increased);
2. Better materials and more precise balancing of rotating elements;
3. Web condensers of various types;
4. Autoleveling which became important with the advent of chute feeding (long-term leveling refers to sliver lengths of 100 meters and longer; mid-term leveling refers to sliver lengths of 1 to 3 meters; short-term leveling refers to sliver lengths of 2.5 to 9.0 centimeters or longer);
5. Multiple lickerins;
6. Improved bearings and tolerances;
7. Stationary non-loading carding plates;
8. More sturdy and precise frames and structures;
9. Real time on-line quality monitors; and
10. Direct current drive motors for ease of speed control.

Figure 5.7. The Marzoli C300 card. (Courtesy Fratelli Marzoli & C. spa)

A view of the Marzoli Model C300 card is provided in Figure 5.7. This card, manufactured by Fratelli Marzoli & C. spa in Italy, is classed as a high production card. It has found acceptance in the industry because of its efficiency and durability.

Shown in Figure 5.8 is a simplified diagram of a typical single cylinder short staple card. The double cylinder (tandem) card is described later in this chapter. At the left is shown the feed roll/feed plate system that feeds the compressed fiber mat into the card. An enlarged view of this system is shown in Figure 5.9. The feed roll turns slowly in the direction indicated by the arrow; the feed plate is stationary. To improve the frictional grip between the mat of fibers and the feed roll, the outer surface of the roll is normally fitted with grooves, blunt projections, or short metallic teeth. The feed roll is held down by springs acting on each end. As indicated in Figure 5.9, the fiber beard is presented to the action of the relatively coarse metallic teeth on the lickerin (taker-in) as it is fed over the nose of the feed plate. Because of the rotational effect of the lickerin teeth, the action at "a" (Figure 5.9) is "combing". Fiber tufts are reduced in size as the tufts and individual fibers are

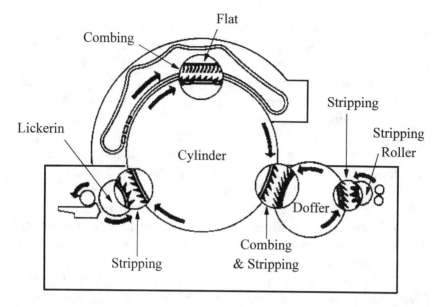

Figure 5.8. Diagram of a typical single cylinder short staple card. (Courtesy Crosrol Ltd.)

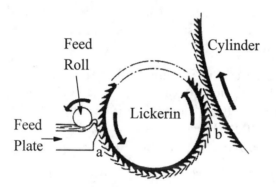

Figure 5.9. Diagram of a typical feed system for the card. (Courtesy Crosrol Ltd.)

caught on the teeth of the lickerin. Trash particles and very short fibers that are not held by the teeth are thrown off by centrifugal force, to be deposited on the floor of the card below the lickerin. This collection on the floor is classed as undercard waste.

The undercard waste is normally removed by a pneumatic suction system located at the feed end (back) of the card. The waste material is caused to be moved into the influence of the suction system by several means, such as a slowly revolving belt, or a series of pulsing air jets located near the sides of the card. In the earlier days this undercard waste was removed manually. The percentage of undercard waste for 100 percent cotton is usually in the range of 1.0 to 2.5 percent. As cleaning of impurities is not as important for synthetic fibers, the undercard waste is set lower for carding such fibers as compared to cotton.

The fibers and small tufts held by the teeth on the lickerin are "stripped" at "b" (Figure 5.9) by the action of the wire teeth on the cylinder, whose surface velocity is as much as 2 times or more the surface velocity of the lickerin wire.

The most critical zone of this system just described is at location "a." When the set distance between the lickerin teeth tips and the nose of the feed plate is relatively close (0.4 mm, 0.017 inch) and the production rate is increased, say to 27 kg/hr (60 lb/hr), excessive fiber damage is likely to occur. Such fiber damage will affect adversely the quality of textile products produced at downstream processes.

An experiment designed at the Institute of Textile Technology in 1993 was conducted using a Marzoli C300S card with several different settings for processing 100 percent cotton. The settings used were as follows:

1. Lickerin speeds at 550 and 700 rpm;
2. Feed plate-to-lickerin spacings at (a) 0.51 mm (0.020 inch), (b) 0.69 mm (0.027 inch), (c) 0.81 mm (0.032 inch), and (d) 1.27 mm (0.050 inch); and
3. Production rates at 27 kg/hr (60 lb/hr) and 36 kg/hr (80 lb/hr).

All other settings of the card remained constant throughout the study. Card slivers at about 4.60 grams per meter (65 grains per yard) were processed through two passes of drawing on the Institute's Rieter RSB 851 draw frame (autoleveling operating during the second pass only). A final weight of 4.25 grams per meter (60 grains per yard) finisher sliver was processed into 0.74 hank roving. All conditions were spun into 20's Ne ring yarn on the Institute's Marzoli NSF2 ring spinning frame.

The overall yarn quality results were as follows:

1. Lower card production rate improved (a) strength, grams per tex; (b) Strength variability, %Vo; (c) Elongation, %; (d) Uster IPI thick places, number; and (e) Uster IPI neps, number.

2. Wide lickerin-to-feed plate spacing improved (a) Elongation, %; (b) Uster IPI thin places, number; and (c) Classimat II long thin places, number.
3. Lower lickerin speed improved (a) Uster 1-meter coefficient of variation, %CV; (b) Strength variability, %Vo; and (c) Uster IPI neps, number.

Multiple Lickerins

The standard card has one lickerin. As the desire for higher productivity increased, several manufacturers of carding machines attempted to increase the fiber opening ability at the feed end of the card by incorporating more than one lickerin. Both two and three lickerins are being used. A drawing showing the arrangement of multiple (3) lickerins is provided in Figure 5.10. In the case of the Truetzschler DK 803 card, the diameters of the three rolls are 175 mm (6.9 inches). The roll that first contacts the fiber beard at the feed roll/feed plate zone is called lickerin no. 1. Lickerin no. 2 rotates in a direction to strip the fibers from lickerin no. 1, and lickerin no. 3 (nearest the cylinder) rotates in a direction to strip the fibers from lickerin no. 2, and at the same time allows the fibers on its teeth to be readily stripped by the cylinder. The speed of the second lickerin is higher than the speed of the no. 1 lickerin, and

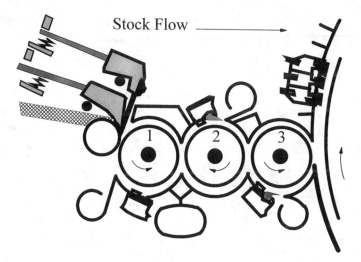

Stock Flow

Figure 5.10. Diagram of the arrangement of three lickerins as used on the Truetzschler DK 803 card.

the speed of the third lickerin is higher than that of the second. Each succeeding lickerin speed is approximately 25 percent higher than the speed of the one before, so that stripping action can occur efficiently. The teeth become finer on each succeeding roll. On the DK 803 card the covering for lickerin no. 1 is composed of needles rather than metallic wire. It is emphasized in the card sales brochure that the needle roll "doubles the maintenance interval in the pre-opening area as compared to single taker-in machines." Metallic wire teeth cover lickerin nos. 2 and 3.

Revolving Flats and Stationary Flats

In the upper part of Figure 5.8, the diagram indicates a moving chain that carries the individual flats slowly over the top section of the rotating cylinder. It had been standard for the movement direction of the flats to be toward the front of the card; on some of the newer cards, however, the direction of flat travel is toward the back. The speed of this movement of the flats is usually about 152 mm (6 inches) per minute. As the surface speed of the cylinder is approximately 10,000 times faster than the flats, the direction of flat movement has no effect on the combing action of the fibers and small tufts that occurs at the interface between the two components. The interspaces between the flexible fillet wire on the flats fill with waste, mainly short fibers, neps, trash particles, and dust. Therefore, it is necessary that this buildup in the flats be cleaned continuously. It is this cleaning aspect of the flats that plays a vital roll as to the travel direction of the flats.

Shown in Figure 5.11 is a cross-sectional view of a flat with flexible fillet wire. With some synthetic fibers, such as polyester and acrylic, it may be desirable not to remove fibers that build up in fillet wire. For this purpose a non-loading metallic wire is mounted on the flat bar in place of the fillet clothing. For carding cotton, however, the fillet wire is commonly used.

The clearance space between the tips of the combing wire on the flats and the teeth on the cylinder surface is usually set at approximately 0.35 mm (0.010 inch) for processing cotton. For synthetic fiber this clearance normally is set at a greater distance, such as 0.43 mm (0.017 inch). As shown in the enlarged view of the wires in Figure 5.8, the direction of the wire teeth is in the carding (combing) attitude.

Figure 5.11. Cross section of a fillet wire flat.

It is customary on the newest cards to utilize stationary carding plates to enhance the total combing action needed, especially when the throughput rate is raised. These carding plates or flats are covered with non-loading metallic wire. Figure 5.12 is a diagram of the Rieter C4 card which shows the positions of station-

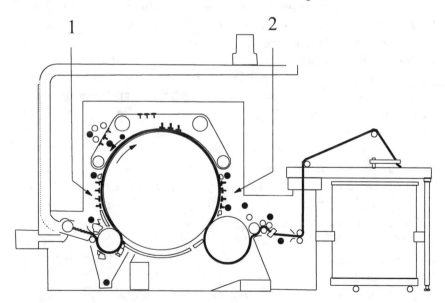

Figure 5.12. Diagram of a Rieter C4 card showing stationary carding plates at "1" and "2." (Note that the travel direction of the revolving flats is toward the back.)

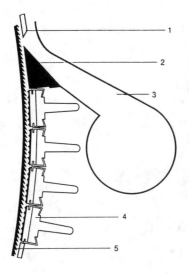

1. Cylinder cover plate
2. Mote knife
3. Suction hood
4. Post carding segment
5. Cylinder

Figure 5.13. The Webclean KR system as used on Truetzschler cards. (Courtesy Truetzschler GMBH & Co.)

ary carding plates. Those at location "1" are commonly called pre-carding plates, and those at location "2" are called post-carding plates. This diagram also shows two stationary carding plates located immediately under and set, with clearance, to the lickerin.

A system of post-carding plates used in conjunction with a mote knife cleaning device is used on Truetzschler cards; it is called "Webclean KR" by the manufacturer. A drawing of the system is shown in Figure 5.13. Short fibers, impurities, and fine dust pass through the gap between the cylinder cover plate and the mote knife, and are removed by a suction system. The post-carding plates smooth out the web held on the card clothing and have a beneficial effect on air flow in this zone.

Another adaptation of stationary carding units is a series of 4 (usually) relatively large curved plates mounted in place of the revolving flats on top of the card. A drawing of a set of cards equipped with stationary tops is located in Figure 5.14. The carding surfaces of these stationary top plates are composed of metallic wire (non-loading) designed specifically for the type of fiber to be processed. As one would imagine, these plates must be removed at intervals for cleaning and inspection. The non-loading feature of the metallic wire used on these plates, and on stationary bars, is

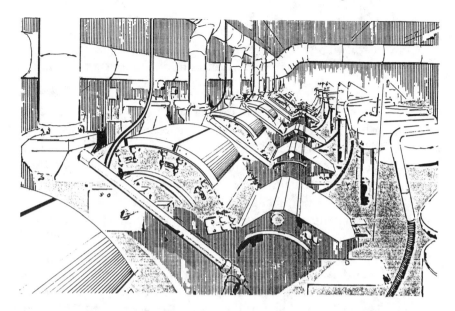

Figure 5.14. Diagram showing cards equipped with stationary top carding plates. (Courtesy John D. Hollingsworth on Wheels, Inc.)

accomplished because the metallic teeth of the wire are relatively short with ample space between.

As no material is removed by stationary tops or stationary non-loading flats (or plates), cleaning (removing) of contaminates cannot be accomplished by these elements. A better choice for use with cotton is revolving flats rather than stationary tops, allowing the removal of flat strips which contain non-lint materials. The fibrous strips contained in the fillet wire act as filters to catch pieces of leaf and stem, seed coat fragments, short fibers, and dust.

Doffer

The position and direction of rotation of the doffer can be seen in Figure 5.8. The essential functions of the doffer are as follows:

1. Transfers a portion of the fibers on the cylinder to its own wire clothing (transfer ratio of fibers from cylinder is only partial, usually in the range of 0.1 to 0.2);
2. Condenses fibers to a heavier mass per unit length as compared to mass on cylinder;

3. Disorients fibers somewhat to give a degree of random orientation for web stability; and
4. Reverses fiber ends to give trailing hooks in exiting sliver.

As the wire clothing enlargement in Figure 5.8 shows, the wire teeth on the doffer are longer and have a more aggressive angle than the teeth on the cylinder clothing. The directions of the front edges of the teeth are such as to create a carding (combing) action at and near the set line between the cylinder and doffer. The setting clearance between the doffer and cylinder is as close as feasible, depending on the mechanical perfection of the structures. A card with excellent mechanical elements can be set (when on a rigid floor) to a tight clearance of 0.10 mm (0.004 inch). A card with poorer mechanical tolerance might have to be set with a clearance as great as 0.15 mm (0.006 inch).

The carding action referred to above is that which occurs when the cylinder wire clothing actually cards the fibers that have been transferred to the clothing on the doffer as the surface of the cylinder approaches the set line and leaves the set line. The surface speed of the cylinder is normally approximately 20 to 25 times the surface speed of the doffer.

The doffer maintains a relatively thick layer of fiber on its surface, even though some of the fibers return to the cylinder as it combs through the fiber ends that are stretched out beyond the wire teeth on the doffer. As mentioned above, the cylinder surface speed is as much as 25 times the surface speed of the doffer. Consequently, fibers lying on 1 meter of cylinder surface pass only 40 mm of doffer surface. In spite of the fact that not all fibers are transferred to and remain on the doffer, the thin layer of fibers on the cylinder surface becomes a relatively thick layer on the doffer.

Web Travel from Stripping Roller to Coiler

Immediately past the doffer and set with a clearance of 0.38 mm (0.015 inch) to the doffer surface is the stripping roll (Figure 5.8). It is covered with intermittent projectors (teeth), and its function is to remove the fiber web from the doffer surface, and to direct the web forward to a pair of crush rolls. Note the direction of rotation of the stripping roller, as shown in Figure 5.8. Normally, there is a brush roller mounted just on top of the stripping roll whose purpose is to clean off any fibers that might become attached to the teeth on the stripper. This clearer roll which turns counter to the direction of the stripper roller is brushed clean when fibers have accumulated on its surface. This is usually a manual operation offering an opportunity for automation. Another view showing the stripper roll and brush roll is illustrated in Figure 5.15. This illustration also shows the squeeze rolls at "4" (crush rolls), the web condenser just past the web bridge at "8," the trumpet, and the calender rolls at "6."

As it is stripped from the doffer, the web is approximately the full width of the card and remains at that width until it exits the

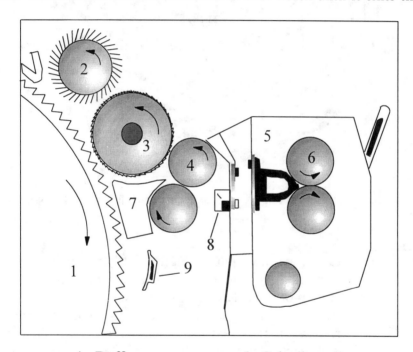

1. Doffer   6. Calender roll
2. Brush roll   7. Suction
3. Stripper roll   8. Web bridge
4. Squeeze roll   9. Guide rail
5. WEBSPEED

Figure 5.15. Illustration of the sequence of elements that guide the fiber web from the card doffer with condensing and calendering actions. (Courtesy Truetzschler GMBH & Co.)

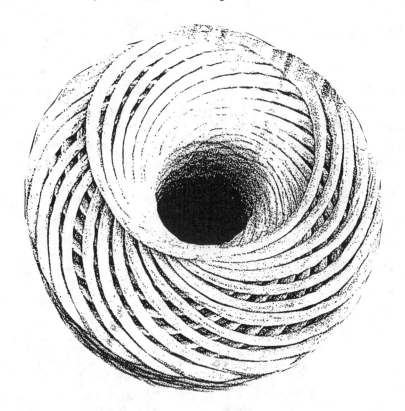

Figure 5.16. Top view of one pattern for coiling card sliver in a can. (Note acceptable spacing between successive coils.)

crush rolls. In the arrangement of Figure 5.15, the web is condensed to a center fiber bundle before entering the trumpet located just upstream of the calender rolls, whose function is to pull the strand through the trumpet and compress/compact the fibers into a tighter strand that is reasonably stable. The two calender rolls are pressed together with coil springs to effect a nipping line that grips the fiber bundle with no slipping. The sliver strand proceeds upward over suitable guide pulleys to enter the coiler that consists of a trumpet guide and a second pair of calender rolls which deliver the finished card sliver through a revolving tube gear into the card sliver can. A normally used coiling pattern is shown as a top view in Figure 5.16. The can dimensions in normal use are 1200 mm (48 inches) in height, with diameters of 760 mm (30 inches), 900 mm (36 inches), or 1000 mm (40 inches).

There are two popular mechanical methods of producing the sliver coiling pattern in the can. In one system the rotating tube gear, which causes the sliver circle in the can build, is mounted in a larger diameter fixed plate. The axis of the tube gear is positioned to cause the sliver circle to be offset relative to the axis of the sliver can, which is positioned by the adjustable location of a metal platform at floor level. The can sits on this platform which is rotated slowly to allow the sliver coils to be laid into the coiling pattern shown in Figure 5.16. This system of coiling is typically called the turntable system, referring to the platform that turns the can.

In the second mechanical method, the rotating tube gear is mounted in a plate that turns slowly in the coiler head. This turning causes the coils to be laid in the coiling pattern. The sliver can sits on the floor and is held in position by a simple holding strap mounted to a curved positioning bar. This mechanism is referred to as a planetary coiler.

The first coil that is positioned into the can actually is laid onto a plastic (or metal) plate at the top of the can. This plate is normally called a piston or piston top. It is held up near the top of the empty can by a coil spring whose function is to keep the top of the coiled sliver near the top of the can regardless of the amount of material in the can. The spring is designed so its compressed force will balance the total weight of sliver resting on the piston. The piston plate has at its rim a vertical skirt approximately 100 mm (4 inches) in height. This skirt fits over the outside diameter of the coil spring which can fit into the cavity under the piston when it is fully compressed with a full load of sliver. It follows that ideally the spring in the can should be designed according to the type fiber being processed as specific gravities of the various fiber types are substantially different. The diameter of the piston used in the can is slightly less than the inside diameter of the can so it can travel down into the can without scraping the can wall. A drawing of a sliver can is shown in Figure 5.17. It is normal to add several coils of sliver above the can rim of a full can. The sliver height extending above the can top is designated the crown. The above considerations are applicable also for sliver cans used at the drawing and combing processes. The graph in Figure 5.18 is useful as a guide for determining sliver can capacity.

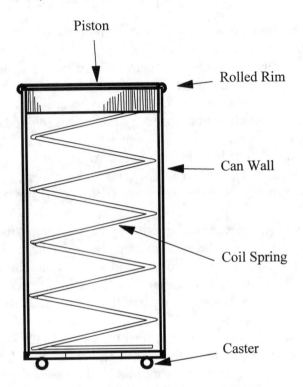

Piston

Rolled Rim

Can Wall

Coil Spring

Caster

Figure 5.17. Cross-sectional drawing of an empty sliver can.

Crush Rollers

Crush rolls are used to pulverize bulky contaminates in fibrous material being carded. Historically, this innovation was incorporated into the mechanism of the woolen card to crush burrs entangled into parts of the woolen fleece. In the middle of the 1950s a similar system was introduced to the short staple carding arena with the purpose of pulverizing cotton material contaminates composed of leaf and stem particles especially. After being pulverized into very small pieces, these contaminating materials literally fell from the cotton fibers, thus causing the ensuing yarn to be cleaner and more lustrous, as well as reducing breaks in the spinning process.

The first such crush rolls made available for cotton (short staple) cards were manufactured by Crosrol Ltd. of Halifax, England. One engineering problem with this system was the design of the mounting for the rolls so that the pressure between the rolls was the same

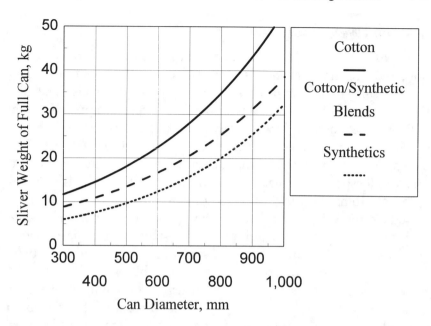

Figure 5.18. Graph relating sliver can capacity to can diameter, with a can height of 1200 mm, floor to rim.

for each position along their contact line when the spring force for that pressure was applied at each end of the top roller. In the case of the Crosrol unit, this problem was solved by a slight cross angle of the rolls; this cross angle was automatically changed as the spring pressure was altered, both lower and higher. There appeared soon on the market several other designs by competing firms that had different solutions to this problem, such as tapered rolls and an additional pressure point at the center of the roll length. All of these crush roll units were arranged so the pressure between the pair of rolls could be changed, and the rolls could be separated slightly when fibers other than cotton were processed.

During the past few years several factors have changed the practical effectiveness of crush roll use. The stickiness of cotton has increased because of greater insect infestation and an increase in the number of seed coat fragments, the oil of which causes fibers to adhere to the crush roll surfaces. As this sticky condition causes defects and downtime at carding when crush rolls are used with pressure, the trend has been a diminution of their use in commercial practice.

Webs and Web Condensers

In older cards, many of which continue in commercial operation, the fibrous web travels from the crush rolls to a collecting trumpet mounted just prior to the front calender rolls. The distance between the crush rolls and the trumpet is approximately 300 mm (12 inches). If a card is not equipped with a crush roll system, the distance is approximately 450 mm (18 inches). The web as it angles into the trumpet is rather delicate and is subject to breaking, especially when production is relatively high, say 45 kg/hr (99 lb/hr) and higher. Web condensers have been developed to reduce the web break rate, as the condenser causes a rope-like flow of fibers into the trumpet. The most popular design of web condenser is shown in Figure 5.19 in which two transverse belts carry the web to the center gap between the two belts, and the heavy rope-like mass then travels toward the trumpet. Another modification of the belt condenser is the direction of the web to one side of the card in such manner to cause the rope-like fiber bundle to exit the single belt at the right side of the card working

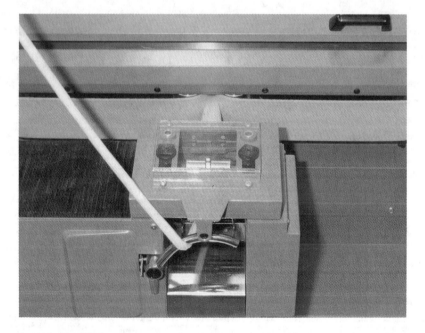

Figure 5.19. Web condenser as used on the C300/S and CX400 cards. (Courtesy Fratelli Marzoli & C. spa)

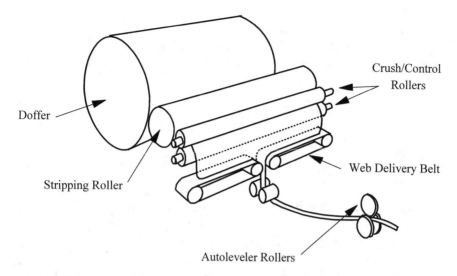

Figure 5.20. Diagram of a web condenser for Crosrol cards. (Courtesy Crosrol Ltd.)

width. Such a system is employed on the Rieter cards, models C4 and C50.

Still another belt web condenser arrangement was developed by Crosrol Ltd. It is shown in the diagram of Figure 5.20. A general feature of web condensers is increased carding efficiency because of fewer ends down and fewer broken selvedges (edges of webs).

Trumpets and Calender Rolls

The main trumpet just behind the front calender rolls is a critical unit in the card, as well as is the trumpet in the coiler. The bore in the main trumpet controls, with the pressure of the calender rolls, the condensing of the sliver. Too much condensing of the sliver can cause coring, which is a hard core in the center of the sliver. This core is approximately the size of a common pencil, and the smooth drafting of cored sliver is practically impossible. Too little condensing will produce a fluffy sliver that will stretch easily and is apt to pull apart causing disruptions in downstream processes. Normal trumpet bore sizes are listed in Table 5.2. These bore diameters do not apply to main trumpets that are used as sensing devices for card autolevelers.

Table 5.2. Normal Bore Sizes for Card Trumpets

Sliver Weight		Main Trumpet	Coiler Trumpet
Grams/Meter	Grains/Yard	Bore Diameter, mm	Bore Diameter, mm
2.83	40	4.0	4.5
3.19	45	4.4	4.9
3.54	50	4.7	5.3
3.90	55	4.8	5.4
4.25	60	5.0	5.6
4.60	65	5.3	5.9
4.96	70	5.6	6.3

Calender rolls are hardened steel cylinders of equal diameters, mounted parallel to each other and pressed together with coil springs. The rolls are mounted very close to the discharge of the main trumpet so that the operating clearance between the trumpet and the surfaces of the rolls is approximately 0.5 mm. The diameters of the front calender rolls on many card models are 75 to 80 mm (2.95 to 3.15 inches). Some models have tongue-and-groove or stepped calender rolls.

The sliver is guided by low friction flanged pulleys as it flows from the front calender rolls upward to the coiler head. An ends-down detector is mounted to sense the presence of the sliver somewhere along the path. If the sliver is not detected, the sensor circuit causes the card feed roll and front rolls, including the doffer, to stop.

Card Drives

For many years the cards were driven by only one electric motor. Figure 5.21 is a drawing representing such a drive arrangement. In this drawing the motor (MP) is connected to the cylinder and lickerin shafts by belting for a direct drive. All the rotating elements in the front zone of the card are interconnected and driven from the lickerin shaft by the pulley "C" which is belted to the transverse shaft that drives gear "A". The speed ratio between gears "A" and "B" can be changed to alter the production rate of the card. As shown, the doffer shaft is connected to both the other rolls in the front zone and to the feed roll at the feed end of the card. Gears can be changed to alter the relative speeds of the front zone and the feed roll so that the draft of the machine can be changed. A change in draft, which usually is in the 100 to 120

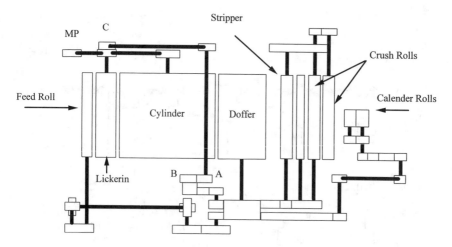

Figure 5.21. Diagram of a single motor card drive.

range, will change the exiting sliver weight. The relative speeds of the calender rolls and doffer must be set to give as low a tension draft as practical to prevent undue stretching of the delicate fibrous web as it flows forward. A typical such draft is 1.35. The can coiler, not shown, is driven by the same gear train that drives the front calender rolls. The total connected drive power for the card arrangement shown in Figure 5.21 is 6 kilowatts.

To facilitate changing the card production rate, draft, and simplifying the control of autoleveling, the drives of the most recent model cards have been changed to utilize additional direct current motors. The diagram of Figure 5.22 shows a multi-motor card drive. Motor no. 1 is an alternating current motor that drives the cylinder, lickerin, and flats. Motor no. 2 drives the feed roll and the chute delivery rolls. Motor no. 3 drives the front zone elements and the can coiler. Motor nos. 2 and 3 are direct current motors for ease of speed control by electronic circuits. The total combined power of these three motors is 4.9 kilowatts. All of the card drives are equipped for a relatively slow thread-up speed which is usually one-third the production speed.

Autoleveling

When cards were fed with fiber laps made on the machine called the picker, or scutcher, the control of weight variability of the sliver produced on the card was associated with the picker lap

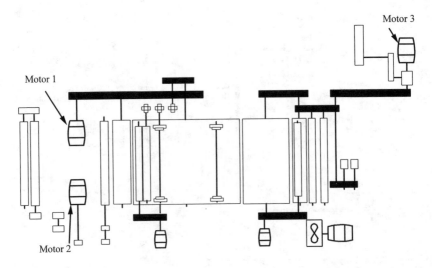

Figure 5.22. Diagram of a multi-motor card drive. (Courtesy Truetzschler GMBH & Co.)

regularity. Thus, there was no weight regulation needed on the card itself. With the advent of chute feeding it was quickly learned that keeping the mat weight consistent was very difficult because of slight changes in fiber frictional properties, changes of temperature in the air used for transporting fiber stock, changes in atmospheric pressure, and other factors. These factors generally caused a relatively slow drift in mat weight referred to as long-term drift. The first long-term levelers that were developed utilized the principle of measuring the sliver bulk exiting the card with resulting corrective changes in feed roll speed (input) to counterbalance a detected change in output bulk. These systems of control did improve the overall consistency of sliver weight. One such attachment is the "Card Control" supplied by Zellweger Uster AG.

A diagram of a pneumatic sensing trumpet is shown in Figure 5.23. As the fibers in the web (or loose rope) are condensed in the two-step trumpet, air is displaced and causes pressure to be generated in the small tube fixed to the exit hole (A) in the trumpet. The tube is connected to a pressure transducer which signals the control system when a change in pressure occurs. More fibers being condensed in the trumpet will cause an increase in pressure; when that occurs, the system will cause the feed roll to slow down slightly. Fewer fibers will cause the opposite reaction. For this control to remain calibrated, it is critical that the fibers remain consistent in size

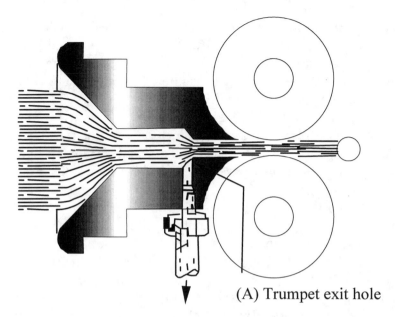

(A) Trumpet exit hole

to pressure transducer

Figure 5.23. Diagram of a pneumatic sensing trumpet positioned at the front calender rolls of the card.

and moisture content. As an example, a change in micronaire of the cotton fiber would cause a shift in sliver weight. The average micronaire of the carded stock must consequently remain constant.

Another popular type autoleveler is one that senses the thickness of the mat as it is compressed under the feed roll. A thicker bulk of mat, which causes the feed roll to be raised, indicates the mat is heavier at that instant. When this occurs, the sensor circuit causes the feed roll speed to be reduced in proportion to the change in height of the feed roll. This method responds to changes in bulk (mass) on the input rather than on the output as described above, and is known in the trade as an open-loop control. The method that senses the output is known as a closed-loop control. This open-loop control can respond faster than the closed-loop control, and this allows corrections to occur within shorter lengths of sliver. Thus, it is referred to as a mid-term leveler, which means the weight of sliver is controlled in lengths approximately 1.0 to 3.0 meters.

An autoleveler as just described is manufactured by Automation Technology Corporation of Easley, South Carolina (USA). It is called Slivertrol and can be retrofitted to most card machines. A drawing of the thickness sensor of this unit is illustrated in Figure 5.24. One of these sensors is mounted on each end of the feed roll. The speed of the feed roll drive is altered in response to the vertical positioning of the feed roll.

Several other autoleveling methods have been used commercially to control card sliver weight including torque sensors on the feed roll, changes in light reflection from the cylinder, and segmented pressure plates to average the bulk of feed stock along the full length of the feed roll.

The more recent developments in autoleveling at the card are short-term levelers which aim at controlling the sliver weight in lengths of 25-mm (about 1 inch) to 200-mm (about 7.9 inches), depending on the type and model. One such device is the Masterleveler supplied by John D. Hollingsworth on Wheels of Greenville, South Carolina (USA). This unit is optional to be fitted on the Hollingsworth Type 2000 card. A schematic drawing of this unit is included in Figure 5.25, together with brief descriptions of its features. Note that the sensing mechanism shown at

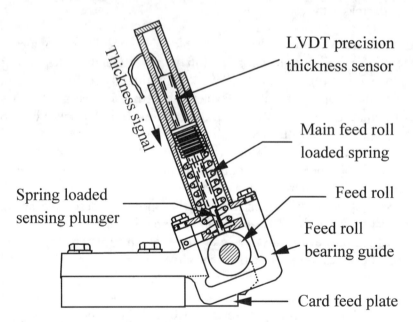

Figure 5.24. Drawing of the thickness sensor of the Slivertrol card autoleveler. (Courtesy Automation Technology Corp.)

Hollingsworth Masterleveler

An optional item on the Type
2000 Card, this leveler corrects
long-term, medium-term, and
short-term irregularities in
material length as short as 0.5 meter.

Features include:
a. Stepless correction response,
b. Simple electro-mechanical control,
c. Self-cleaning measuring system,
d. Control system operative at slow,
   fast, and piece-up speeds, and
e. Correction range of +/- 25%.

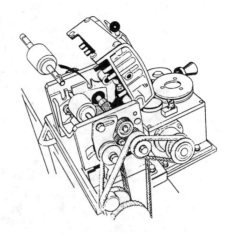

Figure 5.25. A schematic drawing and listing of features of the Masterleveler. (Courtesy John D. Hollingsworth on Wheels, Inc.)

top right is a tongue-and-groove roll pair. The drafting rolls are normal fluted rolls which are driven in response to the sensor rolls, giving a weight correction range of ± 25 percent.

Another card short-term leveling unit is produced by Myrik-White, Incorporated located in Durham, North Carolina (USA). Strain gauges are used to sense the pressure against the main trumpet, and a roller drafting system mounted immediately past the trumpet responds to the pressure signal from the trumpet plate. This mechanism can be retrofitted to a card that has a front drafting unit, such as the Marzoli Model CX300.

Another carding machine that embodies the short-term leveling principle is the "Sliver Machine" produced by Draper-Texmaco of Spartanburg, South Carolina (USA). The patented draft system, as shown in the photograph of Figure 5.26, utilizes tongue-and-groove drafting rollers which respond mechanically and electronically to pressure sensors located on the trumpet mounting plate. The main feature of this unique system, according to the manufacturer, is the production of a card sliver which may require no drawing processes before spinning.

Autodoffing

Until recent years the full cans of card sliver were doffed manually "by the clock". The operator was instructed to take out the

Figure 5.26. Photograph of the drafting unit on the Sliver Machine (Courtesy Draper-Texmaco Corp.)

full can and install an empty can under the coiling unit at a time interval to be regulated by a normal large face clock clearly visible to the operator. This method of doffing had certain drawbacks caused by imprecise doff times occasioned by unavoidable delays of the operator and card production interruptions for ends down. Thus, the sliver yardage fed into the cans varied a great deal. To correct this large variation the card manufacturers introduced autodoffing to the carding machine. These mechanisms, one of which is illustrated by the photograph in Figure 5.27, are important assets in increasing carding efficiency. A meter counter driven by the front calender rolls signals the unit to actuate at the total length preset on the counter. Air operated pistons push the full can from the coiler position and install an empty can ready to be filled; the counter is reset automatically. Such a system supplies full card sliver cans with much lower variability in sliver lengths between cans.

Figure 5.27. An autodoffing system on the C1/3 card. (Courtesy Rieter Machine Works, Ltd.)

Metallic Wire Clothing

Of all the individual components of the card, wire clothing has the greatest influence on quality and productivity. Criteria for choosing wire type include material being processed, type and design of card, allowable cylinder speed range, production rate required, and overall quality requirements. Exhibited in Figure 5.28 is a drawing illustrating the geometry of typical metallic wire. To make wire configurations for specific purposes, most of the dimensional distances can be changed. For example, a wire to increase the number of points per square inch would have reduced "a1", "T", and "h3" dimensions. The carding angle "b" is altered depending on the fiber to be processed. In the drawing of Figure 5.28 the carding angle is shown as a positive angle which is a nor-

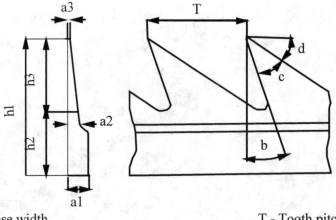

a1 - Base width

a2 - Tooth thickness at the root

a3 - Tooth thickness at the tip

h1 - Overall height

h2 - Height of the base

h3 - Depth of the tooth

T - Tooth pitch

b - Carding angle

c - Tooth apex angle

d - Tooth back angle

Figure 5.28. Typical geometry of metallic carding wire (clothing).

mal arrangement for cotton. For synthetic fibers which have higher frictional drag against the steel wire, the carding angle for best control is typically negative. The carding angle fixes the position of the tooth leading edge compared to a line perpendicular to the base of the wire. Normal wire tooth carding angles are given in Table 5.3.

After the teeth have been cut and the broaching burrs have been brushed, the teeth are hardened. The wire feed stock is fed into the continuous operation of broaching to shape the wire, for debur-

Table 5.3. Metallic Wire Tooth Carding Angles

Wire Location	Carding Angle Range, Degrees
Lickerin (Cotton)	+5 to +15
Lickerin (Synthetic)	0 to −10
Cylinder (Cotton)	+10 to +30
Cylinder (Synthetic)	+5 to +15
Doffer	+20 to +40

ring, and for hardening. The graph in Figure 5.29 gives tooth hardness numbers from the tooth tip to the tooth base. It is apparent that the tip of the tooth is the hardest. When the wire is ground to put a sharp edge on the tooth, the wire becomes progressively softer. Consequently, ground teeth will not stay sharp during production as well as the original teeth before grinding. This fact is shown clearly by the graph in Figure 5.30, which shows that the nep level in card sliver rises more quickly after grinding when compared to the original tooth condition of new wire. This reduction in quality and durability can be explained by the reduced hardness of the ground teeth and by the shape of the tooth tip, which departs from the sharp point of new teeth to the flat plateau of the ground teeth.

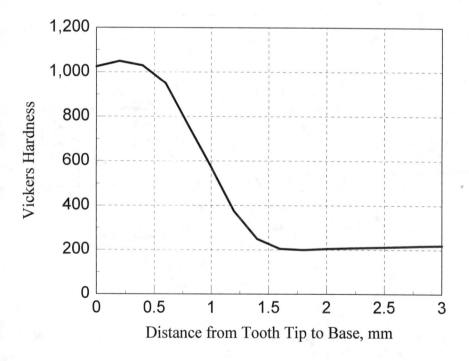

Figure 5.29. Graph showing typical wire tooth hardness at various distances from tooth tip to tooth base.

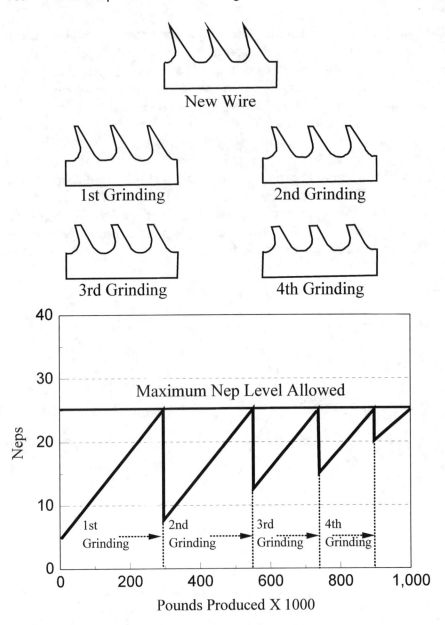

Figure 5.30. Typical grinding cycle of card cylinder wire in relation to nep quality and pounds of stock throughput.

The current practice of maintaining good sharp cylinder wire on the cards is to reduce the number of grinding cycles before new wire is installed. In some incidences mill management has decided to eliminate grinding altogether, opting for replacement when quality benchmarks are threatened.

Each wire manufacturer designates each wire type of specific dimensions with an identifying code number so it can be duplicated for continuing use. This is helpful in reordering and establishing continuity in the carding process.

The results of a test conducted by a member of the Institute staff in a plant environment are listed in Table 5.4. The yarn tested was 20's Ne cotton rotor yarn spun on an Autocoro machine in an Institute member textile plant. The two carding and spinning conditions shown are "Control" and "New". For the control condition, the combing roll in the spinning machine was covered with OS21 wire type which was the normal wire used for cotton spinning. The wire on the card cylinder was a wire with a carding angle of zero, commonly used for both cotton and polyester fibers. For the new wire condition, the wire on the spinning machine combing roll was code AB21 which is slightly different from the OS21 wire. On the card cylinder for this condition, a cotton wire with a positive carding angle was used. The results of the yarn test were clearly superior for the new wire condition. This test and many others justify the optimization of card wire for each particular carding situation.

Metallic wire normally is installed on the lickerin, the cylinder, and the doffer. The use of flexible fillet wire on revolving flats for carding cotton and intimate blends of synthetic fibers and cotton is the accepted practice. The useful life of wire can best be associated with the amount of fibrous stock processed. A guide for wire life expectancy is included in Table 5.5.

Table 5.4. Effect of Card Cylinder and Spinning Machine Combing Roll Wire on 20's Ne Cotton Rotor Yarns

Yarn Property	Wire Used	
	Control Combination Card Wire & OS21 Combing Roll Wire	New Cotton Card Wire & AB21 Combing Roll Wire
Uster %CVm	16.7	15.2
IPI Thins	55	18
IPI Thicks	320	128
IPI Neps (+280%)	52	13

Note: All comparisons significantly different @ 95% confidence level.

Table 5.5. Typical Wire Life Expectancy

Wire Location	Pounds of Fiber Processed
Lickerin	350,000
Cylinder	1,000,000
Doffer	Until Damaged
Flats (Fillet)	4,000,000

These numbers are for use as guides only. Many factors are involved in determining wire life, such as type fiber(s) being carded, production rate throughput, amount of trash (contaminates) in the cotton, and frequency of cleaning. Normal commercial practice is to change lickerins when the tips of the teeth on the lickerin wire become slightly worn. Some operators grind the cylinder wire once or twice to extend its life to some extent. Wear does not occur rapidly on the doffer wire, but the teeth are relatively weak (easy to bend) because of their height. Small clumps of fibers sometimes get caught between the cylinder and doffer causing damage to the doffer wire. Fillet flat wire must be ground frequently to maintain sharp tips on the wire points.

Point density of the mounted wire is determined by multiplying the number of teeth per linear inch of wire by the number of turns (lays) per inch of width of the wire installed on the rotating element such as the lickerin or cylinder. Normal ranges for wire point density are listed in Table 5.6.

Fiber Hooks Formed at the Card

An important characteristic of card sliver is the hooked nature of the fibers in the sliver. A thesis study by a graduate student, Robert J. McInnish, conducted during the academic year 1962-1963 at the Institute of Textile Technology, revealed that major hooks are formed in a large percentage of the fibers at the lickerin

Table 5.6. Wire Point Densities, Normal Ranges

Wire Location	Density Ranges	
	Points/Sq. Centimeter	Points/Sq. Inch
Lickerin	3.7 to 7.4	24 to 48
Cylinder	62.0 to 139.5	400 to 900
Doffer	54.3 to 68.2	350 to 440
Flats (Fillet)	41.9 to 89.3	270 to 576

teeth, and these hooks remain in the fibers as they are formed into sliver. These hooks are in the leading direction as they are formed by the harsh action of the wire teeth on the lickerin surface and remain in the leading direction until the fibers are transferred to the doffer wire where they reverse direction. Thus, the major hooks in the fibers are in the trailing direction as the sliver exits the card. The presence of these hooked fibers in card sliver makes it impractical to spin yarn directly from card sliver, especially in the spinning systems where roller drafting occurs. The downstream drawing processes remove a large percentage of these hooked fibers, and this function of the drawing process is quite important.

The combing process removes hooks by combing the fibers to a straight condition. This process is more effective in straightening the fibers from their hooked condition than is the drawing process.

Waste Removal and Vacuum Cleaning at the Card

The removal of non-lint contaminants from the fibers being processed is one of the card's primary functions. Such removal is accomplished at the lickerin as heavy particles are thrown out by centrifugal force; at the revolving flats which become loaded with fibers to act as a filter for short fibers, trash particles, and dust; and by general blowout from the lickerin and cylinder. The waste removal percentages for effective cleaning of cotton at these particular zones are in the following ranges:

1. Lickerin removal (undercard) at 1.75 to 2.25 percent;
2. Flat strip removal at 1.50 to 1.75 percent; and
3. Blowout (fly and dust) at 1.00 to 1.25 percent.

These waste removal percentages can be reduced when less cleaning is desired. Screens and covers set under the lickerin can effectively alter the quantity of undercard waste removed. Flat wire can be chosen to hold more or less strip load, and for 100 percent synthetic fiber a metallic non-loading flat might be the wise choice.

On all modern cards the blowout is captured inside cabinet enclosures and is removed by adequate suction systems. The drawing in Figure 5.31 illustrates by the darkened sections the suction system on the Truetzschler DK 740 card. The contaminated air exiting the screens and grids near the top of the chute is shown to be

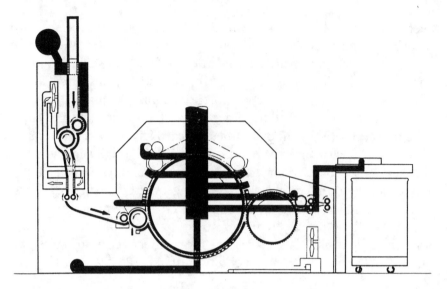

Figure 5.31. The continuous suction system on the DK 740 card, as shown by the shaded components. (Courtesy Truetzschler GMBH & Co.)

captured and sucked away through the circular duct connected to an exhaust system. In addition to blowout slots and covers on the card, an exhaust line is positioned near the floor of the card at the back to suck the undercard waste and to transport it up through the main vertical suction duct. A small fan is located near the floor at the front of the card to blow all the undercard accumulations toward the waste extraction orifice.

Shown clearly in Figure 5.31 is a screen under the front half of the lickerin extending forward under the bottom sector of the cylinder to near the set line between the doffer and cylinder. At both sides at the front edge of this screen, as well as at the side spaces between the doffer ends and the card frame, lint tags tend to accumulate. As is the case with all lint tags, they continue to enlarge until pulled away by the flow of the stock stream. This occurrence of being pulled into the stock causes slubs to be introduced into the sliver. Such slubs are devastating to downstream efficiency and product quality. Even now it is difficult to clean out these tags before slubs occur. Similar tags tend to develop at the ends of crush rolls, on crush roll clearer blades, on condensing belts, and above and near the main front trumpet. Thus, there still remain many opportunities for improved automatic systems that will eliminate tagging at the front end of the carding machine.

Tandem Carding

Over a hundred years ago, carding cotton fibers twice was common practice in Lancashire, England. The practice was cumbersome because the sliver from the first carding step had to be formed into a lap in order to feed it to the second card. This idea was revived in about 1961 at Swift Spinning Mills, Inc. (USA) by Otis B. Alston in an investigation to improve the cleaning of cotton. Two conventional cards were joined together in developing the Duo-Card. The web from the doffer of the breaker card (first card) was transferred to a doffer cylinder which replaced the feed and lickerin section of the finisher card (second card). Even with an increase in production, the nep levels in the web and the appearance of the subsequent yarns were improved by use of the Duo-Card. Such cards were set up in only a few plants because of the complexity of changeover and relative high cost of maintenance.

Crosrol Ltd. of Halifax, England, developed a more precision tandem card, initially called the Crosrol Varga Tandem Card. Its introduction was in the 1970s, timed with the advent of rotor spinning, which is seriously affected by unclean cotton slivers. A later model is illustrated in Figure 5.32, and a drawing showing important elements is located in Figure 5.33. The production rate of this card is in the range of 27 to 55 kg per hour (60 to 120 lb per hour).

Note in Figure 5.33 that the Crosrol tandem card incorporates two crush roll units. Also, the finisher card is fed with the relative thin web produced by the breaker card, enhancing the carding efficiency of the finisher card. Dust is further reduced by the suction plenums applied to both sections of the card.

The results of a test conducted at Texas Tech University to compare single versus tandem carding in preparing 100 percent cotton stock for rotor spinning led to the following conclusions:

1. The breakage rate incurred when rotor spinning from single carded stock was always greater than when spinning from tandem carded stock.

Figure 5.32. Crosrol tandem card. (Courtesy Crosrol Ltd.)

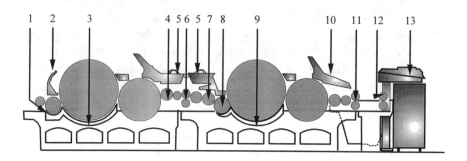

1. Fiber retreiver
2. Dust & waste extraction
3. Under screens
4. Stripping unit
5. Dust & waste extraction
6. Crush rolls
7. Transfer rolls

8. Taker-in
9. Under screens
10. Dust & waste extraction
11. Stripping unit & crush rolls
12. Calender rolls
13. Planetary coiler

Figure 5.33. Diagram of Crosrol Varga tandem card elements. (Courtesy Crosrol Ltd.)

2. The improvement in spinning performance arose because tandem carding was effective in removing materials which were not cotton fiber.
3. Finer yarns could be spun from tandem carded sliver as a result of improved cleanliness.
4. Tandem cards extracted more dust from the cotton, permitting yarns to be spun for longer periods between rotor cleanings without deterioration in yarn quality, particularly yarn strength.

The tandem card offers many advantages such as good cleaning efficiency, low nep count in card sliver, low ends down at spinning, and improved yarn appearance. Disadvantages are high initial cost of the card, higher power consumption, and somewhat higher maintenance cost.

Another source of tandem cards is John D. Hollingsworth on Wheels, Inc., Greenville, South Carolina (USA).

Monitoring of the Carding Process

There are several important qualities of card sliver that cannot be detected except by sample testing in a laboratory. Sample testing is done "after the fact", so much product can be produced before unsatisfactory quality is detected. On-line monitoring of quality characteristics, therefore, is much preferred. Zellweger Uster AG, of Uster, Switzerland, supplies such a system that can be applied to several textile processes, including carding. It is known in the trade as Sliverdata. Ultrasonic sensor trumpets are installed on all sliver producing units, including the card, to measure mass in sensitive length of approximately 2 cm at very high sliver speeds of about the neighborhood of 800 meters (about 875 yd) per minute. A diagram of the trumpet and a schematic of the electronic circuit are included in Figure 5.34.

Sliverdata as applied to sliver producing machinery supplies both on-line production and quality information. Production data recorded include amounts of sliver produced, frequency and duration of long stops, and reasons for long stops as coded at the machine entry stations. Quality data recorded include sliver weight variation, short-term mass irregularity, heavy places (slubs), mean

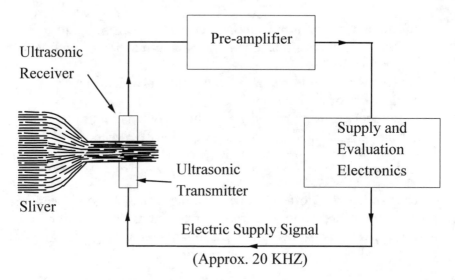

Figure 5.34. Diagram of sensing trumpet with schematic of electronic circuit of Sliverdata monitor. (Courtesy Zellweger Uster AG)

weight deviation, and spectrograms. In addition, alarm systems can be actuated by limits set on production and quality deviations, even to the point of stopping the machine. One of the latest additions to the quality record is the slub count in sliver.

Critical Factors for the Carding Process

In this section a number of critical factors that influence quality of the carding process are considered.

Card Mat Openness

The higher the degree of openness of the fibrous mat fed to the card, the smaller the tuft size in the mat. Consequently, the fibers are easier to separate at the lickerin. Also, the draft of the mat as it is condensed under the feed roll is more stable. When the fibers are thoroughly separated prior to the cylinder, more effective carding takes place between the cylinder and flats; hence, fiber alignment, mass uniformity, cleaning, and nep removal are enhanced.

In a recent plant study of rayon fiber carding, it was found that nep removal and Uster %CVm were significantly improved when the card mat was more open. The results are listed in Table 5.7.

Table 5.7. The Effect of Mat Openness on Rayon Carding Characteristics

Measurement	Mat Openness Index (ITT	
	145	115
Card Nep Removal Efficiency, %	91	77
Card Sliver Evenness, Uster %CVm	6.1	6.8

In a thesis study completed in 1994 at the Institute of Textile Technology, graduate student Jeff S. Frye determined that the following sliver and yarn quality characteristics were significantly improved when the Openness Index of the cotton mat was increased from 112 to 135:

1. Card Cleaning Efficiency,
2. Yarn Single-End Strength,
3. Yarn Short-Term Evenness,
4. Yarn 50-Meter Evenness,
5. Yarn Appearance Grade,
6. Uster IPI Thins (–50%),
7. Uster IPI Thicks (+50%),
8. Uster IPI Neps (+200%),
9. Classimat Minor Defects, and
10. Classimat Long Thin Defects.

Yarns were 20's Ne of 100 percent cotton, spun on the ring system. In many plants the average ITT Openness Index of card mats is significantly below the 140 target. Table 5.8 contains openness data generated in recent plant studies.

Card Mat Weight

The weight of card sliver being produced on groups of cards processing the same product in a mill location is generally well controlled. However, the weight of the mat being fed to these same cards is not well controlled. This leads to quality inconsistencies caused by variations in speed of the feed rolls and variations in draft in the cards. Typical findings in textile plants are listed in Table 5.9. Those data show the relatively large variation that exists from plant to plant. It is difficult to specify exactly which mat weight is optimal; however, cards are best suited for drafts between 100 to 120. There are mat related details that should be controlled including the following:

Table 5.8. Card Mat Openness Data from Recent Plant Studies

Plant	Fiber Type	Average Card Mat Openness Index
1	Cotton	100
2	Cotton	117
3	Cotton	110
4	Cotton	125
5	Cotton	141
6	Cotton	132
7	Polyester/Cotton	135
8	Polyester/Cotton	155
9	Polyester/Cotton	145
10	Polyester/Cotton	117
11	Polyester/Cotton	138
12	Polyester/Cotton	141
13	Polyester	146
14	Rayon	145
15	Rayon	122

1. Mat Weight Variation—Even more important than the average mat weight is the consistency of mat weights from card to card. Every yarn plant monitors sliver weight variability, and it is expected that the weight variation be no more than 1.2 to 1.5 percent. Very few plants monitor how much the weight varies in the mat from card to card, and that variation will have some effect on weight variation, cleaning, nep removal, and alignment of fibers in the card sliver. Mat weight variation in textile plants typically is between 8 and 12 percent. Those plants that have started monitoring this variation have reduced weight variation in card mats to below 3 percent. The results have been noticeably fewer stops at carding and better mass uniformity. Card mat weight and uniformity can be calculated by hand, or with the Institute's SMART-MAT program on the computer.

2. Mat Tension—The tension between the chute delivery rolls and the feed roll to the card should be enough to keep the mat from buckling, but not so high as to pull thin places in the mat. If the mat "buckles" up when tufts are removed, then tension needs to be increased.

3. Chute Run Times and Spanker Plate Rates—To provide the most consistent stock from centimeter to centimeter or inch to inch in the mat, chute feed rolls that are not continuous feed

Table 5.9. Mat Weight (Under Feed Roll) and Card Draft Data from Recent Plant Studies

Plant	Fiber	Mat Weight Under Feed Roll		Card Draft
		gm/m	oz/yd	
1	Rayon	1000	32.3	157
2	Rayon	403	13.0	98
3	Cotton	772	24.9	122
4	Cotton	481	15.5	104
5	Cotton	425	13.7	100
6	Cotton	490	15.8	98
7	Cotton	679	21.9	112
8	Cotton	465	15.0	101
9	Cotton	806	26.0	126
10	Polyester/Cotton	682	22.0	132
11	Polyester/Cotton	319	10.3	74
12	Polyester/Cotton	471	15.2	95
13	Polyester/Cotton	394	12.7	86
14	Polyester/Cotton	499	16.1	117
15	Polyester/Cotton	620	20.0	145
16	Polyester	484	15.6	94

should run as much as possible, preferably over 85 percent of the time. In addition, chutes with spanker plates should have the spanker rate equal to or higher than 60 beats per minute. Slow spanker rates cause thin places to be pulled into the mat. Inch-to-inch variability in the card mat translates into yard-to-yard and meter-to-meter variation in card sliver, which can be directly related to count variation in the yarn.

Lickerin Zone

The lickerin is the location where a great deal can be done to help or hurt the quality of the card sliver. Effective separation of the fibers allows the cylinder to do its job, but over aggressiveness can damage fibers, which will cause downstream problems with ends down and yarn quality. Lickerin settings depend on mat openness, throughput rate of the card, and the characteristics of the fiber being processed. In most cases lickerin settings are set overly aggressive rather than under aggressive. The best indicator of whether the lickerin is damaging fibers is by checking fiber mean length and short fiber content in the card mat and card sliver. An increase of more than 2 percentage points of short fiber signals damage is occurring. If damage is occurring check the following:

1. Feed plate-to-lickerin setting—In most cases settings tighter than 0.43 mm (0.017 inch) for cotton and 0.56 mm (0.022 inch) for synthetics are too close. However, this setting is dependent upon carding rate, so for high speed carding (above 27 kg/hr or 60 lb/hr) feed plate-to-lickerin settings will likely have to be increased even further.
2. Lickerin speed—The appropriate lickerin speed depends on the wire and the openness of the stock entering. For 6.2 pt/cm^2 (40 pt/inch^2) wire, lickerin speeds of approximately 900 to 950 rpm are common with lickerins of 23.5 cm (9¼ inches) diameter. These speeds result in surface velocities of 665 to 701 m/minute (2180 to 2300 ft/minute). For lickerins with larger or smaller diameters than 23.5 cm (9.25 inches), the speed should be set to give equivalent surface speeds.

Waste Removal Uniformity

The amount of undercard waste removed at carding influences the weight of card sliver, efficiency of cleaning, and nep removal capability. The waste removed from each card in a set has been found to range as high as 100 percent. Waste removal must be maintained consistent by careful adjustments of the screens, mote knives, and carding plates associated with the lickerins.

Flat Strip Weight

Most of the short fibers and dust removed at the card are removed in the flat strips. On cards that remove strips, it is important not only for the cards to remove consistent levels of flat strips from card to card, but also that the strip is sufficiently heavy. A good target for flat strip weight is 30 grains/strip. Light strips usually indicate either cylinder-to-flat settings that are too open or the percentage plate at the front of the card is set too close. Normally, the percentage plate setting is one mm or 0.040 inch. If the percentage plate is not open enough, some of the trash and short fiber loosened by the cylinder and flats will be held on the cylinder instead of being released in the strips. It is not uncommon to see flat strips ranging from 19 to 35 grains/strip in the same card room.

Cylinder and Flat Wire

Cylinder wire is critical for nep removal, fiber parallelization, and mass uniformity. The Card Room Manager appreciates the value of sharp wire for quality. In addition to wire sharpness, wire point density and speed of the cylinder are very important. Two critical issues that make wire density so important are use of finer denier synthetic fibers and lower micronaire cotton.

Too often nep levels become unacceptable when finer fibers are carded. In the majority of these cases the cylinder wire is suited to coarser fibers. When polyester finer than 1.0 denier is being processed, 110 to 140 points per square centimeter or 700 to 900 points per square inch (depending on card speed) should be on the cylinder wire. Similarly, if low micronaire cotton is being processed, the number of points must be sufficient to handle all the fibers.

Just as important as the cylinder are the flats. Dull flats can prevent neps from being removed, but loaded flats can be just as ineffective. Fibers and trash packed between the flat wire prevent the flats from penetrating the stock on the cylinder. Dull flat wire is corrected by grinding. Loaded flats can be corrected by proper brushing.

Doffer Wire Loading

Loaded doffer wire causes holes to be produced in the card web. These holes generate small slubs and neps in the card sliver. Trash and/or fiber accumulations can cause loading. Doffers should be inspected frequently, at least once every 8 hours, and any loading should be brushed free.

Tension Draft on Web and Sliver

Tensions on the web between the doffer and the main calender rolls and tension on the sliver between the main calender rolls and coiler calender rolls are caused by the total positive drafts (speed ratios) set for these zones. Tension that is too high in either of these locations has an adverse effect on the short-term evenness (Uster %CVm) of the sliver, because uncontrolled fiber slippage is likely to occur. If possible, the draft on the web should be less than 1.35. The primary drawback to lower tension is floppiness of the web, causing ends down. Too little tension should be avoided, but overcompensating with a web tension that is too high will hurt mass evenness.

Similarly, too much tension on the sliver going to the coiler can hurt short-term mass evenness. In many carding operations, this tension draft is set on the high side, making it easier to thread the sliver at start-up. The tension draft in this zone should not be set higher than 1.05.

Slub Generation

The two most common causes for slub generation are lint accumulations and excessive pressure at the main calender rolls. Slubs that are caused by lint accumulations on cards with visible open webs can be seen by very careful observation of the open web as it approaches the main trumpet, but this method of detection is very inexact. Slubs caused by excessive calender roll pressure cannot be seen by looking at the sliver in its path to the coiler. The Institute Sliver Analyzer, as described below, is useful in determining slub level in card sliver. Also, the Zellweger Uster Sliverdata Monitoring system can supply data on slub counts.

Fiber accumulations can develop at several places on the card. A very common location is the nose of the screen just below the set line of the cylinder to the doffer. Another location for lint accumulations is in the space between the ends of the doffer and card frame. Some cards tend to have lint buildup on the top of the doffer. During the operation of the carding machine, these lint tags grow in size to the point where some of the tag is caught into the material flow where it will produce the slub. Even though a good suction (vacuum) system will reduce some of these lint tags, it will continue to be necessary to clean the card at frequent intervals. One frequency that is commonly used is every 8 hours.

Lint tags that occur at the ends of the stripping roll and crush rolls cause rough web selvedges, a condition that is referred to as "snowballing." This condition causes both slubs and torn webs which usually cause ends down. The most successful strategy in prevention of slub generation is keeping the card clean.

The Sliver Analyzer is a very simple tool for the analysis of defects/slubs in sliver. It utilizes a drafting zone similar to that found on a draw frame and a lighted box. The lighted box is mounted so that the face of the box is nearly perpendicular to the floor and the drafting unit delivers stock down over the lighted surface.

As is shown in the schematic diagram front view of Figure 5.35, the drafting unit delivers a fiber web which feeds down in front of the light box. When the fibers are backlit, it is possible to easily detect defects in the sliver. Slubs will appear as dark spots in the fiber web. Slubs will also have a detectable mass which can be judged by pressing lightly against the slub with a finger or thumb.

Because the Sliver Analyzer is a subjective measurement tool, it is suggested that sliver samples from trials be viewed back to back with standard sliver in order to make accurate and consistent classification of the two slivers. An important consideration is to ensure that different observers classify slubs using the same criteria

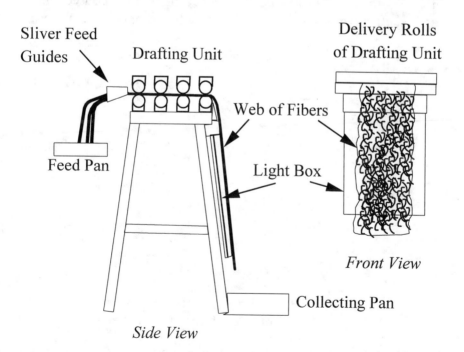

Figure 5.35. Schematic diagrams of the Sliver Analyzer.

(such as: size, darkness, mass). Therefore, it may be useful to place a web containing the minimum size slub on a black background and use this as a reference for different observers. An accurate assessment of the level of slubs can be made by using 225 grams (about 0.5 lb) of sliver.

Production Rate and Sliver Weight

The performance of cards is a function of production rate and mass being processed. In most cases the performance is compromised when the throughput gets too high. Modern cards perform well at production rates up to 36 kg per hour (80 lb per hour) for cotton and 45 kg per hour (100 lb per hour) for synthetics. Numerous studies have shown the effect of carding production rate on quality as illustrated in Figures 5.36 and 5.37. Essentially, all data presented in the literature indicate that lower throughput improves quality. Less emphasis has been placed on the influence of sliver weight on quality. As sliver weight is increased, more mass of fibers must be processed on the cylinder and, consequently, carding effectiveness is decreased. A safe benchmark for maximum

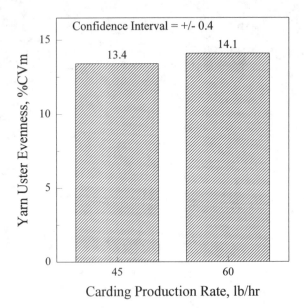

Figure 5.36. Graphical representation of the effect of card production rate on the evenness of 38's (Ne) 100 percent combed cotton yarns.

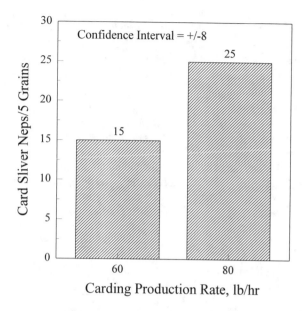

Figure 5.37. Graphical representation of the effect of card production rate on the number of neps measured in the resulting 80 grains/yard cotton slivers.

sliver weight for the production of good quality results is 4.60 grams per meter or 65 grains per yard. In trials conducted on the Model 300CS Marzoli card at the Institute, a greater deterioration in cotton sliver (Uster %CVm, neps, and alignment) resulted when sliver weight was increased from 4.25 to 5.67 grams per meter than when production rate was increased from 27.3 to 36.4 kg per hour (60 to 80 lb per hour).

Quality Benchmarks for Card Slivers

Quality expectations for most high quality carding operations are given in Table 5.10. When card sliver properties meet these quality criteria, both downstream processing and product quality should be excellent. Naturally, there will be some special carding situations that will not be required to meet these quality levels.

Manufacturers of Carding Machines

Until about 1980 many of the carding machines in commercial operation were termed "rebuilt high production" cards. This

Table 5.10. Quality Benchmarks for Card Slivers (Grains per yard ranges and variabilities apply to 65 grains/yard sliver)

Property Measurement	Desired Results
Weight of 100-Meter Lengths	± 0.11 grams/meter or 1.5 grains/yard; 1.2% variability maximum
Weight of 3-Meter Lengths	± 0.23 grams/meter or 3.25 grains/yard; 2.5% variability maximum
Short-term Evenness, Uster %CVm	4.0% maximum
Uster Spectrogram Periodicities	Zero
Uster Inert Test (3-Meter Lengths)	10% Range; 2.5% variability maximum
Nep Count, AFIS test	Cotton, 62 neps per gram maximum Synthetic, 15 neps per gram maximum
Coarse Trash, MTM test	Cotton, 0.1% maximum Poly/Cotton Blend (50/50), 0.050% maximum
Sliver Analyzer Defects	Cotton, 1.40 per kg or 3 per lb maximum Synthetic, 0.45 per kg or 1 per lb maximum
Crown Height on Full Cans	30.5 cm or 12 inches maximum
Cans with No Breaks	85% minimum

meant that old cards had been modernized by machine companies that were so equipped to upgrade the mechanism to allow higher cylinder/lickerin speeds for substantial increases in productivity. Today, however, the tendency is to modernize the carding process by replacing all machinery with the latest models of newly built machines. The most popular suppliers of cards in the USA are listed in Table 5.11.

Table 5.11. Suppliers of Popular Carding Machines for the USA Market

Company	Model	Home Country
Crosrol Ltd.	Mark 5	England
Draper-Texmaco	Sliver Machine	USA
John D. Hollingsworth on Wheels, Inc.	2000	USA
Fratelli Marzoli & C. spa	CX400	Italy
Rieter Machine Works, Ltd.	C50	Switzerland
Truetzschler GMBH & Co.	DK 760; DK 803	Germany

The Drawing Process

In an ideal world the process of drawing, which is typically used after carding and combing, would not be necessary. The draw frame corrects for weight variation and fiber misalignment that is introduced by the card. Similarly, the comber introduces weight variation in the form of piecings that the drawing process subsequently helps to reduce. Over the years as the card, comber, and draw frame have been improved to deliver sliver with more uniform mass per unit length, the number of drawing processes has been reduced, especially when the sliver is being used in rotor spinning. However, for ring spinning, air jet spinning, and fine count rotor spinning (finer than 24's Ne), drawing is critical to provide the necessary fiber alignment in the sliver to achieve satisfactory yarn strength and defect levels. In the future, though, as attempts are made to shorten the yarn manufacturing process, drawing will be a likely candidate for elimination.

Fiber Alignment in the Drawing Process

In the drawing process, fibers are moved past each other in an attempt to align the fibers without breaking the continuous sliver strand formed at the previous process. The movement is achieved by using several pairs of rollers, as shown in Figure 6.1, which run progressively faster from the beginning to the end of the process. As the fibers move past each other, the hooked ends are straightened.

An example of how fibers are straightened by using rollers is illustrated in Figure 6.2. Two pairs of rollers are shown with an arrow indicating the direction of fiber movement. The rolls labeled set "A" are moving at a surface speed of 100 meters per minute, and set "B" rolls are moving at a rate of 500 meters per minute

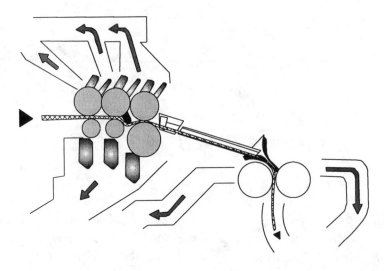

Figure 6.1. Schematic of the 3-over-3 drafting system of the RSB 851 draw frame. (Courtesy Rieter Machine Works, Ltd.)

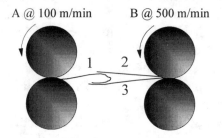

A @ 100 m/min   B @ 500 m/min

Figure 6.2. Illustration of the fiber aligning mechanism at the drawing process.

(m/min). To simplify the situation, just three fibers are considered instead of the thousands that usually are present at any given instant. In this illustration, fiber 1 is moving at 100 m/min because it is held by roll set "A", but fibers 2 and 3 are moving at 500 m/min because they are controlled by roll set "B". When the faster fibers come in contact with fiber 1, the tendency will be to pull the fiber, and the hooked end will become aligned with the other fibers. The ratio in surface speed between the two pairs of rolls influences how much improvement in alignment will occur.

The mechanism of moving fibers at a faster rate through the machine also causes the fiber bundle to become lighter. In fact in the previous example illustrated in Figure 6.2, the mass per meter

Figure 6.3. View of multiple ends of sliver fed into the draw frame. (Courtesy Rieter Machine Works, Ltd.)

would be approximately five times lighter after roll set "B" (compared to before roll set "A") because the roll speed was five times faster. This reduction in mass is called actual draft. In a typical drawing process, the draft is between 5 and 9.

Because the weight per unit length is reduced at drawing, more than one sliver is fed into the machine to offset this situation. In a typical plant six to eight slivers would be fed into the draw frame simultaneously as shown in Figure 6.3. The multiple slivers being fed are called doublings. For example, if six slivers were being fed into the draw frame, there would be six doublings. Doublings are helpful for providing a mixing at drawing that helps to reduce mass variation in slivers produced at the previous process by averaging out the heavy and light sections of the sliver. More simply put, if a sliver is lighter than normal, but it is doubled with five other slivers, the impact of that light sliver is lessened.

There is a useful equation for predicting the impact of doubling on the long-term weight uniformity of drawn fiber assemblies. The approximate change in weight variation is estimated by the following equation:[6.1]

$$\text{Weight \%CV}_{\text{out}} = \frac{\text{Weight \%CV}_{\text{in}}}{\sqrt{\text{\# doublings}}} + 0.2 \quad (6.1)$$

If six doublings of 60 grains per yard (4.25 grams per meter) card sliver are fed into drawing, and the machine is set up with a draft of 6.5 (ratio of surface speed between back and front rollers), then the delivered sliver weight would be determined by the following equation:[6.2]

$$\text{Delivered Weight} = \frac{\text{Weight Fed}}{\text{Draft}} \quad (6.2)$$

$$= \frac{6 \times 60 \text{ gr/yd}}{6.5} \text{ or } \frac{6 \times 4.25 \text{ g/m}}{6.5}$$

$$= 55.38 \text{ gr/yd or } 3.92 \text{ g/m}$$

This calculated weight is only approximate because the relationship between actual draft and mechanical draft is influenced by roll pressures, tension on the sliver prior to and after the drafting rolls, and fiber characteristics. Consequently, the above equation(6.1) is used as a starting point. If the input weight is known, the delivered weight is calculated based on the draft gears set up on the machine. The actual weight must be measured and could likely be different by up to two percent.

Purpose of Drawing

The drawing process orients the fibers in the input sliver through drafting that occurs between several pairs of rollers. Drafting reduces the weight per unit length of the sliver, thus requiring several slivers to be fed. This doubling of slivers helps to blend out inconsistencies that occur in the individual slivers produced at the previous process.

As a result of the drafting and doubling that occur at drawing, the purpose of the draw frame is to accomplish the following:

1. Straighten the fibers and make them parallel to the central axis of the sliver;
2. Improve short-term, mid-term, and especially long-term evenness in the sliver;

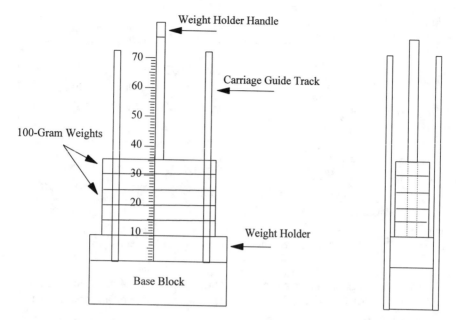

Figure 6.4. Drawing of the ITT Bulk Density Tester, used for sliver fiber alignment measurement.

3. Improve consistency of the product delivered to spinning for lower yarn variation; and

4. Produce the proper weight sliver for the following process.

Fiber Straightening

The degree to which fibers are straightened, or aligned, in the drawing process is important, but not often quantified in the yarn manufacturing plant environment. Two procedures that are used to quantify the alignment include bulk density and drafting force analysis.

The bulk density tester is shown in Figure 6.4 and quantifies how much the sliver will compress under load. As fibers become more aligned, they will compress more. In this test several cuts of sliver are placed in a 20.3-cm (8-inch) trough. The height of the sliver bundle is measured after 200, 400, and 600 grams of weighting are applied. An alignment index is then calculated based on the total volume and the weight of the sliver. The entire procedure for determining a sliver alignment index is documented in Appendix I.

Fiber Alignment Index

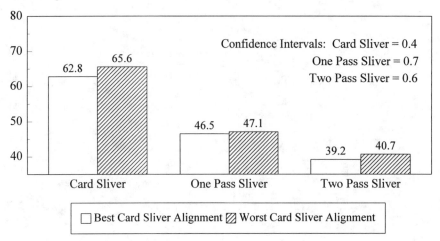

Figure 6.5. Graphical illustration of the effect of number of drawing passes on fiber alignment index with a lower index meaning better alignment of the fibers. (Barnes, 1996)

As shown in Figure 6.5, the alignment index of test slivers improved (smaller number indicates better alignment) as the slivers were drawn. For card slivers the improvement in alignment at the first passage of drawing was approximately 27 percent, and an additional 14 percent improvement was achieved at the second drawing passage (Barnes, 1996). Because the majority of alignment improvement occurs in the first drawing passage, some rotor spinners are now using only one passage of drawing.

Drafting force is also used to quantify alignment. The two instruments currently available to measure drafting force are the ITT Draftometer and the Rothchild Cohesion Tester. The drawing process helps to reduce the drafting force of sliver as shown in Figure 6.6. In turn, lower drafting force has been shown to help improve strength and evenness in rotor and air jet yarns.

Evenness of Drawing Sliver

The distance between the rolls on the draw frame must be adjusted to suit the length of the fiber being processed. Unfortunately, not all the fibers are the same length, so the rolls must be set to handle the longest fibers in the bundle. The fibers that are

Drafting Force, grams/grain/yard

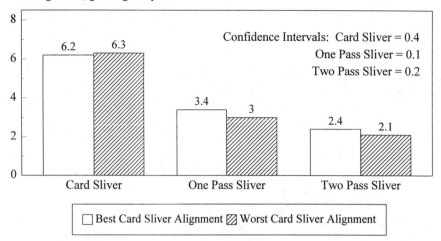

Figure 6.6. Graphical illustration of the effect of number of drawing passes on the force required to draft 100 percent cotton slivers. (Barnes, 1996)

shorter than the roll settings will not be under control of either set of rolls for a given period of time. These floating fibers tend to cause unevenness in the strand, known as drafting waves, because they tend to group together with other short fibers and come through the draft zone in a group or wave. The more short fibers that exist and the higher the draft, the more pronounced the drafting wave will be.

On evenness testers the magnitude of the drafting wave can be monitored by use of the spectrogram that denotes repeating defects. As shown in Figure 6.7, the drafting wave between 5 and 18 inches on the drawing sliver spectrogram is much higher than it is on the card sliver spectrogram, even though the short-term mass variation (%CVm) is better in the drawn sliver. With roller drafting, a drafting wave will exist corresponding to each zone. Total draft, draft distribution, roll spacings, roll pressures, and pressure bar settings can all be set to control these drafting waves so that they do not have a detrimental effect on yarn defects or fabric appearance.

Despite the spectrogram often getting worse at drawing, the short-term, mid-term, and long-term evenness are often improved. Improvement in long-term evenness is most often accomplished

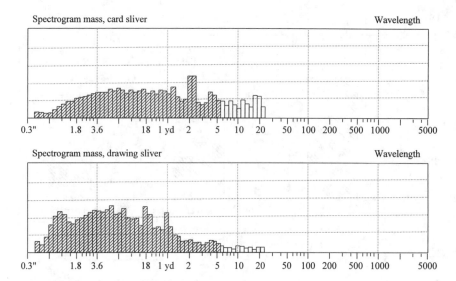

Figure 6.7. Spectrograms of card sliver and drawing sliver.

through the help of sliver doublings. Improvement in short-term evenness (%CVm) is less likely to occur, especially if autoleveling is used at the card and if the short-term variation of the input sliver is already low. However, 1-meter, 3-meter, and 10-meter mass variation improvements are expected. Table 6.1 contains evenness data on several card slivers and the corresponding evenness after one and two passes of leveled drawing.

For the short-term leveled (Hollingsworth) and mid-term leveled (Truetzschler) card slivers, the short-term evenness was not improved at drawing. However, for each card sliver, 1-meter and 3-meter mass variations were improved at the first passage of drawing. In some cases minor improvements were recognized in the second drawing passage as well. It is interesting to note how similar the three slivers were after two drawing passages.

Common Draw Frames in USA Industry

The draw frames currently used in the USA textile industry are divided into two categories: leveled and non-leveled. Leveled draw frames contain mechanisms to measure the mass of sliver entering the machine and electronically send a signal to the draft zone to cor-

Table 6.1. Influence of Drawing on Short-Term and Mid-Term Mass Variation

Measurement	Truetzschler DK 760	Crosrol Mk5	Hollingsworth 2000
Uster %CVm in:			
Card Sliver	2.93	5.46	2.98
Once Drawn Sliver	3.17	3.29	3.07
Twice Drawn Sliver	3.13	2.94	3.03
1-Yard %CV in:			
Card Sliver	1.77	3.47	0.75
Once Drawn Sliver	0.56	0.53	0.43
Twice Drawn Sliver	0.55	0.51	0.56
3-Yard %CV in:			
Card Sliver	1.60	2.83	0.50
Once Drawn Sliver	0.42	0.37	0.32
Twice Drawn Sliver	0.35	0.31	0.39

rect for off-weight situations. Draw frame leveling is quickly becoming standard within the industry. Currently, leveled draw frames produced by Cherry, Hollingsworth (also Saco Lowell), Rieter, Vouk, and Zinser are being used in USA plants. Truetzschler and Howa also have autoleveled frames in trial applications. Rieter, a provider of many draw frame models over the years, has a new autoleveled frame to soon be available which is designated as the D-30 model.

Common unleveled draw frames currently used in the USA industry are as follows:

1. Zinser 720;
2. Rieter DO/2, DO/6;
3. Rieter SB 51, SB 851, SB 951;
4. SacoLowell DF-11; and
5. Vouk SH2

Technical details of selected draw frames are listed in Table 6.2.

Main Parts of the Draw Frame

Today's draw frames are usually built with one or two deliveries. Single delivery frames are more flexible and efficient, whereas two delivery frames have definite initial cost and floor space advantages. A common draw frame in today's industry is the Rieter

Table 6.2. Technical Details of Popular Draw Frames in the USA Industry

Manufacturer	Model	Drafting System	Maximum Delivery Speed (m/min)	Auto-leveler	Number of Deliveries
Cherry	DX-500	5-over-4	500	Yes	2
	DX-800	5/4	800	Yes	2
Howa	DFK	3/3	500	Yes	2
	DFH	4/3	800	Yes	2
Hollingsworth	DF-11	4/4	800	No	2
	DJ-11	4/4	800	Yes	2
Rieter	DO/2	3/5	350	No	2
	DO/6	3/3	550	No	2
	SB51	3/3	600	No	1
	SB851	3/3	800	No	1 or 2
	SB951	3/3	900	No	1 or 2
	RSB851	3/3	800	Yes	1
	RSB951	3/3	900	Yes	1
	D-30	4/3	1000	Yes	1
Truetzschler	HS 900	3/3	900	Yes	1
	HSR 1000	3/3	1000	Yes	1
Vouk	SH2/D-E	3/4	600	Yes	2
	SH802/D	3/4	800	No	2
	SH802/D-E	3/4	800	Yes	2
Zinser	720/2	5/3	380	Yes	2
	720/2	4/3	500	Yes	2

RSB 951, which is shown in Figure 6.8. The main parts of a draw frame are the creel, drafting system, sliver condensing section, coiler, and suction system.

Creel

The cans behind the draw frame are arranged in a manner that is most practical for the floor space available in the plant. The two most common arrangements are the in-line creel, which has all of the cans lined up straight, and the nested creel, which has the cans arranged in a rectangular group. On single delivery machines, the nested creel is used exclusively, whereas on a dual delivery machine the in-line creel is more popular. The two types of arrangements are illustrated in Figure 6.9.

Figure 6.8. View of the RSB 951 draw frame filling rectangular cans. (Courtesy Rieter Machine Works, Ltd.)

Another option for the creel is whether the sliver will be pulled out of the can by the drafting rolls or whether power driven rolls above the cans will be used to help get the slivers out of the can. The power driven creels also incorporate a stop motion to detect

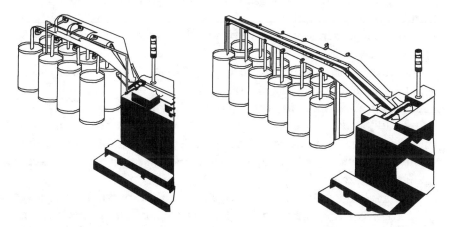

Figure 6.9. Illustrations of a nested creel arrangement (left) for a single delivery frame and an in-line creel arrangement (right) for a dual delivery frame. (Courtesy Rieter Machine Works, Ltd.)

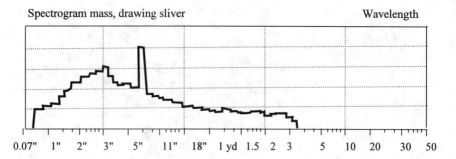

Figure 6.10. Spectrogram containing a drafting roll periodicity.

sliver breaks. If the stop motions do not have a sliver in between the two metal rollers, the machine will not operate. Power driven creels are used mainly with in-line creels, but some plants with rectangular creels also incorporate power driven creels to avoid sliver stretch.

Drafting System

Every model draw frame has some feature that is slightly unique in the drafting zone. The most basic arrangement is a two-zone system with three top rollers and three bottom rollers, forming two draft zones. The back zone is typically called the break draft zone, and the front zone is called the main draft zone. Draw frames are also available with 4-over-4, 4-over-3, 5-over-3, and 3-over-4 roller arrangements, all of which have been proven to provide acceptable results when set appropriately.

In today's draw frames, the top rollers are rubber coated and typically called cots. The bottom rollers are steel with fine flutes, usually parallel with the axis of the roll. The cots are usually between 70° and 83° Shore hardness, depending on the location in the draft zone and pressure applied. Usually, an anti-static treatment is applied to the cots, and in most cases ultra-violet treatment is used to prolong the life.

Damage to cots, bottom steel rolls, or to the bearings and gears used to drive the cots can be detected through the spectrogram provided by the evenness tester. Wear problems generally cause a periodicity on the spectrogram, similar to that shown in Figure 6.10.

Another critical component in the draft zone is the pressure bar. The pressure bar is situated in the main draft zone, helping to guide floating fibers as they move through this zone. As a result, the pressure bar serves to reduce the main drafting wave that appears on the spectrogram. The pressure bar acts essentially like another low pressure nip.

Sliver Condensing Section

The flat fiber web exiting the draft zone must be formed back into a sliver. In the condensing section, the web is gathered in a condensing funnel and is then passed through a sliver tube as shown in Figure 6.11. It is compressed further as it passes through a small hole in a condensing trumpet from which it is pulled by a set of calender rolls and delivered through a coiler tube and into a can.

Figure 6.11. View of the draft system of the RSB 851 draw frame showing the web condensing plate and tube on the right. (Courtesy Rieter Machine Works, Ltd.)

The act of condensing the sliver at the trumpet is critical for providing sufficient fiber-to-fiber cohesion to hold the sliver together at the next process. However, if too much condensing occurs, the draw frame will run poorly and/or thick places will be formed in the sliver.

Coiler Section

The coiler gear below the calender rolls lays the sliver into the rotating can after it passes through the coiler tube. The coiler tube consists of many shapes, sizes, and materials depending on the fibers used, speed of the draw frame, and sliver weight. The lay of the coils in the can depends on the following:

1. Speed of the coiler tube,
2. Speed of the bottom turntable, and
3. Alignment of the turntable to the coiler tube.

The diameters of most finisher drawing cans are between 450 mm and 600 mm (18 to 24 inches), and some breaker drawing cans are as large as 1000 mm (40 inches).

Suction System

As the fibers move through the drafting zone, dust and lint are dislodged from the fibrous mat into the air. The purpose of the suction system within the draw frame is to remove these particles from the air so that they do not redeposit on the drawing rolls. Collection of fibers on the roll surface quickly accelerates into greater and greater accumulation until a fibrous lap-up exists around the rolls. Lap-ups first cause unevenness to exist in drafting and eventually cause the machine to stop.

The suction system creates a vacuum by using a fan that pulls the dislodged dust and lint away from the draft zone, through a filter, and into a waste collection unit either at the machine or in a centralized filter house. For 100 percent cotton, it is expected that the draw frames will perform poorly if suction is less than 75 mm H_2O static pressure. Some modern draw frames require up to 125 mm H_2O pressure.

Critical Factors for Optimal Drawing

The draw frame can be divided into four major zones for discussion of optimization. These are the creel zone, drafting zone, coiling zone, and autoleveling zone. The major functions of each zone are as follows:

1. Creel Zone—Slivers should be delivered to the draft zone with the appropriate tension to avoid stretch while keeping the slivers taut. Also, excessive abrasion of the fibers should be avoided.
2. Draft Zone—The fiber bundle should be reduced to the appropriate weight while maximizing fiber alignment and avoiding creation of defects.
3. Coiling Zone—Sliver should be laid in the can to maximize the amount in the can while allowing for removal of the sliver without tangling and scuffing.
4. Autoleveling Zone—Autoleveler should be set up to provide a sliver with the most consistent short-term, medium-term, and long-term mass consistency.

The critical control factors for achieving optimal performance in each zone are discussed.

Creel Zone Optimization

The critical factors in the creel zone include the following:

1. Total mass fed,
2. Tension selection,
3. Creel height,
4. Stop motion function, and
5. Housekeeping.

Total Mass Fed

Perhaps the most critical decision for quality purposes is how many ends of sliver will be fed into the creel. The mass fed depends upon the number of doublings and the weight of each doubling. In the drawing process a trade-off exists regarding number of doublings. The more doublings that exist, the better the blending and consistency of the product; however, excessive mass in the draft zone makes it difficult to improve the short-term mass evenness. Consequently, between 6 and 8 ends are usually fed into the draw frame as a compromise. The more consistent the mass of the incoming slivers, the more likely it is that 6 ends will be preferred.

As a rule of thumb, the total mass at drawing should be kept below 38.3 g/m (540 gr/yd) for breaker or first pass drawing and 34.0 g/m (480 gr/yd) for finisher drawing. For drawing prior to combing, less than 28.3 g/m (400 gr/yd) is desirable.

Tension Selection

Tension selection is critical in the drawing creel. On a day-to-day basis, once tension is set up, there will likely be no reason to change it. However, with changes in fiber type, there is often a need to adjust tension to avoid buckling of the incoming sliver (tension too low) or stretching of the slivers (tension too high).

Creel Height

Creel height is often not considered to be important, yet when the creel is too low, there is a higher likelihood for sliver to tangle as it comes out of the can. If sliver tangling is recognized in the creel, then either the sliver coiling in the previous process should be adjusted to provide more space between the coils, or the creel should be raised to provide extra space for the sliver withdrawal from the can. Operators often compensate for tangling sliver by moving the sliver can to the side of the creel until the can becomes half empty in order to provide this extra space between the sliver can and the creel guides. This procedure is not appropriate because it causes extra abrasion of the sliver on the side of the can that can increase yarn defects.

Stop Motions

Stop motions exist on the creel to detect sliver breakages. Over time these systems can become faulty, so routine checks on their functionality must be made. If one end of sliver is missing, the resultant sliver weight could be 12 to 16 percent light if no autoleveler is on the frame. On the newer draw frames which have autolevelers, corrections can be made to keep the mass per unit length of the sliver fairly consistent if a stop motion fails and if one fewer sliver is fed into the machine. However, the difference in total draft to achieve the same weight causes the alignment (draftability) of fibers in that sliver to be different for the next process. As a result, the final product will likely have different characteristics. Thus, even on autoleveled draw frames, functioning stop motions are essential. In most operations stop motion functionality is checked each shift.

Housekeeping

Finally, housekeeping is important at the creel. Anything that looks wrong probably is wrong. Thus, attention to fiber tags on sliver guides and removal of trash on the tray holding the slivers will go a long way toward minimizing thick places in the yarn.

Draft Zone Optimization

The draft zone is the heart of the draw frame. Critical control factors in the draft zone include the following:

1. Roll spacings,
2. Draft distribution,
3. Roll pressure,
4. Speed,
5. Pressure bar setting,
6. Cot condition, and
7. Suction.

Roll Spacings

When optimizing a draft zone, the first thing to assess is roll spacings. In an ideal situation all of the fibers would be the same length, and the rolls could be set just wider than the fiber length for maximum control of the fibers without breakage. Unfortunately, in the real world the fiber processor must deal with a distribution of fiber lengths caused by normal variation in length and the presence of hooks.

When setting the roll spacings, it is desired that the distance between the nips (contact points) of the rollers be just wide enough to let the longest fibers "grow". If the roll spacings are too tight, the longer fibers will break by being held by two sets of rolls turn-

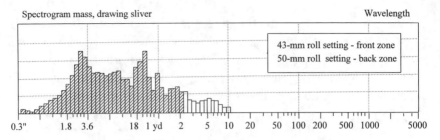

Figure 6.12. Example spectrogram for sliver with excessive drafting wave.

ing at different speeds. If the roll spacings are too wide, too many fibers will "float" in the draft zone and mass consistency will deteriorate due to the loss of fiber control (high drafting wave).

To optimize roll spacings, the following information must be known:

1. Longest fiber length of incoming and delivered sliver (2.5 percent length on the Uster AFIS Tester or 1.0 percent length on the Peyer AL-101);
2. Short fiber percentage in incoming and delivered sliver;
3. Spectrogram of sliver with the current setup; and
4. Inside nip-to-nip spacing of both drafting zones.

If the longest fiber length does not increase in the drawn sliver, if the short fiber content increases, or if any point on the spectrogram vertically exceeds the third line (see Figure 6.12), an opportunity exists to evaluate roll spacings.

As a starting point, it is worthwhile to compare the longest fiber length of the incoming fibers to the inside-to-inside spacing of the rollers. The best way to get the true nip-to-nip roll spacing is by using carbon paper to get an impression, and then measuring the distance between roll impressions.

Because more hooks exist in card sliver, more distance must be provided to allow for fiber "growth" in the first drawing process. As a result, a good starting point for first passage drawing is as follows:

1. Front Zone Spacing = Longest fiber length in fed sliver + 3 mm to + 5 mm; and

2. Back Zone Spacing = Longest fiber length in fed sliver + 6 mm to + 7 mm.

During the first drawing process many hooks are removed. Thus, less space beyond the longest fiber length is needed for second and third pass drawing. For finisher drawing, a good starting point is as follows:

1. Front Zone = Longest fiber length in fed sliver + 1 mm to + 3 mm; and
2. Back Zone = Longest fiber length in fed sliver + 4 mm to + 6 mm.

When roll spacings are evaluated, which should be anytime the fiber significantly changes (i.e., cotton crop change, synthetic fiber merge change, new blend, new carding), a few roll spacings should be tried and evaluated based on the following:

1. Fiber growth (increase in fiber length);
2. Spectrogram height;
3. Short-term evenness;
4. Yarn strength;
5. IPI thick places;
6. Classimat minor defects;
7. Yarn 1-meter, 3-meter, 10-meter evenness; and
8. Spinning performance.

The condition that provides the best fiber growth and spectrogram will likely be the best setup, but sometimes tight roll spacings will look better in sliver than they really are. Consequently, it is always advised to check the drawing setups relative to their yarn performance. The setup that provides the best fiber control will provide the best yarn 1-meter, 3-meter, and 10-meter evenness. Short-term evenness (%CVm) in the yarn does not always differentiate the preferred drawing setup. Yarn strength, Classimat minor defects, and IPI thick places are most useful for detecting tight roll spacings. If the roll spacings are too tight, yarn strength will decrease while Classimat minor defects and thick places will increase.

Draft Distribution

Another critical factor at drawing is draft distribution. In the previous section it was noted that the amount of mass fed into the draw frame influences short-term evenness. With more mass, more draft is required. This draft must then be distributed among the draft zones. Typically, a small amount of the draft (1.2 to 2.0) is provided in the back zone. This preparatory draft is called break draft. In a two-zone drafting system the remaining draft is provided in the main draft zone. At this point it should be noted that the drafts are multiplicative. Thus, if the total draft is 8.0, and the break draft is 2.0, then the main draft would be 8.0 ÷ 2.0, or 4.0.

The appropriate level of break draft is determined by fiber control and downstream processing. Usually, lower break drafts help sliver evenness; however, alignment usually suffers with low break drafts. Thus, a compromise must be established. The best indicator of whether the draft distribution should be evaluated is the spectrogram from the evenness tester. Figure 6.13 contains spectrograms from sliver produced at different break drafts. In this case higher break draft caused the evenness to be worse. When

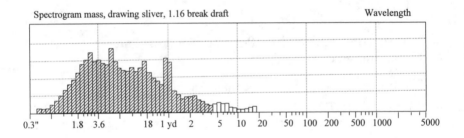

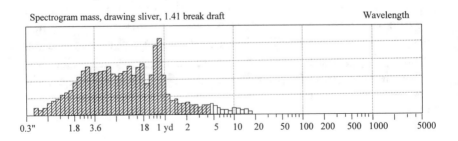

Figure 6.13. Spectrograms for 100 percent cotton slivers processed with different break drafts at drawing.

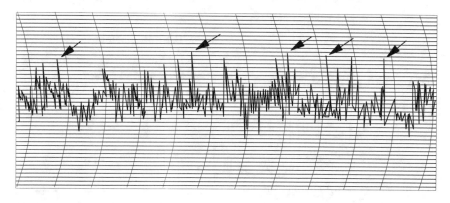

Figure 6.14. Example of a Uster strip chart showing spikes that can be caused by low break draft.

evaluating break drafts, a break draft that is too high usually is indicated by a higher hump (or humps) on the spectrogram between 10 inches and 1 yard. A break draft that is too low usually is detected by spikes on the Uster strip chart as shown in Figure 6.14.

On modern draw frames it is typical to find the following trends regarding break draft:

1. Breaker drawing requires a higher break draft than finisher drawing.
2. Synthetic fibers require higher break drafts than cotton.
3. Fine denier synthetic fibers require higher break drafts than coarse deniers.
4. Higher micronaire cottons require lower break drafts than do lower micronaire cottons.

Pressure Bar

The purpose of the pressure bar, as mentioned previously, is to help guide the fibers that are not under control of a drafting roll. The depth of penetration of the pressure bar can be set, typically by altering the number of 1-mm washers on which the edges of the pressure bar rest. In some cases changing the pressure bar depth has an impact on sliver quality. Particularly when 100 percent cotton is used, pressure bar depth is critical.

0.5 to 1.0 mm clearance

3.0 +/- 0.5 mm distance

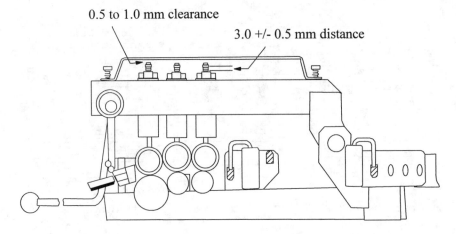

Figure 6.15. Illustration of measurements indicating the top roll pressure setting on the RSB draw frame. (Courtesy Rieter Machine Works, Ltd.)

Roll Weighting

Whether roll weighting is applied mechanically (with a spring) or pneumatically, the consistency of pressure on the rolls is extremely important. Pneumatically supplied systems usually have a pressure gauge that provides an indication of roll weighting, while mechanical systems often require a measurement to assess the setting. For instance, on the Rieter RSB machines a measurement of the rod above the bearing blocks is taken to assess roll weighting as shown in Figure 6.15. If the measurement is outside of specification, quality will likely suffer. A good visual check to evaluate the roll pressure situation on any machine is a nip impression using carbon paper. If the roll weight is insufficient, the impression will be faint.

Cot Condition

Cot condition is critical for drawing quality. If the cots are not cleaned frequently (usually once per day), buildup of fats, oils, and waxes or fiber finish will cause poor evenness to result. Also, cots that become too hard or too small in diameter will lose the effectiveness of their nip and cause quality to suffer. Most cots today are between 70° and 83° Shore hardness.

Suction

Suction is critical for draw frame cleanliness. If suction is below 75 mm of H_2O, the operators have difficulty keeping the drafting zone clean.

Coiling Zone Optimization

After the sliver exits the draft zone, it is condensed through a funnel and a trumpet and is then laid into a can in circular coils. Some factors that influence the quality of the coiled sliver include the following:

1. Tension between the front rolls and the calender rolls,
2. Trumpet size, and
3. Coiler speed.

Tension

The rule on tension is the lower that it can be set and run efficiently, the better. If the tension is too low between the front roll and the delivery rolls, excessive front roll lap-ups and/or condenser clogs will occur. However, if the tension is too high, sliver evenness will suffer. As a rule of thumb, the following guidelines can be used for front roll-to-calender roll tension:

1. Carded Cotton at 1.01 to 1.02;
2. Combed Cotton at 1.00 to 1.01;
3. Polyester/Cotton at 0.99 to 1.00; and
4. 100 percent Synthetic at 0.98 to 1.00.

Figure 6.16 is a presentation of the results from two trials that involved reducing the front end tension at drawing from 1.02 to 1.00.

Trumpet Size

Before the sliver bundle enters the can, it must be condensed. The amount of condensing by the trumpet influences whether the sliver will have enough cohesion to be pulled out of the can at the next process, or too much cohesion for appropriate fiber movement at the next process. The size of the hole selected for the trumpet depends on the sliver weight with a starting point recom-

Sliver Uster %CV

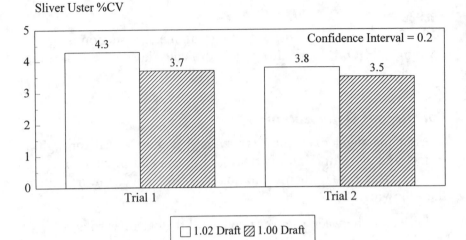

Figure 6.16. Graphical illustration of the influence of front roll-to-calender roll tension draft on the short-term evenness of 50/50 polyester/cotton slivers.

mended in Table 6.3. For combed cotton the trumpet size should be approximately 0.2 mm tighter than that used for carded cottons. For synthetic fibers the size should be 0.2 mm larger than that used for carded cottons.

Coiler Speed

The coiler speed of the draw frames should be chosen so that excessive tension is not placed on the coils. Some signs that typically indicate excessive coiler speeds include:

1. Hairy sliver;
2. Periodic defects on the spectrogram corresponding to the diameter of the coil; and
3. Dirty slubs deposited on top of the sliver can when the draw frame stops.

The speed of the coiler can be measured with a tachometer on many draw frames. In most cases the optimal speed is determined by continually slowing the coiler until the sliver begins to overfeed slightly, causing kinks. Then, the coiler speed is barely increased to lessen the kinks.

Table 6.3. Recommendations for Trumpet Size (Bore) Based on Sliver Weights for Carded Cotton

Sliver Weight Delivered		Trumpet Bore	
Metric Count	Grains per Yard	Millimeters	Inches
0.16	88	4.6	0.18
0.18	78	4.4	0.17
0.20	70	4.2	0.16
0.22	64	4.0	0.16
0.24	59	3.8	0.15
0.26	54	3.6	0.14
0.28	50	3.4	0.13
0.30	47	3.2	0.13
0.32	44	3.0	0.12
0.34	41	2.8	0.11

Autoleveling Zone Optimization

The introduction of the short-term, open-loop autoleveler has done more to improve yarn quality than any other refinement in recent history. In particular, count variation, strength variation, Uster evenness, 1-meter through 50-meter %CV, IPI thins, and Classimat minor defects and long thin places have been positively influenced by draw frame leveling. In addition, the autoleveler has allowed rotor spinners to eliminate the second drawing passage in some cases. As shown in Figure 6.17, the autoleveler consists of a bulk measuring head, which is usually a set of tongue-and-groove rolls, that sends a signal corresponding to mass changes to a servomotor that changes the draft of the main draft zone. The draft is adjusted by altering the speed of the back and middle rolls at the same rate, keeping the front roll speed the same at all times. As a result, the break draft remains constant and the main draft changes. Since the ratio of back roll speed to front roll speed is being altered, the total draft also changes.

Since the autoleveler is an electrical system, it must be calibrated. The critical calibrations include the following:

1. Target weight setting,
2. Response time setting, and
3. Intensity setting.

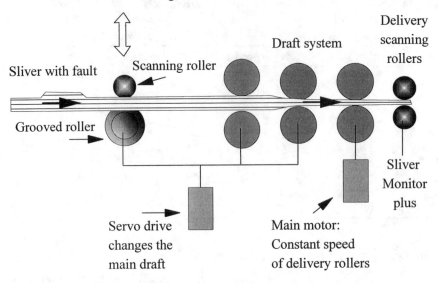

Figure 6.17. Schematic of the autoleveler on the RSB draw frame. (Courtesy Rieter Machine Works, Ltd.)

Target Weight Setting

In any autoleveling situation, it is first important to set up the frame with the autoleveler off. In this case set the draft for a delivered weight as close as possible to the target. Then, with the autoleveler turned on, adjustments can be made with the sliver weight adaption knob to fine-tune the autoleveler to the exact desired weight through precise adjustment of the draft. Periodically, the target weight setting must be checked to ensure it is still calibrated correctly.

Response Time Setting

When an open-loop autoleveling system is used, a time delay exists between the time a measurement is made and when it must be corrected in the draft zone. The response time setting determines the length of this delay. If the timing is incorrect, variation will be made worse, particularly in 1-meter and 3-meter lengths. Two checks can be made to assess the timing.

The most common timing check involves running 200 meters of sliver at several different timing settings (usually 5 settings) and comparing the spectrograms from the evenness testers. The setup that provides the lowest spectrogram height between 35 and 51 cm is preferred. An example comparison is shown in Figure 6.18.

The quick test for autoleveler timing involves removing an end from the creel while the draw frame is running. Then, the sliver is tested on the evenness tester. If the evenness tester reveals that the mass gets heavy for a moment and then is correct, the timing is too quick. If it gets light for a moment, the timing is too slow.

Intensity Setting

Autoleveler intensity refers to the ability of the leveler to correct the appropriate amount. In other words, if the incoming mass shifts 2 percent, the autoleveler must be set to correct it 2 percent. In some cases the autoleveler is set up to correct at the appropriate time, but when it corrects, it corrects the wrong amount. The intensity is analyzed by producing 200 meters of normal sliver. Then, one extra end is added to the creel and 200 meters are produced. Next, one too few ends are fed in the creel and 200 more meters are produced. Twenty non-successive 1-meter weighings are made on each condition and the average sliver weights for each are plotted. If the average sliver weights do not deviate from

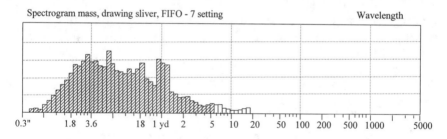

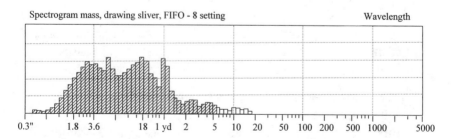

Figure 6.18. Spectrograms comparing slivers produced with different autoleveler timing settings.

Sliver Weight, grains/yard

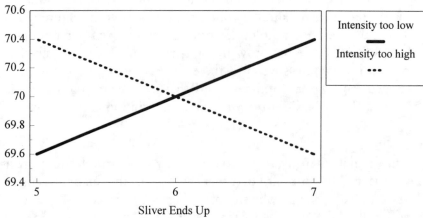

Figure 6.19. Graph used to evaluate draw frame autoleveler intensity setting.

the sliver weight of the normal doubling by more than 0.5%, the intensity is satisfactory. For a 4.25 g/m (60 gr/yd) sliver this means that the tolerance for satisfactory intensity is ± 0.02 g/m (± 0.30 gr/yd). If the intensity is not satisfactory, the slope of the graph, shown in Figure 6.19, will indicate whether intensity should be increased or decreased. A positive slope is indicative of undercorrecting, so the intensity should be increased. A negative slope indicates that intensity is too high.

Documenting Quality of Drawn Sliver

The quality of drawn sliver can be documented through various techniques. Reflecting on the purposes of drawing being to align the fibers; improve short-term, mid-term and long-term evenness; and improve consistency; perhaps it is best to characterize quality assessment relative to each of these purposes. For instance, to assess alignment it would make sense to evaluate these characteristics:

1. Fiber growth—Increase in mean length and/or upper 2.5 percent length after drawing;
2. Bulk density—Using the bulk density tester, an alignment index can be determined for each draw frame sliver; and
3. Drafting force—Using the ITT Draftometer or Rothchild Co-

hesion Tester, a drafting force can be determined that relates to how well the fibers are aligned.

To assess short-term, mid-term, and long-term mass uniformity, it makes sense to evaluate the following:

4. Evenness, %CVm—Remembering that lowest is not always best if it is achieved through roll spacings that are too tight;
5. 1-meter and 3-meter %CV—Indicate mid-term evenness, which is strongly influenced by the number of doublings and autoleveler calibration;
6. 10-meter %CV—Determined using a reel and scale; and
7. Spectrogram—Indicates periodic faults caused by damaged cots, bearings, gears, pulleys, or steel rolls. Also, indicates loss of fiber control in the form of drafting waves.

To assess potential damage incurred to the fibers at the drawing process, evaluations should be made on these items:

8. Short fiber content change—The short fiber content based on the Uster AFIS tester should be lower after drawing, and;
9. Evenness Diagram—Spikes on the evenness tester that exceed the 15 percent line in finisher sliver should be fewer than 0.23/g (0.5/lb).

Benchmarks for Finisher Slivers

The benchmarks in Table 6.4 should be targeted for finisher drawing sliver quality.

Table 6.4. Finisher Sliver Benchmarks

Measurement	Cotton	50/50 Polyester/Cotton
Sliver Weight, %Vo	<0.5	<0.5
Evenness, %CVm	<3.2	<3.0
1-Yard %CV	<0.7	<0.7
3-Yard %CV	<0.5	<0.5
Alignment Index	<40	<40
Drafting Force, g/gr/yd	<3.0	<3.5
Change in Short Fiber Content (Card Sliver to Finisher Sliver)	-1 percentage point	-1.5 percentage points
Change in Upper 2.5% Span Length (Card Sliver to Finisher Sliver)	+1.5 mm	+2 mm

Combing

Cotton sliver as produced on the carding machine contains some small trash particles, small quantities of neps, and short fibers which are defined as being less than 12.7 mm (0.5 inch) in length. Typically, cotton card sliver produced from normal Upland cotton contains approximately 10 percent (by weight) of these short fibers. In addition, the individual fibers in card sliver are not well aligned longitudinally and most of them are hooked on one end and many are hooked on both ends.

The purposes of the combing operation are as follows:

1. To separate long fibers from the shorter fibers with the longer fibers processed into "combed" sliver and shorter fibers processed into "noil" (waste);
2. To straighten and align longitudinally the fibers in the combed sliver; and
3. To remove the noil consisting of short fibers, trash particles, and neps.

Stronger and more even yarn can be spun from combed sliver as compared to "carded" sliver. In combed sliver the fibers have a more uniform length with a longer average staple. Further, yarns spun from sliver with straighter and more parallel fibers have a distinct sheen or luster.

Fine yarn counts, 40's and finer, are usually spun from combed cotton when uniformity of mass and appearance are required for the end use. When strength is of critical importance, even coarser yarns require combed cotton. Many knitted products are made with combed cotton to obtain best appearance and hand.

Cotton used for fine combed yarns is normally of the longer varieties, such as Peru Pima, American Egyptian, or American Pima, with staple lengths in the range of 30 to 38 mm (1³⁄₁₆ to 1½ inches). Medium and coarse combed yarns are made from shorter cottons with staple lengths from 27 to 30 mm (1¹⁄₁₆ to 1³⁄₁₆ inches).

The amount of noil (waste) removed by the combing machine can be in the range of 5 to 25 percent depending on the end use of the yarn. The quality of the yarn improves only slightly by the process of "scratch" combing which takes out as little as 5 percent noil. With removal of up to 9 percent noil, the yarn may be called half combed. For the largest proportion of combed yarn, the noil removed is in the range of 10 to 15 percent, whereas in the case of fine combing, it may be as high as 25 percent.

Cotton combers operate intermittently because the short length of the fiber does not allow the use of a continuous method of combing. One end of the cotton fringe (beard) is combed by a circular comb or half lap, while the other end is combed with a single row of needles called the top comb. After both ends of the fringe have been combed separately, the separated fringes are then re-united by a piecing system.

As the main function of the combing process is to remove short fibers, it is desirable for the material fed to the comber to have the fibers reasonably parallel so that the longer fibers will not be lost in the noil waste. For this reason it is customary to draft the card slivers and prepare a lap which is subsequently fed to the comber.

Definition of Terms

Comber
The machine consisting of six or eight heads which removes short fibers, trash, and neps from a fibrous fringe developed by the pulling apart of a lap sheet of fibers.

Combing Cycle
A complete movement of the combing mechanisms from feed through detaching actions and back to feed.

Comber Lap
Calendered blanket of cotton fibers 267 or 300 mm (10½ or 12 inches) wide, weighing approximately 60 grams per meter for the

266 mm wide lap, rolled into a cylinder weighing up to 25 kg (55 lb) for the 300 mm wide lap.

Cushion Plate
The oscillating plate over which the lap is fed and which, together with the nipper knife, holds the feed end of the lap after detaching and during combing.

Detaching
The movement that separates the lap sheet to form a web of fibers held by the nipper knife and cushion plate.

Detaching Roll Travel (forward)
Total movement of the detaching roll in a forward direction.

Detaching Roll Travel (return)
Total return movement of the detaching roll.

Detaching Rolls
Two pairs of gripping rolls that rotate forward and back intermittently, holding and moving the combed web for a net forward travel.

Detaching Time
Index number at which the fringe of the combed web enters the gripping line (nip) of the rear set of detaching rolls.

Half Lap
A curved arc plate with needles (or metallic teeth) which fits onto the rotating combing cylinder and which actually combs through the fringe of fibers held by cushion plate and nipper knife.

Nipper Setting
Distance between the cushion plate and back bottom detaching roll when the nipper unit is in its extreme forward position. (Ecartement)

Nipper Knife
The vertical plate, grooved at the bottom, which bears down to hold the lap sheet firmly to the forward end of the cushion plate.

Noil Percentage
Weight of noil removed divided by the sum of weight of sliver plus weight of noil times 100.

Overlap
Return of the web less the advance of the web.

Piecing
The movement that places the combed web, held in the nipper, onto the retracted web held by the detaching rolls.

Timing
Index number at which the detaching rolls start their forward motion.

Top Comb
Straight comb with approximately 26 needles per centimeter (66 needles per inch), held vertically and moved up and down to comb the forward section of the detached web from above.

Separation
The movement that separates the combed web, held in the nip of the detaching rolls, from the uncombed lap sheet.

Super Lap Machine
Lap forming machine that normally develops three fibrous webs, each made by drafting predrawn slivers and sandwiched together for calendering and winding; manufactured by Whitin Roberts Company, USA.

Unilap Machine
Lap forming machine that normally develops two fibrous webs, each made by drafting either 12 or 16 predrawn slivers and sandwiched together for calendering and winding; manufactured by Rieter Machine Works, Ltd., Switzerland.

Web Advance
Forward movement of the detaching rolls less their return movement.

Brief History of Comber Development

Several cotton combers were developed in the nineteenth century, but the only one to have real success was invented in 1845 by Joshua Heilmann of Mulhausen, France. His comber was first exhibited at the International Exhibition of 1851. The next successful comber was that introduced by J.W. Nasmith, an Englishman, in 1902. Some of the disadvantages of this early machine were corrected by Nasmith by a redesign in 1928. A swinging nipper mechanism and larger diameter combing cylinder permitted a much heavier lap to be processed. Production was increased and production costs were reduced. Further improvements in the detaching mechanism made it possible to comb fibers much shorter than those for which the Heilmann comber was suitable. Most modern combers are based on the 1928 Nasmith principle, shown in the drawing of Figure 7.1.

The first two-sided comber was developed by John Hethering and Sons, Ltd. of Manchester, England. Patents were issued in 1854 and later in 1883. Saco Lowell (USA) started in 1934 to build combers using this two-sided system with six laps on each side. The Saco Lowell Model 56 which was introduced in 1955 operated at a rate of 140 nips (revolutions or combing cycles) per minute. On these combers the two sides are driven by a common motor. However, each side can be adjusted separately so that it is possible to comb different cottons on the two sides at the same time. The working mechanism of these two-sided combers is shown in Figure 7.2.

Comber Lap Preparation

The objective of lap preparation is to produce a lap which will serve as the best supply package for the comber. Experience and experiments have led to certain criteria for lap quality including the following:

1. Lap thickness and mass should be even both longitudinally and laterally.
2. Fiber hook direction must be leading as the lap enters the comber. Most of the fibers in card sliver have large ("majority")

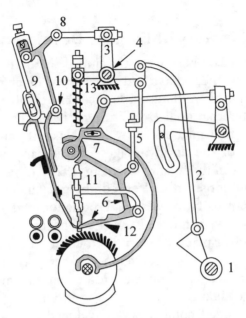

The oscillating main shaft (1) through connecting rod (2) moves lever (3) around point (4). Lever (3) connected with rod (5) swings nipper frame (6) which oscillates around pivot (7) and through connection (8) the top comb holder (9), which later swings around point (10). Nipper knife (11) is pressed onto cushion plate (12) during the combing cycle by spring (13). The pressure is released and the nipper opens when detaching due to the upwards movement of the left end of lever (3).

Figure 7.1. Diagram and explanation of components of the 1928 Nasmith comber.

hooks which are trailing as the sliver exits the card. This means that two separate processes are needed between carding and combing in order to present leading hooks to the combing zone.

3. The total amount of draft in the two processes between carding and combing should be in the range from 8 to 12, unless careful trials on a particular process indicate a slightly higher draft to be desirable.

The fibers in card sliver with hooks at the trailing end as the sliver exits the card have what are called majority hooks. The

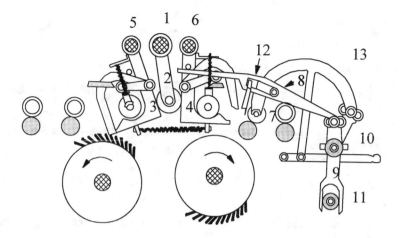

Nippers (3 and 4) swinging around shafts (5 and 6) are actuated by rocker shaft (1) and crank arm (2). Top comb (12) carried by lever (8) receives its back and forth movement from oscillating shaft (10). Top detaching roll (7) carried by curved lever (13) rolls around the bottom detaching roll. This motion is caused by cam (11) which swings lever (9).

Figure 7.2. Diagram and explanation of the working elements of the Saco Lowell Model 56 comber.

number of hooks is reduced by drafting. The hook direction is reversed at each separate process. As a result, after two successive drafting operations following the card, the majority hook will be leading going into the comber.

There have been several different two-step lap preparation methods utilized throughout the years. These methods included the draw frame followed by a sliver lap machine, the sliver lap process followed by the ribbon lap machine, and the draw frame followed by the lap winder.

In 1950 Whitin Machine Works (USA) introduced an improved method of lap preparation by using a draw frame followed by the newly developed Super Lap machine. This machine had three heads, each fed by 16 or 20 draw frame slivers which pass down-

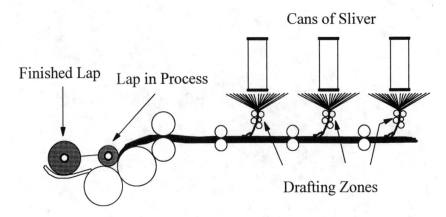

Cans of Sliver

Finished Lap   Lap in Process

Drafting Zones

Figure 7.3. Schematic illustration of the Super Lap machine. (Courtesy Whitin Roberts Co.)

ward through a vertically arranged 2-over-3 roll drafting system as shown schematically in Figure 7.3.

The sheets (webs) delivered by the vertical drafting heads are superimposed on a smooth stationary table; the combined webs are drawn through calender rolls and then wound into laps. The lap doffing and spool changing attachment automatically stops the machine, ejects the full lap onto the tray, feeds an empty lap spool into place, and restarts the machine. The spool magazine holds as many as 7 empty spools, and the adjustable receiving tray holds 4 comber laps. The machine stops automatically in case the lap tray is fully loaded.

The Super Lap machine is available in two models, G1 and H1. The first builds 300 mm wide laps and the second 267 mm wide laps. A view of the Super Lap machine is provided in Figure 7.4. The production of the Super Lap machine is approximately 320 kg (705 lb) per hour at 100 percent efficiency when processing laps up to 85 grams per meter (1200 grains per yard). The net weight of full laps is approximately 15 kg (33 lb).

Lap preparation by Rieter Machine Works (Switzerland) includes three distinct methods as follows:

1. Sliver Lap, Ribbon Lap combinations for long cottons of more than 30 mm (1³⁄₁₆ inches) in staple length;
2. Draw Frame, Ribbon Doubling machine combination for short cottons of 27 mm (1¹⁄₁₆ inches) and shorter; and

Figure 7.4. Super Lap machine. (Courtesy Whitin Roberts Co.)

3. Draw Frame, Unilap machine (newest) combination for cottons of all staple lengths. The Unilap machine was introduced in the early 1990s.

The Sliver Lap machine Model E5/20 is shown in Figure 7.5. It is fed with 20 to 24 card slivers and has a draft range of 1.20 to 2.24. Laps weigh up to 25 kg (55 lb). Production rate is 430 kg (946 lb) per hour.

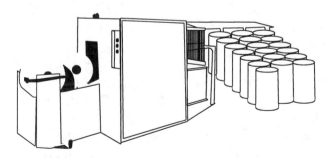

Figure 7.5. Illustration of the Sliver Lap Machine E5/20. (Courtesy Rieter Machine Works, Ltd.)

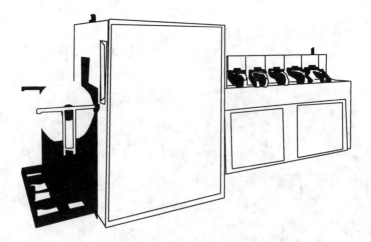

Figure 7.6. Illustration of the Ribbon Lap Machine E5/40. (Courtesy Rieter Machine Works, Ltd.)

The Ribbon Lap machine Model E5/40, shown in Figure 7.6, is fed by six laps made on the Sliver Lap machine. A draft of 6 is applied, and the individual webs are laid superimposed on each other on the table. The full lap weighs 25 kg (55 lb), and the production rate is 430 kg (946 lb) per hour.

A layout of the Ribbon Doubling machine Model E5/30 is shown in Figure 7.7. Twelve to sixteen predrawn slivers are fed to each of three drafting heads which produce drafted webs to be sandwiched together on a smooth table. Draft range is 1.0 to 7.9. Full laps weigh 35 kg (77 lb) and production rate is 500 kg (about 1100 lb) per hour.

A drawing of the Rieter Unilap machine is shown in Figure 7.8. Twelve to sixteen predrawn slivers are fed on two overhead tables into two vertical drafting units whose discharged webs are superimposed and fed into the lap forming unit. This machine can process all cottons from 18 to 50 mm ($^{11}/_{16}$ to 2 inches) in staple length. Draft range is 1.20 to 2.24. Lap weight delivered is up to 80 g/m (1128 gr/yd) and the lap is 300 mm (11.8 inches) wide. Net lap weight is up to 25 kg (55 lb) and production rate is 430 kg (946 lb) per hour.

On the Unilap machine, the web emerging from the drafting mechanism is guided over deflector plates and moved by calender rolls to the winding rollers where the lap is constructed. The lap ejector sequence is started as soon as the counter signals that the

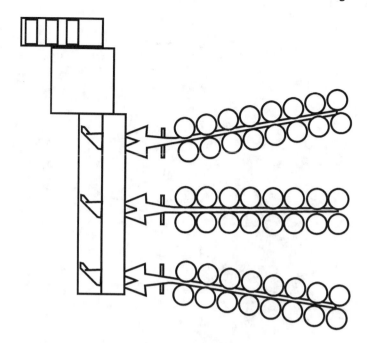

Figure 7.7. Layout of the Ribbon Lap Machine E5/30. (Courtesy Rieter Machine Works, Ltd.)

pre-selected lap length is reached. The machine stops and the batt is severed between the calender rolls and the point of winding with the aid of an auxiliary motor. The lap is ejected with the end of the lap in the correct position for piecing on the comber. A weight correction during lap building is not necessary. The amount of pressure exerted on the lap is automatically controlled in order to compensate for the gradually changing geometric conditions. Thus, there is no practical difference between the inner layers and the outside layers of a full lap. Measuring the evenness in mass of 1-meter lengths taken sequentially on laps with net weights of 25 kg (55 lb), with a batt weight of 80 grams per meter (1129 grains per yard), gives a variability of less than 0.5 percent coefficient of variation.

In Figure 7.8 the Rieter Servolap system for the automated transport of laps to the combers is shown as item 5 where either 4 or 8 full laps are automatically loaded onto an overhead rail system. This "train" of laps is then moved on the rail by a motor operated crane which pulls the train of laps to the comber which sig-

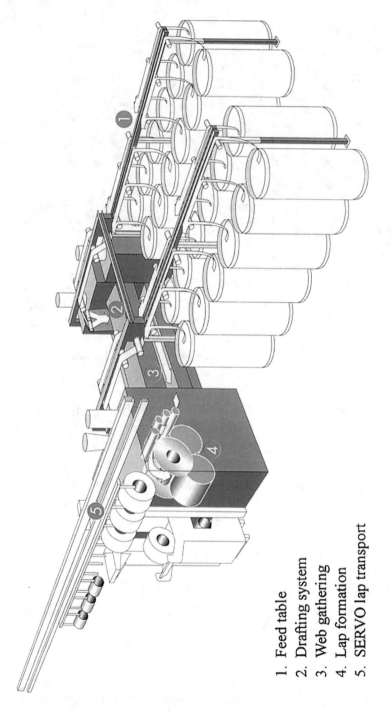

1. Feed table
2. Drafting system
3. Web gathering
4. Lap formation
5. SERVO lap transport

Figure 7.8. Drawing of the Unilap machine. (Courtesy Rieter Machine Works, Ltd.)

nalled for laps. Another arrangement of the Servolap system is shown in Figure 7.9.

Combing Cycle and Comber Mechanisms

The objectives of combing, as listed in the Introduction to this chapter, are accomplished when the comber lap is converted to sliver as the cotton fibers proceed through the comber, being processed by the major comber components known as: (a) feed roll, (b) cushion plate, (c) nipper knife, (d) half lap, (e) top comb, (f) detaching rolls, (g) condensing trumpet, and (h) drafting unit. Six of these components are shown in the drawing of Figure 7.10. The combing of the cotton fibers is achieved during an intermittent cycle which is accomplished in the following manner:

1. Beginning the combing cycle at "feeding", the lap sheet is fed into the comber between the feed roll and a smooth cushion plate. A second plate (nipper knife) moves down to fix the fibrous sheet between the fork at the bottom end of the knife and the contoured front end of the cushion plate. This feeding

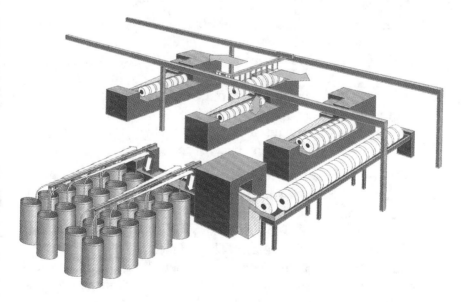

Figure 7.9. Drawing of a typical layout of the Servolap transport system. (Courtesy Rieter Machine Works, Ltd.)

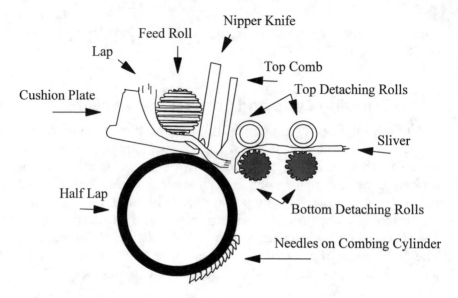

Figure 7.10. Drawing of the components of the typical comber.

occurs *after* detaching, called "reverse" feed. "Forward" feeding refers to the timing which causes feeding *while* detaching.

2. Fibers protruding from the lap beyond the nipping point of the cushion plate and the nipper knife are combed by the passage of the needles (or teeth) on the half lap, fastened to the "cylinder", which revolves and carries away the short fibers, neps, and impurities combed out of the lap fringe. This waste (noil) is removed from the needles of the half lap by a revolving brush mounted just below the cylinder and is deposited (usually by suction) at the back of the comber.

3. A second comb (top comb) with one row of needles is inserted into the fiber fringe from above after the half lap comb has cleared the fringe, and just as "piecing" is occurring by the reverse travel of the detaching rolls and the forward motion of the cushion plate. The connected (pieced) web is pulled through the top comb as the detaching rolls rotate forward to move the combed web toward the web pan. As the web is pulled through the top comb, neps, some short fiber, and entanglements (not removed by the half lap combing) are

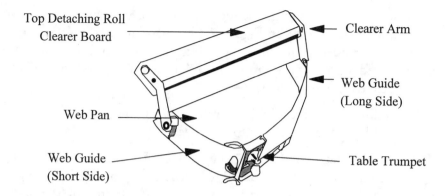

Figure 7.11. Illustration of a comber web pan.

removed. In this manner the continuity of the combed web of fibers is maintained.

4. The top comb is withdrawn upward and the half lap (bottom comb) operates once more, the fringe having been advanced slightly by the feed roll to present a new, short section of the lap to be combed. Thus, the combing cycle is completed.

5. The combed web is delivered to the web pan by the detaching rolls, and thence to the table trumpet located to the side as shown in Figure 7.11. The condensed web is pulled through the trumpet by calender rolls that deliver the combed sliver to the table where slivers from all combing heads are laid side by side and flow to the drawbox located in the head end of the machine. The sliver delivered from the drawbox, which has a draft range of 5 to 12, is coiled into a can which is doffed automatically on modern combing machines.

The bore diameter of the drawbox sliver trumpet varies accordingly to the sliver weight as shown in Table 7.1.

Table 7.1. Drawbox Sliver Trumpet Bore Diameters

Sliver Weight	Bore Diameter
55 grains/yard	0.130 inch
60 grains/yard	0.140 inch
65 grains/yard	0.145 inch

Illustrations showing the sequential moving of principle combing elements are shown in Figure 7.12. There are two methods of setting the sequence in which detaching and feeding takes place, and this sequence has an important influence on noil removal. The two methods are as follows:

1. Feeding after detaching, or "reverse feed"; and
2. Feeding while detaching, or "forward feed."

Today's modern combers can be set for either method. The reverse feed sequence is (a) combing by the half lap, (b) detaching, and (c) feeding. The forward feed sequence is combing by the half lap and feeding while detaching. For combing short fibers with low noil removal, the forward feed is preferred; and, for longer fibers and better quality combing with higher noil removal, the reverse feed is preferred.

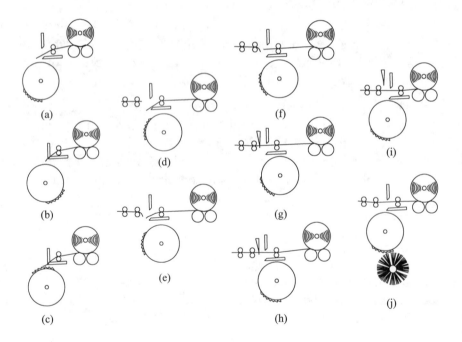

Figure 7.12. Sequence of movements of the combing elements: (a) feeding, (b) nipping, (c) half lap combing, (d) retracting of the nipper knife, (e) piecing initiation, (f) piecing continues, (g) top comb facilitates piecing, (h) top combing, (i) retracting of the top comb, and (j) new feeding.

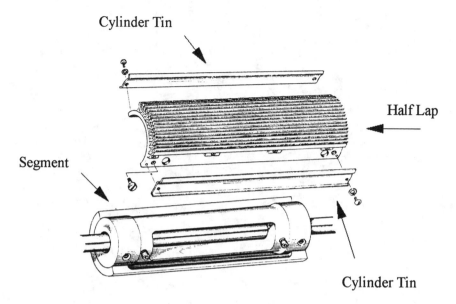

Cylinder Tin

Half Lap

Segment

Cylinder Tin

Figure 7.13. Drawing of a combing cylinder and half lap.

A typical combing cylinder and half lap comb are illustrated in Figure 7.13.

High Production Combing Machines

Hollingsworth Model CA Comber

The Hollingsworth Model CA comber was introduced at the Southern Textile Exposition in Greenville, South Carolina, in October of 1966. The drawing in Figure 7.14 illustrates the two-sided feature of the Model CA. The Model CA comber is a two-sided machine with twelve combing heads, six on each side, and it delivers two slivers, one on each side. Two sliver cans are filled simultaneously.

The head end has two parts, the drawbox cabinet and the main gear box. The totally enclosed gear box is located between the drawboxes in the top of the combing section. The two drawboxes in the top of the cabinet are shown in Figure 7.15. The cabinet also contains the calenders, coilers, turntables, and counter.

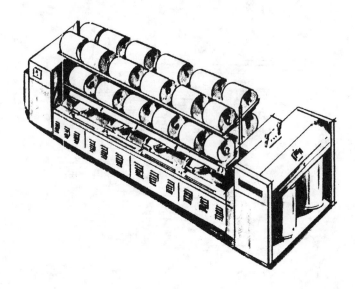

Figure 7.14. Drawing of the Model CA comber that was used to introduce the machine in 1966. (Courtesy John D. Hollingsworth on Wheels, Inc.)

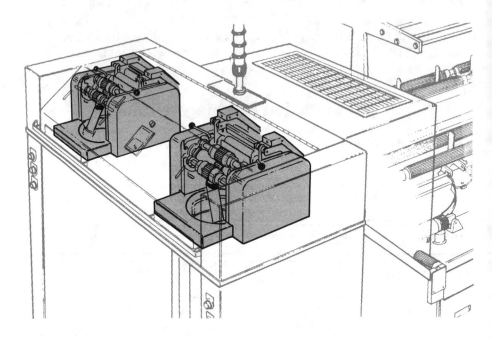

Figure 7.15. Model CA comber drawbox cabinet with twin drawboxes. (Courtesy John D. Hollingsworth on Wheels, Inc.)

The latest Hollingsworth comber Model CA-1D has been improved over the Model CA. The new design facilitates automation of lap handling as shown in Figure 7.16. Other features include automatic doffing, taller cans, larger diameter laps, and higher production capability.

Rieter Model 7/6 and Model E70R Combers

The basic comber supplied by Rieter Machine Works of Wintertur, Switzerland, is the Model 7/6. This comber has been well accepted world wide. It is a single sided machine with eight combing heads, doubled and drafted to deliver a single combed sliver into an automatically doffed sliver can. A view of this comber with a section of the "Servolap" lap transport system is exhibited in Figure 7.17. Features of this machine include high speed nipper rate, large laps, and individual monitoring of each combing head.

The Model E70R comber is the Model 7/6 comber as described above with the additional mechanism to change laps and to piece

Figure 7.16. The Model CA-1D comber with automatic lap handling. (Courtesy John D. Hollingsworth on Wheels, Inc.)

Figure 7.17. Model 7/6 comber with laps being creeled by the Servolap transport system. (Courtesy Rieter Machine Works, Ltd.)

laps automatically. The series of sketches in Figure 7.18 indicate the movements associated with the automated lap changing and piecing procedure. These features are referred to as "Robolap."

Marzoli Model PX2 and PX80

Fratelli Marzoli & C. spa is an Italian textile machinery manufacturer supplying their products world wide. The latest combers are Models PX2 and PX80. These are technologically the same except for the automatic lap piecing capability of the PX80. The combed slivers delivered from eight heads are laid side by side on the conveyor table which guides them into the drawbox. One

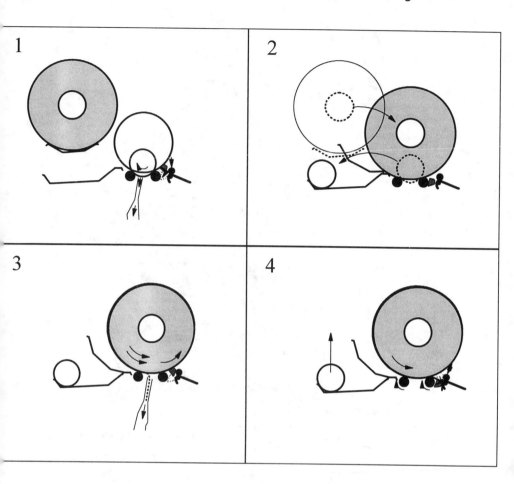

Figure 7.18. Sequence of actions of the Robolap function on the E70R comber: (1) presentation of lap end for piecing, (2) lap change, (3) preparation of the start end for piecing, and (4) piecing and start-off. (Courtesy Rieter Machine Works, Ltd.)

sliver is delivered from the drawbox into a single revolving can. The automatic can doffer is pneumatically driven.

These machines which can operate at speeds up to 350 nips/ minute are available for either 267 mm (10½ inches) or 305 mm (12 inches) wide laps; maximum lap diameter is 600 mm (24 inches). A cross-sectional drawing of the comber showing the combing head section is provided in Figure 7.19, and a sketch of the comber is shown in Figure 7.20.

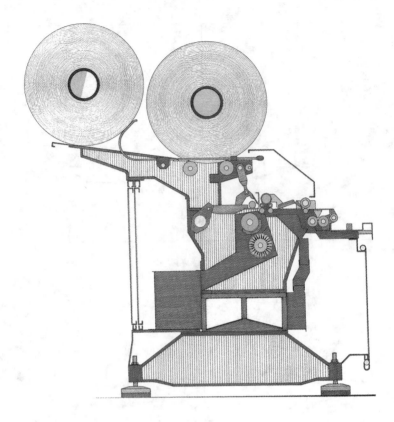

Figure 7.19. Cross-sectional drawing of the PX2 comber. (Courtesy Fratelli Marzoli & C. spa)

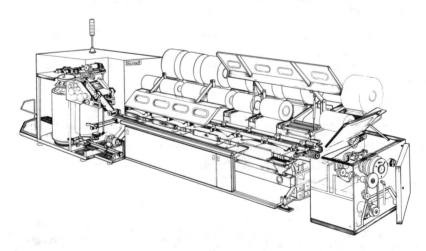

Figure 7.20. Drawing of the PX2 comber (Courtesy Fratelli Marzoli & C. spa)

Other Suppliers of Commercial Combers

Other cotton comber manufacturers are as follows:

1. Howa Machinery, Ltd. of Aichi, Japan;
2. Hara Shokki Seisakusho, Ltd., Yawata, Japan; and
3. Vouk Macchine Tessili, spa, Gorizia, Italy.

Comber Setting and Elements

There are a number of important considerations and settings for best comber operation, most of which are listed below:

1. Total draft in lap formation;
2. Lap weight, grams per meter;
3. Lap weight per meter coefficient of variation;
4. Lap splitting tendency;
5. Feed distance, type feed;
6. Detachment setting;
7. Cushion plate-to-half lap setting;
8. Top comb penetration depth;
9. Noil removal consistency;
10. Speed, nips per minute; and
11. Ambient conditions.

It has been the practice over the years to creel new full laps into the comber on every other combing head, so that full laps and half full laps, or half full laps and near empty laps were being fed simultaneously. This was done because there was troublesome weight drift of the laps from full to empty. The latest lap forming machines, however, are designed with tension compensation so that this is no longer necessary for good sliver weight control. These new machines control lap weight, meter to meter, to no greater than 0.5 percent coefficient of variation.

The speed setting in nips per minute depends on the type machine being set. In most all cases the mechanical speed limit is somewhat higher than that which will give the best quality and efficiency. There is no way to determine the optimum speed setting except by trials.

Splitting of the lap blanket as it is fed into the comber generally is a result of too much draft in lap preparation and/or incorrect ambient conditions control. Drafts as listed in the early section of "Comber Lap Preparation" (8 to 12 range) are sufficiently low to avoid excessive parallelization which tends to aggravate splitting.

For good combing efficiency, a 52 to 58 percent relative humidity and a 27 to 29 degrees Celsius (80° to 84°F) temperature have been found to be satisfactory.

Half Laps

The half lap has 14 to 30 needle bars which are fastened by screws onto the surface of the half lap segment, or are pushed into milled slots in the segment. The angle of the needles to a radius of the cylinder is smallest at the first row and is gradually increased. Generally, the angles vary from 40 to 45 degrees. For progressive combing action, the needles of the first row are coarse, widely spaced, and longer. The needle rows gradually become finer, closer spaced, and shorter toward the last row. Relatively little combing is done by the first rows; their main function is to penetrate the beard, straighten the fibers, and remove fine trash. The combing is performed by the finer and more closely set needles. Different suppliers of half laps offer different needling specifications for coarse, medium, and fine combing.

More recently the Nitto Unicomb half lap was introduced. It has rows of saw-tooth type teeth instead of needles. The shape of the teeth is such that the noil of fibers is easily removed by the circular brush; thus, the Unicomb half lap does not need additional cleaning, has a long service life, and eliminates the necessity of re-needling. The Model NUC 1 Nitto Unicomb is available in three types: (a) the 1300 series for removing about 10 percent noil, (b) the 1500 series for removing 12 to 18 percent noil, and (c) the 1700 series for 20 percent or higher noil removal. Experience has led to the rule of thumb that the Unicomb is preferred up to 30's Ne yarn count with low noils, while the conventional needle half lap is preferred for finer (higher and lighter) counts with higher noil removal.

Self-Cleaning Top Comb

Staedtler & UHL Nadelsysteme of Germany has developed and patented a self-cleaning top comb. This pneumatic top comb was first demonstrated at the ITMA-1991 Exhibition held in Hanover,

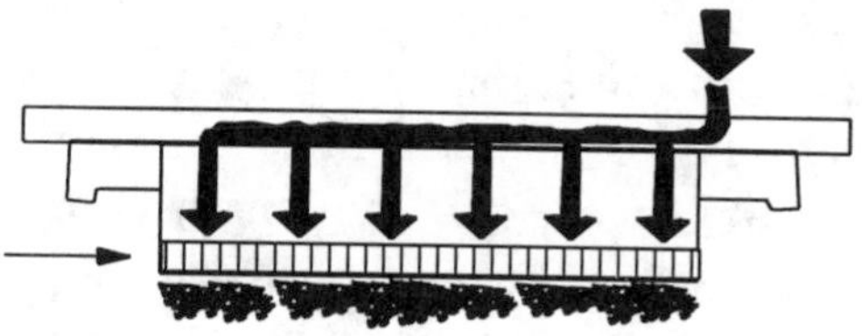

Compressed air in

Top comb ⟶

↓ Air, fiber, and debris out

Figure 7.21. Illustration of the Staedtler self-cleaning top comb.

Germany. This device provides puffs of air during each combing cycle that flow through the comb to prevent short fibers and trash from loading between the pins in the comb. A diagram of the cleaning principle is shown in Figure 7.21.

An experiment conducted by the Institute of Textile Technology staff revealed that these combs do not load as do the standard combs, which usually require manual cleaning on a two-hour cycle. In these trials yarn spun from combed sliver produced with the self-cleaning top combs, using air pressure of 6 bar, contained 10 percent fewer Classimat minor defects. At lower air pressure of 4 bar, the IPI (Uster) defects were reduced 20 percent and Classimat defects were reduced 50 percent on the 38's Ne yarns spun from the combed slivers. It was concluded that the pneumatic top comb reduced comber downtime (for cleaning top combs) significantly and improved yarn quality.

Test Methods at the Comber

For acceptable comber operation several tests should be made. These include lap weight, timing and speed, length of feed, noil removal percentage, top comb setting, and can builds.

Laps can be checked by cutting 1-meter (or 1-yard) lengths and weighing. The lap weight should normally be between 58 and 64 grams per meter (825 and 900 grains per yard) for a 267-mm (10.5–inch) width. For a 305-mm (12-inch) width, lap weight should be between 67 and 74 grams per meter (950 and 1040 grains per yard). Higher weight laps result in more noil removal and fewer neps; lighter laps result in less efficient short fiber removal.

Timing should be set to manufacturer's specification. Speed in nips per minute can be timed with a stop watch; it will generally be set 10 to 15 percent lower than the maximum mechanical speed.

The production rate of the comber will depend on the noil removal percent, the speed, and the length of feed. An increase in the length of feed will increase noil removal. The feed length will depend on the length of cotton; however, a feed length between 5 and 8 mm (0.20 and 0.30 inch) should be maintained.

To check noil removal percentage, the following procedure should be followed:

1. Run the comber for one full doff.
2. Collect all noils removed for the one doff cycle.
3. Weigh noils.
4. Weigh sliver in can.
5. Calculate noil percentage as in this equation: (7.1)

$$\% \text{ Noils } = \frac{\text{Weight of Noils}}{\text{Weight of Sliver} + \text{Weight of Noils}} \times 100 \quad (7.1)$$

Allowable maximum ranges in noil removal are 2 percentage points head to head on the same comber and 0.5 percentage points between combers.

The setting of the top comb should be at a depth that results in 30 to 50 percent of the total noil to be removed by the action of the top comb. Check and set the top comb with this procedure:

1. Determine the total noil removal as outlined above.
2. Remove top combs.
3. Determine the total noil removal of the comber running *without* the top combs.
4. Calculate the percent of noil removed by the top combs.

Top Comb Removal % =   (7.2)

$$\frac{\text{Noil Weight with Top Comb} - \text{Noil Weight without Top Comb}}{\text{Noil Weight with Top Comb}} \times 100$$

The coiling of sliver in the can should be set to prevent the sliver from being scuffed by the inside surface of the can wall. The spring in the can, below the can piston, should have a compressed force such that the spring is completely compressed when the can is filled to the can top. Crowns above the top of the can should be limited to less than 150 mm (6 inches) to prevent sliver damage during handling and transport.

There is a tendency for combers to become very linty, especially around the detaching rolls and at the top comb. Cleanliness and general mechanical condition are critical for producing good quality sliver. The brushes that remove the noils from the half laps must be properly set and in good condition. These brushes should be examined once per week. Top combs should be cleaned at least on a 4-hour cycle. Lint accumulations should be cleaned out at the same frequency.

During piecing, too much or too little overlap will adversely affect short-term mass evenness. The piecing length is highly dependent on the fiber length and the amount of reverse rotation of the detaching rolls.

Quality Benchmarks for Combed Sliver

As pointed out earlier in this chapter, the quality of combed sliver depends primarily on the noil percentage removed during combing. For a slight upgrade of yarn quality a low noil percentage is normal, the noil removed being near 5 or 6 percent. The term used for this type combing is scratch combing. At the other end of the noil removal range used for fine count yarns of superior quality, the noil percentage can be as high as 25 percent. By far the largest amount of commercial combing in the USA is in the noil removal range of 10 to 15 percent.

The quality benchmarks for this medium range of combed sliver are listed below:

1. Short fiber content (<12.7 mm) at <3%;
2. Trash and nep count (AFIS) per gram at <35;
3. Fiber alignment (bulk test) alignment index at <35;
4. Uster normal mass coefficient of variation at <4%;
5. Sliver Analyzer defects per pound at <2; and
6. Three-meter weight variability (overall) at <1.2%.

Chapter 8

Ring Spinning

Introduction

The "spinning" of short staple fibers, primarily cotton and linen, has a very long history. The first method was probably a hand method of both drafting and winding. An early method of spinning is illustrated by the drawing in Figure 8.1. The spinning wheel was the first attempt to mechanize the process; it was in use in India prior to the 14th century when it was brought to Europe. The Saxony Wheel, invented in 1533, used a "flyer" invented by Leonardo De Vinci.

The first significant development was the use of rollers to draft out the fiber bundle; this mechanism was patented by Lewis Paul in about 1731. James Hargreaves developed the "Spinning Jenny" in 1764 which used rotating spindles to insert twist. Richard Arkwright was the first to use the idea of roller drafting practically, and built the first spinning machine in 1769. Samuel Crompton built a machine called the "Spinning Mule" in 1779. Thus, the credit for developing modern spinning is given to four Englishmen: Paul, Arkwright, Hargreaves, and Crompton.

In 1828 an American by the name of John Thorpe of Providence, Rhode Island, patented a ring concentric with the spindle, and in 1830 Jenks of Pawtucket, Rhode Island, is credited with inventing the traveler rotating on the ring. The ring-traveler system was perfected by the end of the 19th century and became the standard manufacturing technology throughout the world for the production of short staple yarns. This system has been called "ring

Figure 8.1. An early method of spinning fibers into yarn.

spinning" since its development. The basic design has not been changed since the 1830s.

In Figure 8.2 is pictured a typical ring frame of the late 1980s. It contained approximately 500 spindles, 250 spindles on each side. Spindle spacing (gauge) could be chosen as 75 mm (2.95 inches), 88 mm (3.46 inches), or 100 mm (3.94 inches), depending on bobbin and ring size desired. Normally, coarse yarns, such as cotton count 5's to 10's (Ne) are spun on larger bobbins with larger diameter rings, typically near 75 mm. Finer count yarns, such as 20's to 60's are spun on smaller bobbins using smaller rings. Older frames than the one shown in Figure 8.2 are creeled and doffed manually; the very latest systems have automated both these operations.

Objectives of the Spinning Process

The objectives of the ring spinning process, as with all other spinning processes, are as follows:

1. To make final size (weight per unit length) of strand;
2. To stabilize strand with twist, or wrap;
3. To make a usable package; and
4. To make desirable yarn characteristics.

Figure 8.2. A typical late 1980s vintage ring spinning frame. (Courtesy John D. Hollingsworth on Wheels, Inc.)

In order to make the final size of the strand (yarn), a precision drafting unit must be incorporated in the mechanism. In the ring spinning system, the strand of fibers is stabilized by the insertion of twist which tends to compress the fibers in the bundle together, and this lateral fiber-to-fiber pressure becomes even greater as the yarn is put under longitudinal tension. In recent years there have been two developments that have changed the nature of the twisted bundle; namely, the wrapping of a strong, light multifilament around the bundle of fibers, and the wrapping by an air nozzle of a portion of the fibers in the strand around the remaining fibers which are in an untwisted bundle. The first system is called "wrap spinning", and the second system is referred to as the "air jet system".

In all spinning systems, the constructed yarn is wound on a package that is suitable for use in the next step toward fabric forming. In the case of a knitted fabric, the next step should be the knitting operation itself in which packages of yarn weighing 1.36

kg to 3.64 kg (3 to 8 lb) are installed directly into the creel of the knitting machine. In the case of the ring system, however, the machine package (bobbin) onto which the ring spun yarn is wound is relatively small because it has to be rotated at high speed for twist insertion. Much more is presented about this aspect later in this chapter. So that the ring yarn packaging is practical for use downstream, the spinning bobbins produced on the ring frame must be taken to an automatic winder for the production of a package of more reasonable size.

The characteristics of the final yarn produced depend, in addition to the spinning system itself, on the fiber(s) in the input stock, the quality of the input strand, the optimization of all the settings of the feed arrangement (creel) and drafting mechanism, and in most cases on the production rate setting. The yarn produced on the ring spinning system is unique in that the fibers are positioned in the twisted yarn bundle in a manner that produces a dense, flexible (soft), and strong yarn. In addition to yarn structure, other positive features of ring spinning are large count range, flexibility to handle most all fibers, and simple machine mechanisms with relatively low maintenance cost. Negative aspects of ring spinning are low productivity per spinning unit (spindle), small package which necessitates an additional process of winding, sensitivity to short fibers, relative high fiber loss, need for the roving process, and high energy usage.

Productivity and Market Share of Commercial Spinning Systems

The established commercial spinning systems in present use for producing short staple yarns are ring spinning, rotor (open-end) spinning, and air jet spinning. Listed in Table 8.1 are the numbers

Table 8.1. Age and Productivity of Spinning Technologies

System	Age (Years)	Productivity per Unit (Meters per Minute)
Ring Spinning	167	18
Rotor Spinning	27	150
Air Jet Spinning	16	350

Table 8.2. Short Staple Yarn Produced in the United States

| System | Percent of Total Weight of Yarn Produced by Year | | | | | |
	1984	1987	1990	1993	1996	2000 (est.)
Ring Spinning	79	69	60	53	44	40
Rotor Spinning	20	26	35	40	45	42
Air Jet Spinning	1	5	5	7	11	18

of years as of 1997 that the systems have been active and the relative productivity of the systems. The market share that these three spinning systems have established in the USA and the predicted market share through the year 2000 are shown in Table 8.2.

The Roving Process

Introduction

Practically all ring spinning machinery utilizes roving as the input material. Roving is a loosely twisted strand of fibers whose weight per unit length is made to allow a draft (reduction in weight per unit length) at the ring frame in the approximate range of 15 to 40. The purpose of the roving process is to reduce the weight of the fiber strand (slivers) for ring spinning.

From time to time a machinery manufacturer has offered to the industry machinery that would spin yarn on a modified ring machine using untwisted sliver as the input material. One such machine was offered by the Japanese company OM, Ltd. in the late 1950s, and several installations were put in place in the USA. This machine was equipped to give mechanical drafts up to 600. After a few years, however, this machinery disappeared from the scene. More recently Suessen introduced its "RingCan" sliver-to-yarn ring spinning machine at the 1991 International Textile Manufacturers Association (ITMA) show in Hanover, Germany. A thesis was conducted by Mark Stone at the Institute of Textile Technology in 1993 to determine the relative quality level of yarns spun on this machine, and the results were very favorable. There have been a few additional designs for systems to ring spin yarn from sliver, but up to the present time roving is considered the universal feed package for ring spinning.

Figure 8.3. The Hollingsworth (Saco Lowell/Platt) Rovematic roving frame. (Courtesy John D. Hollingsworth on Wheels, Inc.)

Description of the Roving Frame

In earlier times there were in use as many as three separate roving processes, one preceding the other. The first machine was called "the slubber", the second machine was called the "intermediate", and the third machine was called the "jack frame". As improvements were made to allow higher drafts on the roving frame, and as the draft of the spinning frame was increased by use of improved drafting systems, the roving process could be effectively accomplished with the use of only one machine. The machine is now referred to as the roving frame, the slubber, or as the fly frame.

The roving frame (Rovematic) shown in the photograph in Figure 8.3 was introduced to the textile industry in the late 1950s. It produced packages of either 305 mm (12 inches height) by 127 mm (5 inches diameter) or 356 mm by 178 mm (14 inches by 7 inches). It had two unique features, the flyer mounting and the Link Belt PIV (pitch infinitely variable) transmission, which permitted speeds as high as 1800 revolutions per minute of the flyer. At the time of its introduction, this speed was considered to be a breakthrough for roving frames.

Typical flyer designs are the top mounted flyer as shown in the drawing at left in Figure 8.4. The drawing on the right shows the unique mounting of the flyers on the Saco Lowell/Platt Rovematic frame. The normal procedure for doffing the full roving packages is to remove completely or raise the flyer on the left so the package can be lifted straight upward. For the Saco Lowell Rovematic system, the full package is "wound down" to the lower position within the fixed flyers, and the full packages are removed by lifting them just above the spindle (which is at its lower position) and pulling them forward to clear the flyers. For the typical flyer which is open at the bottom, the rod on which the flyer is mounted

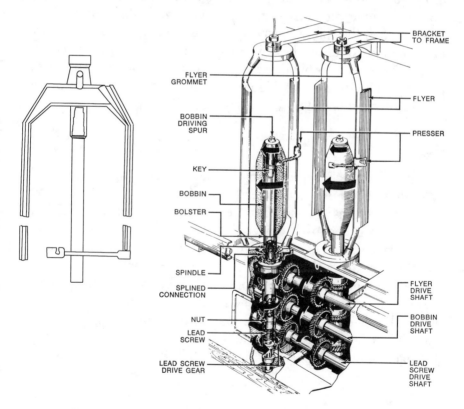

Figure 8.4. Illustrations of a typical flyer on the left and of the Saco Lowell "fixed" flyer on the right. (Courtesy John D. Hollingsworth on Wheels, Inc.)

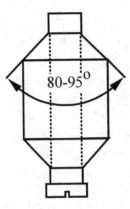

Figure 8.5. Profile of a full roving package.

moves up and down to position the coils onto the package, which does not move vertically during the package build. On the Rovematic frame, the spindle carrying the package moves up and down in a manner to allow the coils of roving to be properly laid onto the package, and to build the tapers at the bottom and top of the package. The diagram in Figure 8.5 illustrates the general profile of a full roving package.

Creel

The input stock to the roving frame is sliver in the weight range of 3.5 to 5.7 grams per meter (50 to 80 grains per yard). Sliver packaged in cans is positioned behind the machine and the sliver ends are threaded over powered "lifter" rolls which turn at the same surface speed as the back drafting rolls. The arrangement for the sliver path in the creel is such that each sliver end is lifted nearly vertically from its can to a turning lifter roll, thus eliminating stretching and rubbing of the sliver. Just behind the draft zone, where the sliver is drafted (attenuated) to its final weight per unit length, is a photoelectric stop motion, or other system, that stops the machine when a sliver breaks and falls downward.

Drafting

The drafting system on the roving frame typically consists of a 3-roll arrangement as illustrated by the drawing in Figure 8.6. The

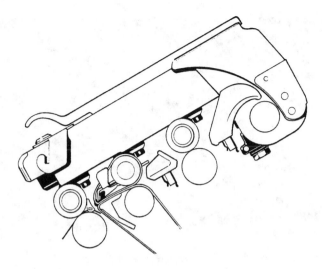

Figure 8.6. Diagram of a typical spring weighted drafting system used on the roving frame. (Courtesy Spindelfabrik Suessen)

bottom rolls are fluted steel and the top rolls are covered with synthetic rubber. When the top pendulum arm is latched down, springs are compressed to apply pressure on the top rolls for good, strong contact with the bottom rolls. Shown in the top view of the drafting system in Figure 8.7 is one pendulum arm in place (left) and one raised showing both the top rolls and the bottom steel rolls. The top rolls attached to the pendulum arm are double ended to control two adjacent deliveries. The latch lever on top of the arm when pushed down fastens the arm to the back mounting rod, and cantilever action causes springs in the arm to compress to generate force on the top roll arbors. The contact pressure lines generated at each of the 3-roll pairs cause the fiber bundles to flow at the surface speed of each of the three rolls, which are set to yield the amount of draft (attenuation) desired.

Shown in Figure 8.6 are a guide trumpet just behind the back roll pair and two guide condensers, one in the back draft zone between the middle roll pair and the back roll pair, and the other condenser immediately behind the front roll pair. The trumpet at the back position guides the sliver input material into the drafting zone, and the guide condensers keep the relatively heavy fiber bundle on the correct path to be drafted. Located in the front draft zone (between the middle and front roll pairs) are top and bottom

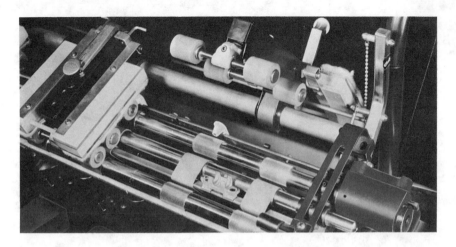

Figure 8.7. Top view of one roving frame drafting system.

aprons, necessary to suppress drafting waves caused by roller drafting. More in-depth discussions of drafting waves are presented in the Ring Spinning section under "Drafting" and in Chapter 6 under "Evenness of Drawing Sliver".

Another type of drafting system is the MagneDraft supplied by Saco Lowell (Hollingsworth) for use on the Rovematic roving frame. It is unique in that the pressure between the synthetic rubber covered top rolls and the bottom steel rolls is generated by magnetized top rolls. A view of this system is shown in Figure 8.8. The back rolls have large interlocking flutes which give a firmer gripping (nipping) hold on the fibers of the input sliver. A similar drafting system is supplied by Saco Lowell (Hollingsworth) for their Spinomatic ring spinning machine. One difference is that the back roll pair for spinning does not have the large interlocking flutes as found on the system supplied on the Rovematic roving frame.

Drafting roll spacing is the distance between the nip lines of successive roll pairs. This distance, as with all roller drafting systems, is adjustable to accommodate the fiber length to be drafted. A general guide to setting the roll spacing (ratch) is to set the space just greater than the longest fibers in the material to be processed. This spacing should be optimized by experimental trials.

Total draft (amount of attenuation) is determined by the surface speeds of the back and front roll pairs. If the front roll surface speed is 10 times greater than the back roll surface speed, the total draft

Figure 8.8. The MagneDraft drafting element for the Rovematic roving frame, utilizing magnetic top rolls.

would be 10. Thus, the weight of the exiting strand would be ¹⁄₁₀ the weight of the input strand. At times there can be a slight difference between the "mechanical draft" (ratio of roll speeds) and the "effective draft" (ratio of input to output strand weights per unit length). As the effective or actual draft is the primary objective, the draft should always be set to give the correct effective draft.

Another important consideration for the draft zone is "draft distribution"; this is the amount of draft in the back zone and the amount of draft in the front zone. The total draft is the product of these two draft ratios. The draft in the back zone is referred to as "break" draft, and it should be as low as feasible so that the largest portion of the total draft will take place in the front zone where apron control of fiber movement is located. Table 8.3 shows suggested roll spacings and break drafts for three different drafting systems in use on the Rovematic frames. In the bottom line are listed break draft factors which are useful in determining break draft ratios when the bottom steel roll speeds in revolutions per minute are known. The middle roll speed divided by the back roll speed times the break draft factor gives the actual mechanical break draft.

Most roving frames are designed to allow a large range in total draft from 4 to 20 and above. However, a more practical range of 5 to 12 has been found to yield a better quality product.

Table 8.3. Typical Rovematic Settings

Roll Zone	Roll Spacings by Drafting System Type					
	Truset (Spring Weighting)		Magnedraft (Magnetic Weighting)		SKF1500 (Spring Weighting)	
	Inches	mm	Inches	mm	Inches	mm
Front-to-Middle:						
Cotton, 1 $^1/_{16}$"	2 $^1/_{16}$	52.4	2 $^1/_2$	63.5	2	50.8
Synthetic	2 $^1/_4$	57.2	2 $^1/_4$	57.2	2	50.8
Middle-to-Back:						
Cotton	1 $^9/_{16}$ to 1 $^5/_8$	39.7 to 41.3	1 $^5/_8$	41.3	2 to 2 $^1/_4$	50.8 to 57.2
Synthetic	1 $^3/_4$ to 2	44.5 to 50.8	1 $^3/_4$ to 2	44.5 to 50.8	2 $^3/_4$	69.9
Break Draft						
Cotton	1.30		1.40		1.10	
Synthetic	1.50 to 1.70		1.70		1.15	
Break Draft Factor	0.975		0.798		0.964	

Wind and Spindle/Flyer Drive

It is essential that the wind of the roving strand be carefully made in a precision manner as illustrated in Figure 8.9. Each wrap of the roving is laid in a precise spacing relative to the previous coil to prevent damage to the loosely twisted strand and to form a uniform surface for the succeeding coils of roving to be wrapped onto as it is being built to a full package. The tapers at both ends of the bobbin are formed by the reduction in the length of the wind traverse after each full layer of roving has been made. Each layer of roving wound onto the package causes an increase in diameter. The surface speed of the package must be regulated to the speed at which the roving strand is delivered from the front drafting rolls. Therefore, the slight increase in diameter makes it necessary for the speed of the spindle, which drives the package, to be incrementally reduced as the package diameter increases. This precision wind requires a rather complicated drive mechanism, as the flyer with its presser guide is maintained at a constant speed.

The incremental change in spindle speed on most roving frames is performed by the use of a belt drive between two cone (tapered)

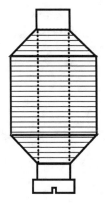

Figure 8.9. Illustration of the precision of the wind required for a roving package.

pulleys as shown in Figure 8.10. The Rovematic model uses a different PIV transmission as described earlier in this section.

On most roving frame models, the wind traverse is accomplished by a vertical up and down motion of the flyer, which is mounted, only at the top, on a vertical shaft that moves in a clearance channel through the spindle. On the Saco Lowell Rovematic, however, the traverse is made by the vertical up and down movement of the spindle and package, as the flyer is in a fixed position. The speed of the wind traverse is set with a change gear that is determined by the size (diameter) of the roving strand being produced.

Stop Motions

The number of deliveries (spindles) on the roving frame can vary depending on the production program in the textile plant. Typically, the number is in the range of 88 to 96. For efficient operation it is crucial that each package be maintained at the same diameter because of the programmed speed changes on the spindles. Consequently, when an end breaks or ceases to operate for any reason, the machine must be stopped quickly, so that only a small length of roving is lost on that package. After the strand is threaded and spliced, the machine is restarted. Stop motions are detectors that are fitted

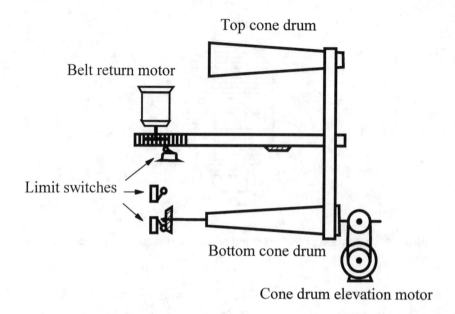

Figure 8.10. Illustration of the belt drive between two cones (tapered pulleys) for some roving frames.

to the machine both behind and in front of the drafting zone. Both mechanical and photo-electric devices are used for this purpose.

Grommetts

As the roving strand enters the top of the flyer, it is rubbed by the inside surface of a grommett which is mounted at the top of the flyer. This frictional contact between the fibrous strand and the grommett imparts "false" twist to the strand in the length between the front roll nip and the grommett. This twist imparts strength in the roving in this zone to allow it to better withstand the slight tension and vibrations that occur. Refer to Figure 8.11 for a view showing several grommetts.

Doffing

The roving frames have yardage counters that are preset to stop the machine after the planned yardage has been made. The full packages normally contain 1.3 to 2.7 kg (3 to 6 lb), depending on the package dimensions being made. Usually, the packages are removed by hand and stacked in a container mounted on wheels to facilitate transport. Empty bobbins are donned onto the spindles,

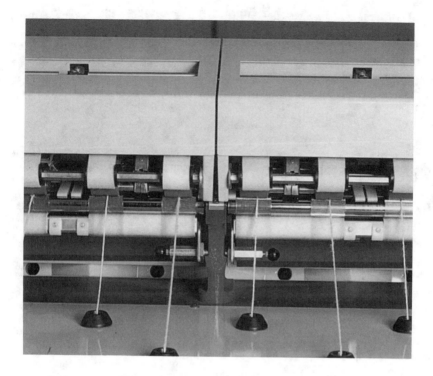

Figure 8.11. View showing several grommetts on a roving frame.

the roving end is attached to each bobbin, and the machine is restarted.

Toward the end of the 1980s the machinery builders began supplying equipment on which the doffing operation was automated. One such machine is shown in Figure 8.12. In addition to the removal of full packages, the unit positions them onto an overhead transport and storage channel that offers automatic transport to the ring spinning machines.

Roving Frame Settings

The trumpet that guides the sliver into the draft rolls should be chosen to properly fit the sliver. If too tight, the sliver will be stretched enough to change the effective draft; if too loose, the sliver will not be guided accurately. It has been determined that the best opening in the trumpet is a horizontal, rectangular slot which positions the sliver in the back nip for good fiber gripping.

Figure 8.12. Roving frame with automatic doffing feature. (Courtesy Fratelli Marzoli & C. spa)

The condenser in the back draft zone should be the same width as the sliver trumpet. The size of the condenser positioned just behind the front roll nip is determined by the roving size (hank) which is the number of 840-yard lengths that weighs one pound. This relationship is also used to determine yarn count on the English yarn/roving numbering system. Table 8.4 shows condenser sizes for a range of roving counts.

Total draft is set to yield the desired weight of roving by the choice of the correct draft change gear. Break drafts and roll spacings are set according to Table 8.3.

Optimum twists per inch in the roving strand are determined by trial and error, or by the use of the Draftometer developed at the Institute of Textile Technology (ITT). The twist level in roving should be sufficient to prevent stretching of the strand in normal handling and in its travel in the spinning creel. Too much twist will cause failures in the spinning draft, usually referred to as "hard ends". A brief description of the Draftometer and the procedure for determining "critical draft" follows. Refer to the sec-

Table 8.4. Condenser Sizes for Several Roving Counts

Hank Roving (Count, Ne)	Condenser Size
0.50 to 0.60	14.3 mm ($^9/_{16}$ inch)
0.61 to 0.75	12.7 mm ($^1/_2$ inch)
0.76 to 1.00	9.5 mm ($^3/_8$ inch)
1.01 to 1.75	7.9 mm ($^5/_{16}$ inch)
1.76 to 2.00	6.3 mm ($^1/_4$ inch)

tion following that discussion for the definition and use of twist multiple.

Determining Critical Draft of Roving Using the ITT Draftometer

Critical draft is the draft at which the dynamic drafting force being applied to the strand becomes highly unstable (stick-slip drafting). Laboratory and plant studies have concluded that the break draft on the spinning frame should be set 10 percent below the critical draft. If the break draft is set too near the critical draft, ends down in spinning will be higher, the resulting yarn will contain many more IPI and Classimat defects, it will have a higher Uster %CVm, and its properties will be more variable than the properties of yarn produced with the proper break draft.

The critical draft is determined by using the ITT Draftometer, a diagram of which is shown in Figure 8.13. The Draftometer uses a simple 2-roll drafting system with a transducer which translates the drafting force into the recordable electrical signal. When testing roving, the recorded drafting force will increase as the draft increases. The tracing of the drafting force should rise to a certain point and then remain stable. When the critical draft is reached, the tracing will become erratic with many peaks and valleys. When the draft becomes so high that it is above the critical draft, the drafting force will be low, yet stable on the tracing.

Shown in Figures 8.14 and 8.15 are tracings of two roving bobbins made from the same finisher sliver, each with a different level of twist. These tracings show that the roving with a twist multiple of 1.48 did not have a break draft where the drafting force was level. Because of the lower level of twist in the roving, there is no

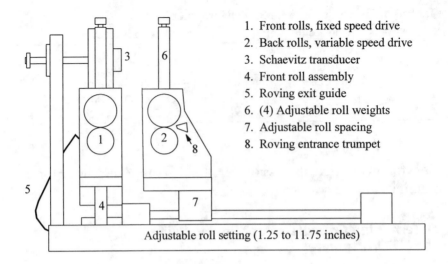

1. Front rolls, fixed speed drive
2. Back rolls, variable speed drive
3. Schaevitz transducer
4. Front roll assembly
5. Roving exit guide
6. (4) Adjustable roll weights
7. Adjustable roll spacing
8. Roving entrance trumpet

Adjustable roll setting (1.25 to 11.75 inches)

Figure 8.13. Schematic diagram of the ITT Draftometer.

break draft which is most suitable for this roving. In the case of the roving with a twist multiple of 1.65, the drafting force is stable until the break draft reaches 1.38. At this draft the tracing becomes erratic. At higher drafts the drafting force is very low and somewhat stable. Therefore, to optimize the quality and performance of ring spinning, this roving should be processed on a spinning machine with the break draft set at 1.25 (10 percent below critical draft).

As was shown in the previous example, the ITT Draftometer is capable of quantifying the way in which a fiber strand (roving or sliver) reacts to drafting force. For roving, the critical draft can be determined and the spinning frame can be set accordingly. Also, as in the case of the roving with a twist multiple of 1.48, the Draftometer can determine whether roving has too little or too much twist. ITT recommends that the critical draft of roving be engineered to 1.40.

The results from testing roving on the Draftometer are extremely valuable for the following:

1. Determining optimum twist level in roving,
2. Determining optimum break draft at spinning, and
3. Evaluation of roving quality.

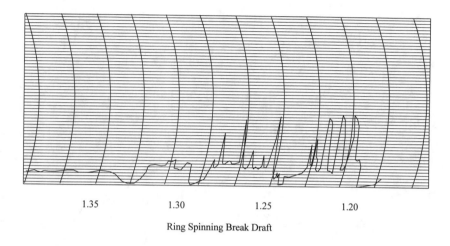

1.35   1.30   1.25   1.20

Ring Spinning Break Draft

Figure 8.14. Draftometer tracing of a 1.00 Hank 100 percent cotton roving with 1.48 Twist Multiplier.

Twist and Production Calculations

A guide to the correct twist in roving is to use the "twist multiple" (TM) constant, which determines the helix angle of the fibers in a twisted roving or yarn strand. The relationship between TM and twists per inch is expressed in the following equation:[8.1]

$$\text{Twists per Inch} = \text{TM}\sqrt{\text{Roving Hank (English Count)}} \quad (8.1)$$

Experience has shown that for certain fiber systems and for certain hank ranges, twist multiples should be consistent for best results. For example, the twist multiple for carded cotton should be between 1.30 and 1.50 for a range of roving sizes from 0.5 hank to

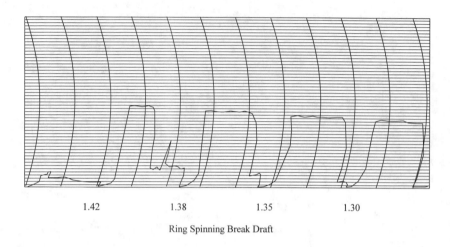

1.42   1.38   1.35   1.30

Ring Spinning Break Draft

Figure 8.15. Draftometer tracing of a 1.00 Hank 100 percent cotton roving with 1.65 Twist Multiplier.

1.0 hank. For combed cotton the twist multiple should be slightly less, 1.20 to 1.30, because the fibers are longer on average. Thus, it can be said that the longer the fibers, the lower the twist multiple.

One method of calculating the twist in roving is to relate flyer speed to the velocity of the strand exiting the front roll. A front roll of 28.58 mm (1⅛ inches) operating at a speed of 300 revolutions per minute will deliver roving at a velocity of 26.9 meters per minute. If the flyer were operating at 1000 revolutions per minute, the roving twist would be calculated by this formula:[8.2]

$$\text{Twists per Inch} = \frac{\text{Flyer RPM}}{\text{Front Roll Surface Speed (Inches/Min)}} \quad (8.2)$$

$$\text{Twists per Inch} = \frac{1000 \text{ Turns/Min}}{26.9 \times 39.37 \text{ Inches/Min}} = 0.94$$

Such a twist level would be appropriate for a 0.52 hank carded cotton roving. It can be observed from this relationship that when more twist is desired in the roving, the front draft roll would be reduced in speed without a change in flyer speed.

Many of the roving frames manufactured today for producing large roving packages are capable of flyer speeds as high as 1300 revolutions per minute. Such speed must be approached cautiously as it can cause both lower quality and a problem known as "surface breaks", which produce excessive flying lint that contaminates adjacent packages while in operation. The main difficulty with surface breaks is that the strand continues to remain intact until it gets to the package surface where all the fibers fly out into the surrounding space instead of being wound onto the package surface; thus, surface beaks. The machine continues to run.

Production rate of the roving frame is calculated using the front roll surface speed and the roving count. Considering the figures used above for calculating twist, the delivery velocity from the front roll is 26.9 meters per minute for the 0.52 hank roving.

On a roving frame with 96 spindles operating at an efficiency of 75%, the production yield would be:

$$\text{Pounds per Hour} = \frac{26.9 \times 39.37 \times 60 \times 96 \times 0.75}{36 \times 0.52 \times 840} = 290.9 \ (132.4 \text{ kg})$$

The meters per minute are changed to yards per hour and applied to all 96 spindles, with a correction for the frame efficiency. Pounds is determined by dividing the yards produced by the number of yards needed to weigh one pound (0.52 x 840). A general formula[8.3] for roving frame productivity in pounds/hour is as follows:

$$\text{Roving Frame, lb/hr} = \text{Front Roll m/min} \times \frac{39.37 \text{ in.}}{\text{m}} \times \frac{\text{yd}}{36 \text{ in.}} \times \frac{60 \text{ min}}{\text{hr}} \times$$

$$\frac{\text{lb}}{\text{Count(Ne)} \times 840 \text{ yd}} \times \frac{\text{Spindles}}{\text{Frame}} \times \text{Efficiency (decimal)} \quad (8.3)$$

Critical Factors at Roving

As with all textile processes, there are a number of critical factors at roving that need consideration and/or optimization for product quality and efficient operation. Listed below are a number of such factors.

1. Creel:
   - Sliver wound in cans without rubbing can wall;
   - Springs and pistons operating properly;
   - No piece cans (cans with many splices) being fed;
   - Cans located directly under lifting rolls;
   - Power creel (lifter rolls) turning;
   - No crossed ends;
   - No rough places on guide rods; and
   - No tags (cleaning).

2. Draft Zone:
   - Sliver trumpet size correct;
   - Good top roll surfaces and pressure;
   - Roll spacing and draft distribution optimum;
   - Proper condenser size;
   - Aprons and condensers well maintained;
   - All surfaces and rolls free of lint accumulation; and
   - Top rolls parallel to bottom rolls.

3. Front Roll to Wound Package:
   - Twist determined by ITT Draftometer;
   - Proper diameter grommet; grommet well mounted with no wobble when turning;
   - Tension, lay, and taper gears correctly selected;
   - Speed of flyer not excessive; use 1000 rpm maximum as a guide;
   - No wear on presser;
   - Bobbin firmly fixed to spindle;
   - No excessive backlash of flyer;
   - Minimum spindle vibration;
   - Tension between front roll and grommet not excessive;
   - Zero surface breaks;
   - Ends down rate not more than one per doff; and
   - No slip or backlash in transmission.

Ambient Conditions

Temperature and relative humidity should be maintained to less than ± 1 ½ points for good operation. A target setting for temperature is 28° Celsius (82° Fahrenheit). For 100 percent cotton material, a relative humidity of 48 percent is suggested; for synthetic fibers and blends, the suggested relative humidity is 43 percent.

Quality Benchmarks

Benchmarks for good quality roving are given below. For certain textile products, these benchmarks can be relaxed somewhat.

1. Uster %CVm <4.5;
2. Count Variation (1-meter weighings) 1.2%;
3. Uster Inert Range <3%; and
4. Ends Down, one per doff maximum.

Machine Suppliers

Popular suppliers of roving frames are as follows:

1. John D. Hollingsworth on Wheels, Inc., USA;
2. Rieter Machine Works, Ltd., Switzerland;
3. Fratelli Marzoli & C. spa, Italy;
4. Howa Machinery, Ltd., Japan;
5. Toyota Automatic Loom Works, Ltd., Japan; and
6. Zinser GMBH, Germany.

The Ring Spinning Process

Description of the Ring Spinning Machine

A cross-sectional drawing of a typical modern ring frame with a Co-We-Mat built-in automatic doffer is illustrated in Figure 8.16. Some key elements of the machine are named in the legend of the figure. The frame is divided into four separate zones for the description that follows. These four zones are the creel, the drafting system, the twisting zone, and the zone for winding onto the bobbin. In addition to the descriptions of these various elements of the ring spinning machine, certain factors that have been found to be critical for an efficient spinning process are discussed.

The Creel Zone

The creel is the superstructure at the top of the frame in which the supply packages (roving) are hung on hangers mounted on channel bars fastened to transverse arms attached to the vertical columns, fixed to the steel supporting frame, referred to as "samsons". These samsons contain the adjustable feet of the frame and are positioned at approximately every 1.2 meters along the frame. The creel of this type is called an "umbrella" creel which is now common on all ring frames. As shown in Figure 8.16, the roving ends are threaded over roving guide rods so that the roving can be guided toward the draft zone. It is critical that the roving guide path be as near as possible the same for each individual end to avoid differential tension (stretching) on the roving. An important quality parameter of yarn is minimum count (size) variability; different amounts of stretching of the roving ends cause count variability to increase.

An experiment conducted at the Institute of Textile Technology showed a difference in count variation that was caused by the vertical position of the roving guide bar that guides the roving as it leaves the roving package. The "high" position of the guide rod was ⅔ the height of the roving package from the bottom of the package, and the low position of the rod was only ¼ the height of the roving package. A 35's count 100% rayon yarn was spun from 1.0 hank roving, first with the roving guide rod set high, and then with the roving guide rod set in the low position (¼ up from the bottom of the roving package). The same roving packages and spinning positions were used in each case. The full bobbins were checked for count by winding off 20 successive skeins for size determination. A typical set of graphs showing the comparison of counts resulting from the twenty skein checks is shown in Figure 8.17. The overall average results showing the average count comparison and the count variability comparison are listed in Table 8.5.

A second critical factor pertaining to the creel zone is vibration. The creel is situated in a manner to magnify any vibration that causes even slight movement in the lower frame mechanism, such

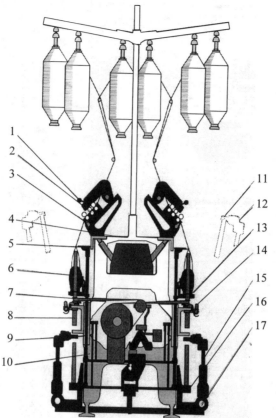

1. Draft unit
2. Bottom steel roll
3. Roller stand
4. Top plate
5. Lappet (pigtail guide)
6. Separator
7. Spindle drive tape
8. Spindle bolster
9. Lifter rod
10. Lifting mechanism
11. Bobbin gripper
12. Doffing bar
13. Yarn cutter
14. Spindle brake
15. Doffing arm
16. Autodoffer main shaft
17. Bobbin conveyor

Figure 8.16. Cross-sectional drawing of a typical ring spinning frame.

as unbalance in the drive motor pulley, or in the main shaft that drives the spindle tapes. Subsequent vibration in the creel accentuates the non-uniform stretching of roving ends in their path from package to draft zone, resulting in increased count variation.

Another inconsistency in the creel operation is the malfunctioning of the roving bobbin hangers. Such a holder, which is fastened

Yarn Count, Ne

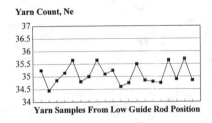

Yarn Samples From Low Guide Rod Position

Yarn Count, Ne

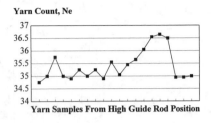

Yarn Samples From High Guide Rod Position

Figure 8.17. The effect of guide rod position in the spinning frame creel on yarn count variation.

Table 8.5 Comparison of Yarn Count and Count Variability Related to Roving Guide Rod Height

Measurement	Roving Guide Rod Height	
	High	**Low**
Average Count, Ne	35.7	35.3
Count Variability, %Vb	1.45	1.25

Note: Comparisons significantly different at 95% probability.

to the longitudinal channels that are mounted to the transverse arms of the creel, is illustrated in Figure 8.18. This model holder incorporates an external brake shown as the vertical curved rod with a smooth weighted section at the bottom. This section rides against the surface of the roving by gravity, and its purpose is to prevent the roving package from unwinding excessively. Another type of brake is known as the internal brake, which has a small amount of friction added to the turning of the ball bearing holder. Both types of brakes perform well when all elements of the holder are in good condition. Excessive wear on the bearing or lint or roving wrapped on the holder can cause differential stretching of the roving ends.

It had been customary for the spinning machine builders to locate some of the roving supply packages in a lower position. There were approximately 20 percent of the packages so positioned, because with the advent of larger roving packages it was difficult to locate all packages in the upper position. This lower positioning in the creel undoubtedly caused increases in yarn count variation because the lower creel roving ends were under different tension than those fed from the upper creel. However, the creels on the newer

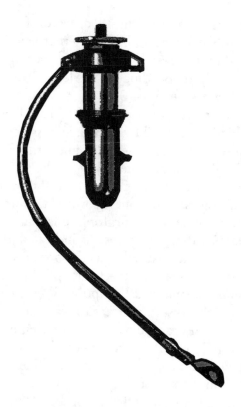

Figure 8.18. Illustration of a roving bobbin holder with weighted brake as used in the ring spinning frame creel.

ring frames have been designed to eliminate this lower positioning. One factor that has helped this latest creel design is the new standard of 406 mm (16 inches) roving bobbin length and 152 mm (6 inches) full package diameter for the roving package build.

The Drafting Zone

All of the elements of the ring frame are important, of course, but from the aspect of yarn evenness and consistency, the Drafting Zone is the most important. The greatest draft on an individual end occurs at the spinning frame. This necessitates a precision control of the fibers in the strand being drafted to suppress the phenomenon of "drafting waves" caused by the imperfect acceleration of individual fibers that exists in all roller drafting systems. It is usual for fibers in the strand to have different lengths, and this situation is referred to as "fiber array". Cotton is a system of varying fiber lengths, usually (depending on variety) 38 mm (1.5 inches) (longest fiber) down to 6 mm (0.25 inch, very short fiber). Even synthetic fibers that are cut to a uniform length exhibit varying fiber lengths which are caused by breakage in processing. Synthetic fibers also usually have a percentage of fibers that are hooked (not entirely straight), these hooks emanating at the carding process.

In Figure 8.19 is a sketch of a pair of typical drafting rolls depicting a flow of a fibrous strand from right to left. Normally, the drafting system contains three sets of drafting rolls, but only two sets are shown for explaining drafting waves. As indicated, the top and bottom rolls on the right are referred to as the back rolls turning at low speed; the roll pair on the left, the front rolls, are turning at high speed. The top and bottom rolls are pressed together with pressure to create a nip which causes the fibers to travel at the surface speed of the rolls. It is usual practice to set the spacing between the rolls (nip points) slightly longer than the longest fiber in the fiber system, so no breakage will occur in the drafting unit. Ideally, the fiber that is just released from the nip of the back rolls should be grabbed instantly by the nip of the front rolls (fast speed). Understandably, this ideal action cannot exist with the shorter fibers. And as these shorter fibers become under the influence of their neighboring longer fibers, which have already gained the high speed of the front rolls, they are accelerated earlier than desired. This early acceleration of the short fibers causes them to accumulate into a heavy mass leaving a lighter mass behind. This drafting wave of higher mass and then lower mass occurs in succession at a wave length of approximately 2½ times the mean fiber length.

Notice in Figure 8.19 the dotted lines which indicate the general position of flexible aprons that are typically applied to the drafting system. Such aprons are driven by the back rolls causing their surface speed to be the speed of the back rolls. These aprons tend to keep the fibers, regardless of their length, traveling at the back roll (apron surface) speed, thus preventing short fibers from being accelerated before they are gripped by the front roll nip. The apron systems are not perfect, but they reduce greatly the drafting wave effect on mass variation.

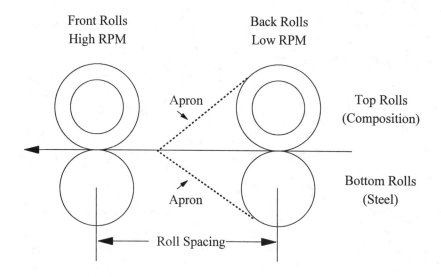

Figure 8.19. Sketch of a simple roller and apron drafting system as used in ring spinning.

There are only two types of apron systems that are now used commercially for ring spinning; these are the short bottom apron system shown in Figure 8.20, and the long bottom apron system shown in Figure 8.21. Both of these figures are cross-sectional drawings of these mechanisms.

For many years the short bottom apron system (Figure 8.20) was the preferred system used in the USA, whereas the long bottom apron system was preferred in Europe and most other countries. Within the past ten years or so, the long bottom apron arrangement has spread into the USA with the installations of the latest machine technology.

Seen at the very top of Figures 8.20 and 8.21 is the mechanism that holds the top rolls in position and aligned with the bottom fluted steel rolls. This mechanism is called the "pendulum arm". The pendulum arm contains clips that are spring actuated to hold the center section of the top roll arbors that accommodate two drafting units, one on each side of the arm. As it is necessary for the top rolls to be pressed down against the bottom rolls to establish a positive grip for the grist of fibers being drafted, the pendu-

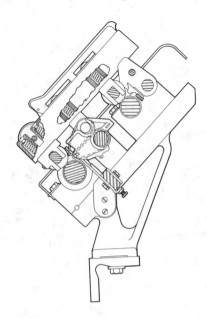

Figure 8.20. A cross-sectional view of the short bottom apron short staple drafting system. (Courtesy Whitin International Ltd.)

lum arm is fitted with springs, usually coil or leaf. The force to actuate the springs to press the top rolls against the bottom rolls is usually cantilever action by the mounting of the pendulum arm to the heavy mounting bar at the top end of the arm. Some earlier designs utilized latch down straps hooked near the front end of the arm just behind the front rolls.

The back top rollers and the front top rollers are covered with a relatively thick tube of synthetic rubber, which provides a cushion type grip on the flutes that have been broached into the surface of the bottom steel rolls. The bottom rolls are driven from one end (both ends on some extra long frames), and the top rollers are turned by frictional contact with the steel rolls. The center top roll is normally a steel roll with a knurled surface that carries the top apron. The top apron, and bottom apron on the short bottom system, are guided at the front (near the front roll nip) with a steel "U" pin that is fixed into a "cradle" that is separate for each spinning position. For a long bottom apron system as shown in Figure 8.21, the front apron guide is a steel bar fixed in the roll stands, which are positioned at every six or eight spin-

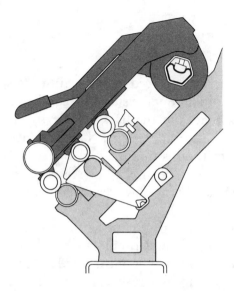

Figure 8.21. A cross-sectional view of the long bottom apron drafting system. (Courtesy Rieter Machine Works, Ltd.)

ning positions. The top apron in this latter system is guided at the front by a spring loaded plate which is fitted and attached to the pendulum arm.

The synthetic rubber covering (cot) for the front and back top rolls is formulated according to the following criteria:

1. Achievement of desired hardness;
2. Resistance to ozone cracking;
3. Resistance to fiber sticking or static attraction that usually leads to roll lapping;
4. Resistance to wear;
5. Production of a fine, smooth ground surface; and
6. Resistance to set marks which can occur when the frame is stopped for extended periods of time with the pressure applied.

As mentioned above, the pressure of the top rollers against the bottom rollers is applied by a latch-down of the pendulum arm causing compression of coil or leaf springs. The total pressure applied on the individual rolls at their center position is approximately as follows:

1. Back roll at 13.5 kg,
2. Center roll at 10.0 kg, and
3. Front roll at 18.0 kg.

The pendulum arm is forced down against the springs mounted in the arm by a latch-down system, by cantilever action through mechanical attachment to the back bar, or by air pressure at the attachment mechanism at the back bar. A convenient feature of the air pressure method is that the pressure on all pendulum arms of the full machine can be applied or released simultaneously. Rieter Machine Works, Ltd. of Switzerland incorporates the air pressure design (Figure 8.21).

One concern of the top synthetic rubber covered top rolls and aprons is wear. The normal running schedule for these spinning machines is 24 hours per day; most textile plants operate 7 days per week. Wear occurs where fibers come into contact with the roll and apron surfaces. The bottom fluted steel rolls are not a problem because they are surface hardened. The roving strand is guided into the drafting system by a roving trumpet mounted just behind the back (feed) rolls (Figure 8.20). This trumpet is fixed to a bar which oscillates sideways at a very slow speed. This movement of the guide trumpet spreads the wear over a centimeter of the roll and apron surfaces.

The primary goal of the back roll (feed) nip is to provide a positive no-slip grip on the roving strand. Usually, the hardness and the composition of the back top roll covering are aimed at this objective.

As the back rolls turn at a relatively slow speed, lapping of fibers around the back rolls presents no serious problem. The front rollers rotate at 20 to 40 times the speed of the back rolls in order to provide the necessary draft or attenuation of the strand of fibers. Lapping at the front rolls can be a troublesome problem. Lapping can be described as the fibrous strand winding around either the top or bottom roll instead of traveling beyond the roll nip to be collected into the suction pipe (or tube) which is adjacent to the front bottom steel roll. In most cases lapping does not occur until the end going into the twisting zone breaks and is expected to be picked up by the suction system. Several causes have been found that tend to increase lapping frequency including dirty or contaminated rolls, surface cuts on the top roll surfaces, low suction on the collection tube, static buildup, and improper finish level on synthetic fibers. Below is a checklist that is helpful when lapping is a serious problem:

1. Finish level on synthetic fiber. Check finish level on incoming bales. Ensure good openness of fibers on blending table.
2. Tinting and/or oversprays can be troublesome.
3. Avoid high relative humidity with a 43 percent maximum.
4. Replace old steel rolls that have corrosion pits or worn flutes and keep steel roll flutes clean.
5. Replace damaged top roll cots, or cots with too small diameter, 27 mm (1.06 inches) to 32 mm (1.26 inches) minimum.
6. Increase buffing cycle of cots and treat cots before use by exposure to ultraviolet light.
7. Clean finish/wax buildup from top front roll cots. Use mild soap and warm water or suitable solvent.
8. Position vacuum suction slots in best location.
9. Keep at least 50-mm (about 2 inches) water gauge suction on end collection system.
10. Experiment with J 490H cot material. (Hardness of cot is probably significant.)
11. Check steel rolls for good electrical grounding.
12. Check static charge on top front roll.
13. Ensure good fiber alignment. If processing a draw frame blend of polyester/combed cotton, predrawing polyester before blending might help. For polyester/cotton intimate blend, three drawing processes might help.
14. Avoid low twist in roving.

There have been several attempts to develop a practical system to stop the supply of input roving to the draft zone when a break at the output occurs. One notable example is the Roving Stop Motion marketed by SKF Textilmaschinen - Komponenten GMBH. This unit can be retrofitted to both cotton and worsted ring frames. Roving stop hardware has not been well accepted in industry, mainly because the economic balance has not been favorable.

A variation of the normal drafting arrangement is the INA V-Draft high drafting unit manufactured by INA Walzlager Schaeffler KG, Germany. A drawing of the unit is shown in Figure 8.22.

As can be seen in that drawing, the back rolls have been repositioned in a manner that increases pressure on the roving strand *after* it has been released by the back roll nip line. This increased pressure against the fibers in the roving causes the fibers to maintain the back roll surface velocity farther into the drafting zone, and this control reduces the magnitude of the drafting wave described previously. The objective of the V-Draft is to allow a higher draft ratio in the back draft zone (back roll-to-middle roll) without impairing yarn quality. Tests made at the Institute of Textile Technology have in fact verified this feature. Of course, a higher draft in the back zone translates to a higher total draft of the system. Such a higher total draft will allow the yarn to be spun from a coarser roving strand, resulting in a cost saving at the roving process.

Critical Factors in the Drafting Zone

The most important part of the ring spinning system is the drafting zone, and several critical factors are discussed in the following paragraphs.

Roving Trumpet

The roving trumpet guides the roving strand into the back roll nip. It is located as close as practical to the back rolls. As mentioned earlier the trumpet is fastened to a transverse bar which oscillates slowly to cause the roving strand to move laterally in order to spread wear on the rolls and aprons. The distance of the side-to-side travel of the trumpet can be altered and should be set in a manner to use as much surface as practical without getting too close to edges of cots (covering on top rolls) or aprons. The surface of the trumpet must be very smooth and of low friction, with a hole (slot) diameter which is large enough to prevent chokes from collecting in the trumpet. Chokes in trumpets cause excessive yarn hairiness and also cause the drafted strand to be lighter in mass.

Roll Spacing

The rolls must be set apart so that the longest fibers in the bundle being drafted are released by the grip (nip) of the back roll pair before entering the grip of the front roll pair. When the draft ex-

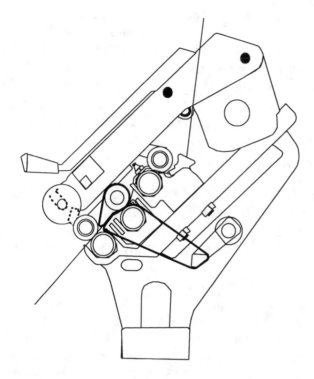

Figure 8.22. Drawing showing a cross section of the INA V-Draft system.

tension of the zone is relatively high (greater than 2.0), the rolls should be set as close as possible; when the draft is relatively low (less than 1.5), the roll spacing can be set farther apart. In the high draft zone where the drafting aprons are located, the discharge nip (pressure point) of the apron pair must be set as close to the forward roll nip as possible for best control of mass uniformity in the resulting yarn.

Roll Positioning

The bottom steel rolls are typically of equal diameters, or very nearly equal diameters, with their centers positioned so they are all in the same plane. Lines drawn through these centers and perpendicular to the plane establish the zero "overhang" for the centers of the top rolls. The back top rolls on all systems, except the INA V-Draft, are positioned with zero overhang, meaning they are exactly above (on the perpendicular line) the bottom roll plane. In many cases the front top rolls are positioned with positive (forward) overhang of about 2 mm. Such positioning can be advantageous as it can allow the apron discharge lips to be set closer to the front roll nip, thus helping mass uniformity in the yarn. Please refer to the section on "Ring Frame Geometry" for further discussion on this point. The middle top roll is usually positioned with zero overhang or a slight negative overhang to allow the apron plane discharge to be properly positioned.

Apron Surfaces and Positioning

Management of the draft aprons is critical for good drafting. The fit of the short aprons, either top or bottom, must be optimal in that the aprons must move steadily without buckling or slipping. The tensioning bar for the long bottom apron must be in good condition. Both the top and bottom aprons must be of the identical surface material which allows good control of the fibers being drafted. The discharge plane of the apron system must be slightly higher (1.5 mm) than tangent to the front roll surface. Another critical consideration is the spacing of the bars (or plates) that are the *front* guides for the aprons. Recommended starting point spacings between the top and bottom bars, which control the clearances for the apron movement and the pressure brought to bear on the fibers, are given in Table 8.6. Notice that these spacings (openings) are associated with the yarn count being spun, and that the space openings are different for the short bottom system (solid tensor) and long bottom apron system (guide plate and guide bar). For best optimization, trials should be run with openings both larger and smaller than the starting point spacings given in Table 8.6.

Draft Considerations

The term "draft" is given to the amount of attenuation, or reduction in mass per unit length, of a textile strand. "Effective draft" is the actual weight per unit length entering the drafting zone divided by the weight per identical unit length exiting the drafting zone, or:

$$\text{Effective Draft} = \frac{\text{Weight per Unit Length Input}}{\text{Weight per Unit Length Output}} \quad (8.4)$$

Table 8.6. Front Apron Guide Bar (Tensor) Openings

	Opening in Millimeters	
Yarn Count (Ne)	Short Bottom Apron (Tensor Pin)	Long Bottom Apron (Guide Bar & Plate)
14's and Coarser	6.0	5.6
15's – 18's	5.5	4.9
19's – 24's	5.0	4.5
25's – 30's	5.0	4.1
31's – 36's	4.5	3.8
37's – 45's	4.0	3.4
46's and Finer	3.5	3.2

Another way to express the amount of draft is by relating the output velocity to input velocity; this ratio of velocities is typically called "mechanical draft", which can be stated as:

$$\text{Mechanical Draft} = \frac{\text{Output Velocity}}{\text{Input Velocity}} \quad (8.5)$$

In many situations the effective draft and mechanical draft are not *exactly* the same, so one must be careful in setting the draft unit to give an anticipated amount of drafting. The mechanical draft is normally calculated from the yarn count desired when an input roving count is known, and the drafting unit is geared to that amount as closely as possible. Then, some yarn is actually spun and the weight of the yarn is measured. If the yarn weight is correct, no change is made. If the yarn count is off of the target, a change is made in mechanical draft and additional checks on yarn count are made.

It is obvious that the roll speeds and diameters must be determined for calculating mechanical drafts. This consideration requires special attention where the thickness of the draft aprons are part of the diameters of the middle rolls. This must be kept in mind when the draft in the back zone (break draft) is being calculated. Further, the diameters of the steel rolls themselves should be carefully measured.

Draft distribution, that is the amount of draft in the back zone compared to the amount of draft in the front zone, is a most important consideration. The aprons that are used to suppress the drafting wave are typically applied to the front zone only; consequently, the greatest amount of draft should occur in the front zone. As described in "The Roving Process" section of this chapter, the best break draft most likely can be determined by use of the Institute of Textile Technology (ITT) Draftometer. The usual procedure would be to set the total draft to give the desired yarn count, and set the break draft according to the Draftometer results. As the total draft is the mathematical product of the back draft times the front draft, the front draft equals the total draft divided by the break draft. For example, suppose a 35's yarn is to be spun from a roving that measured 1.05 hank (yarn count), and the Draftometer check on the roving suggested a break draft of 1.32. The total mechanical draft would be estimated by this equation:[8.6]

$$\text{Total Mechanical Draft} = \frac{\text{Nominal Yarn Count (Ne)}}{\text{Nominal Roving Hank (Ne)}} \quad (8.6)$$

$$\text{Total Mechanical Draft} = \frac{35}{1.05} = 33.3$$

$$\text{Back Draft} = 1.32 \text{ (Draftometer Result)}$$

The expected main or front draft is then calculated by this formula:[8.7]

$$\text{Front Draft} = \frac{\text{Total Draft}}{\text{Break Draft}} \quad (8.7)$$

$$\text{Front Draft} = \frac{33.3}{1.32} = 25.25$$

Synthetic Rubber Cots on Front and Back Top Rolls

The synthetic rubber coverings (cots) fitted on the front and back top rolls of the ring frame 3-roll drafting system have to perform several functions. On the back top roll, the cot material must provide a firm grip on the roving strand to prevent any slippage of the fibers at the grip (nip) line. It should provide good wear properties and resist surface contamination from cotton waxes and synthetic fiber finishes. Hardness, as measured by the Shore Durometer A Scale, is normally in the range of 75° to 80°. This back roll usually runs at relatively slow speeds; consequently, lapping at the back line of rolls is very unusual. The functions of the front top roll cots are similar to those of the back rolls except for lapping resistance and hardness. As the front roll pair turns at a much higher speed than the back roll pair, lapping can be a serious problem. For this reason the front roll cots are formulated to reduce static charge buildup. Further, tests show that softer cot material, such as 60° or 65° Shore, give more even yarn with fewer IPI thick and thin places. A negative feature of the softer cots is the increased buffing cycle frequency. Table 8.7 contains certain yarn data for 38's combed cotton comparing front top cots with Shore hardness of 60° and 75° after 800 hours of operation.

Vacuum Tube

On all spinning frames there is a slotted flute tube, or a suction pipe, positioned as close as practical to the front roll discharge. The purpose of this suction tube is to collect fibers that might shed off from the strand of fiber exiting the front roll nip, and to collect all the strand after a break occurs. The collected fibers are usually deposited in a filter box at the foot end of the frame. The fiber collection in the filter box can be removed manually or automatically.

Table 8.7. Comparison of Certain Properties of Yarns Spun with Different Front Top Roll Cot Hardness Measurement, Shore Durometer A-Scale

Yarn Property	Cot Hardness, Shore A-Scale	
	60° Shore	75° Shore
Yarn Count, Ne	37.7	38.2
Variability, %Vb	1.6	1.9
Break Factor	2461	2560
Single-End Strength, g/tex	15.3	15.5
Variability, %Vo	8.9	10.1
Single-End Elongation, %	**5.2**	4.6
Variability, %Vo	**2.1**	3.8
Uster %CVm	14.3	14.2
IPI Thins per 1000 m	4	7
IPI Thicks per 1000 m	73	67
IPI Neps per 1000 m	28	27
Uster Hairiness Index	**5.0**	5.2
Classimat A1 Defects	60	79
Ends Down/M Spindle Hr	10.2	11.6

Note: Boldface type indicates significant difference @ 95% probability.

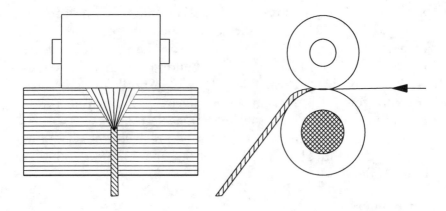

Figure 8.23. Drawing showing the triangle of fibers at the front roll exit zone of a ring spinning position.

Fibers Exiting the Front Drafting Rolls

The configuration of the fibers exiting the front roll nip is in the form of a flat ribbon as indicated in Figure 8.23. The twist in the yarn emanates in the "balloon" of yarn formed around the spindle as it rotates. This twist propagates upward through a pigtail guide and up toward the front roll exit nip. It is critically important that this twist *does not* travel all the way to the roll nip, but stops at the apex of an inverted triangle of fibers that occurs because of the wrap of fibers on the front bottom roll. This triangle causes the edge fibers to be under tension and the fibers near the center of the triangle to be loose. This condition allows the fibers to enter the yarn structure in a manner to produce a relatively dense structure with the majority of fibers having one end bound into the yarn center, and the other fiber end near the yarn surface. This fiber positioning is called "migration", which is the main reason that ring spinning yields a yarn with strength that is greater than yarn made by competing systems. The length of the so called "spinning triangle" depends on the geometric construction of the spinning machine.

Twisting Zone - Ring Frame Geometry

Both the twisting and winding zones are illustrated by the drawing in Figure 8.24. A more detailed drawing showing relative posi-

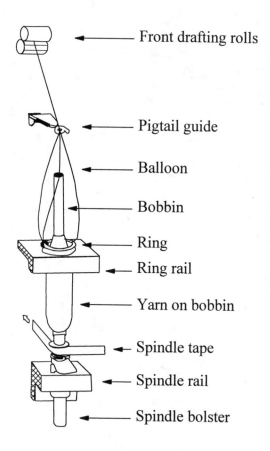

Front drafting rolls

Pigtail guide

Balloon

Bobbin

Ring

Ring rail

Yarn on bobbin

Spindle tape

Spindle rail

Spindle bolster

Figure 8.24. Drawing of the ring spinning frame twisting and winding zones.

tions of major machine elements and important angles is presented in Figure 8.25. The angle designated as "A" of the 3-roll drafting zone is determined by the angle of the roll stands which are mounted on roller beam "Z". This angle, together with the mounting position of the stands on the roller beam and the location of the pigtail guide "F", establish the spinning triangle as described above. At "C" is the angle between the yarn reach "E" and the axis line of the spindle. For best spinning performance, the balloon angle "D" should equal the spinning angle "C". Where these two angles are equal, the yarn will be relieved of tension against the pigtail guide during part of each rotation of the bal-

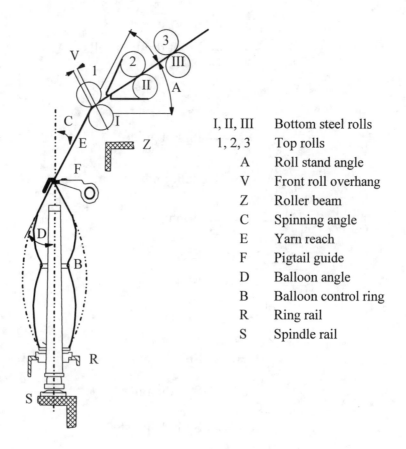

I, II, III	Bottom steel rolls
1, 2, 3	Top rolls
A	Roll stand angle
V	Front roll overhang
Z	Roller beam
C	Spinning angle
E	Yarn reach
F	Pigtail guide
D	Balloon angle
B	Balloon control ring
R	Ring rail
S	Spindle rail

Figure 8.25. Ring frame geometry.

loon which occurs with each rotation of the traveler. The tension release at the pigtail thread guide allows a higher percentage of the twist in the balloon to propagate into the reach of yarn above the guide, thus giving the best performance with regard to ends down rate. The solid balloon lines show the shape of the balloon envelope when a balloon control ring "B" is in place. The broken line depicts the balloon envelope without a control ring. This drawing shows the position of the ring rail "R" at the very beginning of the "doff", the term given to the period of time during which the empty bobbin is completely filled with yarn.

When the spinning position is operating, the twist in the yarn strand between the front delivery rolls of the drafting zone and the bobbin onto which the twisted yarn is wound originates in the balloon as depicted in the drawing of Figure 8.26. Although twist is actually inserted into the region between the pigtail thread guide and the traveler, the yarn between the front roller nip and the thread guide (reach) and that wound onto the package (bobbin) is also twisted. Therefore, twist must have passed both the thread guide and the traveler. As indicated in Figure 8.26, the twist flow at the thread guide is opposite to the yarn flow, whereas at the traveler, the twist flow is in the same direction as the yarn flow.

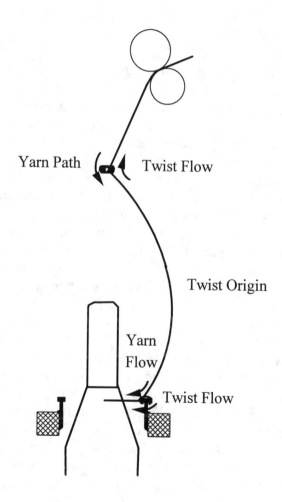

Figure 8.26. Illustration showing twist insertion in ring spinning.

The thread guide and traveler offer resistance to the passage of twist because of frictional contact. Photographic studies have shown that the twist above the thread guide is approximately 95 percent of the twist in the section of yarn in the balloon in the dynamic state when the spinning angle and the balloon angle (see Figure 8.25) are equal. In the same studies it was found that the twist in the yarn wound on the bobbin is approximately 92 percent of the balloon twist. Thus, there is more resistance to twist flow at the traveler.

Ring Spinning Winding Zone

Bobbin Build

The yarn travels from the front drafting rolls down through the pigtail guide and on down to the traveler which rotates around the stationary (typically) ring and guides the yarn onto the bobbin to form the yarn package. Obviously, the traveler in its function as yarn guide must move gradually up and down to position the yarn uniformly over the whole bobbin, and allow the wound-on yarn to build up the diameter to the point which gives adequate clearance between the yarn package and the rotating traveler. The position of the ring determines the plane of the traveler. Consequently, the ring rail ("R" in Figure 8.25) must be moved in accordance with the desired winding plan. The ring rail movement is actuated by mechanical linkages, including cables or flexible bands, caused to oscillate by a cam drive at the gear end (head end) of the machine. The ring rail is a continuous channel in sections, and covers the entire length of the frame except for the two ends where there are no spinning deliveries. The weight of the ring rails, one on each side, is counterbalanced by either weights or springs.

The relative speed of the ring rail up and down is fixed by the cam contour; this difference in up and down speeds typically is 3 to 1. The sketch in Figure 8.27 is an illustration of the yarn build (wind) on a package with a "filling" wind. The lines drawn at the top taper (cone) section of the package indicate the difference in these up and down speeds. In the slower speed direction, the lay of the yarn onto the package is called "wind", and in the higher speed direction, the lay of the yarn onto the package is called "bind". This principle of winding has been developed to produce

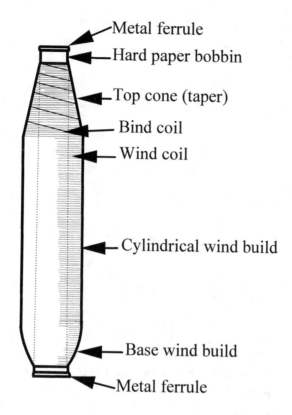

Metal ferrule

Hard paper bobbin

Top cone (taper)

Bind coil

Wind coil

Cylindrical wind build

Base wind build

Metal ferrule

Figure 8.27. Drawing illustrating the build of yarn as wound on a ring spinning bobbin.

a compact and stable yarn package, and to yield highly efficient withdrawing of the yarn during high speed winding, the next step in the processes after ring spinning.

The ring rail speed cycle of fast in one direction and slow in the return direction can be changed by turning the actuating cam on its drive shaft. Traditionally, the cam has been mounted to have the faster speed on the downstroke because this setting ensures less sloughing at the winding process which follows spinning. However, when the coils per inch are adjusted according to the graph in Figure 8.28, the cam can be mounted to have the slower speed on the downstroke without inducing sloughing at winding. Experience has shown that ends down at spinning are slightly lower when the downstroke is slow. In some cases with coarse yarns, the ends

Total Coils Per Inch, Up + Down

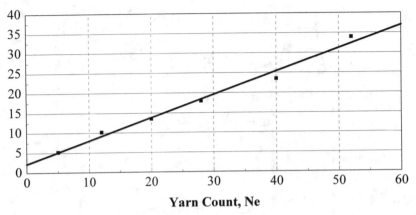

Yarn Count, Ne

Figure 8.28. Suggested maximum coils per inch (up + down) for each full ring rail stroke.

down are as much as 25 percent lower when a change of the ring rail movement from down fast to down slow is made.

With further reference to Figure 8.27, the stroke of the ring rail, which is represented by the tapered section at the top of the package, is normally slightly longer than the ring diameter. This stroke length (the total movement up or down for each stroke) remains the same during the cylindrical and tapered wind sections. During the base wind section, however, the stroke is shorter and gradually lengthens to the stroke for the main portion of the package build. The average position of the rail with respect to the package (bobbin) gradually changes upward during the doff (the building of the full package). This rate of upward movement, which is changed by the movement of a ratchet wheel on the build mechanism, is set to yield a package diameter approximately 3 mm (nominally ⅛ inch) less than the inside ring diameter.

The speed of each stroke of the rail is determined by the speed of the cam rotation which is adjusted by a change gear in the cam drive gear train. The higher the revolutions per minute of the cam, the fewer the coils per inch of ring rail stroke. This speed of the ring rail stroke is governed by the yarn count being spun. The graph in Figure 8.28 is the suggested number of average coils per inch on the yarn package for the up plus downstroke, based on yarn count. Remember that the ratio of wind to bind coils is about 3 to 1.

The total distance traveled by the rail during a whole doff is called "traverse". There is an allowance of approximately 19 or 20 mm (nominally ¾ inch) for clearance between bottom of traverse to bottom of bobbin when added to clearance between top of traverse and top of bobbin. These clearances allow for slight differences between deliveries for the fit of bobbins onto the spindles, and give a few minutes safety if the automatic stop at the end of doff fails to function quickly. Bobbins are made of hard impregnated paper with metal ferrules at each end for protection of the tube. Bobbins made of plastic are available also and are tapered slightly to fit the taper on the spindle blade. There are several different designs for fixing the bobbin to the spindle. One is as mentioned above for the inside (hole) taper in the bobbin to match the slight taper on the spindle. Another design causes the top of the bobbin to come in contact with spring loaded protrusions on the spindle tip. Still another is to have a magnet ring at the bottom ferrule location to hold the bobbin firmly at the bottom. For all arrangements it is critical that there is zero slippage between the bobbin and spindle, and that there is very little vibration of the bobbin as it spins at speeds up to 20,000 revolutions per minute and higher.

Doffing the Ring Frame

At the end of the doff when the packages are full of yarn, the spinning machine is stopped; this is usually done with an electrical switch mounted in such a manner that the ring rail actuates the switch when the rail reaches its top traverse level. Before the full packages of yarn are actually removed from the spindles (doffed), the ring rail must be moved to its very bottom position while spindles are still turning so that the yarn is wound helically down the full bobbin, and with a few turns of yarn around the spindle below the bottom bobbin ferrule. This operation is called "bear-down", and it can be done manually or automatically. The full packages are removed, by hand or automatically, and empty bobbins are installed on the spindles (donned). Since the yarn is left threaded through the traveler and held by the wraps around the spindle, the machine is ready to be started as soon as the thread boards, on which the thread (pigtail) guides are mounted, are rotated down into running position. Just as the machine is started, the ring rail is actuated up and down over a small stroke to make certain the yarn is firmly attached to the base wind on the bobbin.

The weight of yarn on the doffed packages depends on both the length of the bobbin onto which the yarn is wound and the package diameter. The range of weights without regard to yarn count is approximately 0.05 to 0.27 kg (0.1 to 0.6 lb).

Yarn Tension at Ring Spinning

The tension pattern in the yarn in the twisting and winding zones is of critical importance. After studying how the ring rail motion moves throughout the build of the yarn package, it is obvious that the yarn balloon changes height considerably during the package construction. It is important to know that when the balloon height is high, the tension in the yarn is relatively high; when the balloon height is low, the tension in the yarn being twisted is relatively low. Measurements have shown that the average tension in the yarn balloon is 1.7 times higher when the yarn is being wound at the bottom of a 25-centimeter or a 10-inch long bobbin (long balloon height) than when it is being wound near the top of the traverse (short balloon height).

Illustrated in Figure 8.29 is the yarn geometry in the lower part of the balloon and between the traveler and the bobbin/package (dotted lines). In this illustration the spindle is turning clockwise as viewed from above. Angle "A-1" is the angle between the traveler-spindle center line and the yarn being wound on the bare bobbin; this angle is called "angle of pull". Angle "A-2" is the angle between the traveler-spindle center line and the yarn being wound on the nearly full yarn package. The "A-1" angle in this drawing is 30 degrees; the A-2 angle is 65 degrees. Thus, during the build of the yarn package, the angle of pull changes significantly. By examining the top view in Figure 8.29, one can see that the only force causing the traveler to move around the stationary ring is the tension generated in the yarn between the traveler and bobbin, or yarn package. Because of the force vectors involved in this geometry, and because the force needed to move the traveler

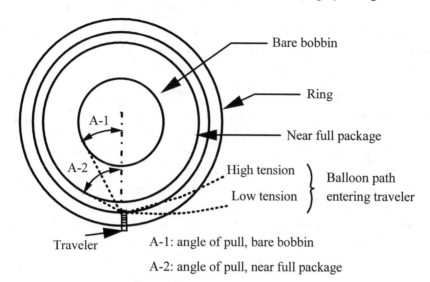

Figure 8.29. Illustration of angles of wind and yarn paths as yarn is wound onto package with the spindle turning clockwise as viewed from above.

around the ring is very nearly constant, the tension in the yarn going on to the bobbin/package is much greater with a small angle of pull (A-1) than with a large angle of pull (A-2). Naturally the tension in the yarn between traveler and package is translated to the tension in the yarn balloon with a slight reduction caused by the friction between the yarn and traveler. A high tension in the balloon causes the yarn to be more upright as it enters the traveler as shown in the Figure 8.29 illustration. The combinations of these angels and tensions have led to the adoption of the filling build or filling wind, as described above, to be used almost universally.

Twist in Ring Yarns

When the spindle turns in the clockwise direction as viewed from the top, the twist direction is called "Z" twist, or regular twist. On most ring spinning machines the spindles can be rotated in a counter-clockwise direction to produce "S" twist, or reverse twist. The angle of the central parts of the letters S or Z indicates the helix inclination of the fibers in the yarn as shown in Figure 8.30. Most staple yarn is spun with Z twist. It has been stated that 95 percent

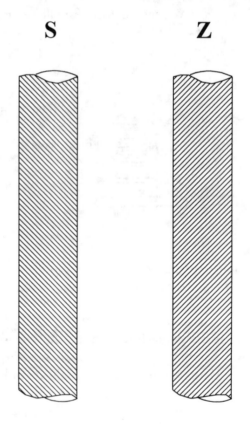

S   **Z**

Figure 8.30. Diagram of the two possible directions of twist (S or Z) in ring spun yarn.

of short staple yarns are produced with regular twist. In many in-stances S twist yarn is made to be plied (2 yarns twisted together) with Z twist yarn to produce a plied yarn with "balanced" twist.

The amount of twist in yarn, usually stated in turns per inch (TPI) or twists per inch, is determined in large measure by end use, and by the type and length of fibers in the yarn structure. The most common procedure for working with twist levels is the use of a constant called "twist multiple", or simply TM. The use of TM re-sults in a constant helix angle of the fiber in the yarn regardless of the count. The relationship is defined in this equation:[8.8]

$$\text{Twist per Inch (TPI)} = \text{TM}\sqrt{\text{Count (Ne)}} \quad (8.8)$$

For a particular count, the larger the TM, the higher the num-ber of turns per inch in the yarn. The following twist multiples are typical for carded cotton yarns:

1. Warp Yarn at 4.25,
2. Filling Yarn at 3.75, and
3. Knitting Yarn at 3.50.

As can be noticed from the above relationship, for a given TM, as the yarn count gets greater (lighter yarn), TPI increases. When the desired TPI for the yarn has been determined, the relationship between the spindle speed and the front roll speed is adjusted by the number of teeth needed in the "twist gear". Even though the traveler turns slightly slower than the spindle, this difference is not accounted for in practice. The unique characteristic of the ring-traveler system is that the traveler lags the spindle speed just enough to wind the yarn delivered by the front roll of the draft zone. As shown in Figure 8.29, the yarn is wound onto differing diameters throughout the package build which causes the traveler to change its speed continuously. The average difference in traveler speed and spindle speed is usually about one percent, and the change in traveler speed is considered of no practical significance.

Ring Spinning Production Rate

The production rate of a ring spinning machine is determined by the speed of the front steel drafting roll and the count of yarn being produced. For example, if the front roll diameter is 27 mm, its speed is 175 revolutions per minute, and the yarn count is 35's Ne, the production rate would be as follows:

$$\text{Production Rate} = \frac{27\text{mm} \times \pi \times 175}{1000}$$

$$= 14.84 \text{ meters per minute}$$

To determine the amount of yarn produced in one hour, on average, one must know the number of spindles on the machine and the spinning frame efficiency. For a 500-spindle frame operating at an efficiency of 95 percent, using the 35's English count yarn, the pounds produced per hour would be as follows:

$$\text{Ring Spinning Frame Pounds per Hr.} = \frac{14.84 \times 60 \times 500 \times .95 \times 1.0936}{35 \times 840 \text{ yd./lb.}}$$

$$= 15.73 \text{ (or 7.15 kg)}$$

A general formula[8.9] for calculating ring spinning position productivity in meters per minute (m/min) is as follows:

$$\text{Ring Spinning Delivery m/min} = \text{Front Roll Diameter mm} \times$$

$$\text{Front Roll rpm} \times \frac{1 \text{ meter}}{1000 \text{ mm}} \quad (8.9)$$

A general formula[8.10] for determining ring spinning frame productivity in pounds per hour is as follows:

$$\text{Ring Spinning Frame lb/hr} = \text{Front Roll m/min} \times \frac{1.096 \text{ yd}}{\text{m}} \times \frac{60 \text{ min}}{\text{hr}} \times$$

$$\frac{\text{lb}}{\text{Count (Ne)} \times 840 \text{ yd}} \times \frac{\text{Spindles}}{\text{Frame}} \times \text{Efficiency (decimal)} \quad (8.10)$$

Ring and Traveler System

As noted in the introduction, the ring-traveler system was developed and patented in 1830, and it has been the heart of ring spinning ever since. The attractive features of this mechanism are its tremendous flexibility and its elimination of the need for expensive and difficult mechanical arrangement in a driven "flyer" system to wind the yarn onto the spinning bobbin while simultaneously inserting twist.

Table 8.8. A Listing of Upper Speed Limit Guides for Travelers

Type Yarn	Upper Limit Meters/Second
Carded Cotton	34.5
Combed Cotton	33.0
Rayon	33.0
Polyester	30.0
Acrylic	28.4
Polyester/Cotton Blends:	
40/60	33.0
50/50	32.5
65/35	31.5

On the negative side, the ring-traveler system has its limitations. One huge problem is the management of the frictional drag of the traveler as it travels on the ring flange at a speed in the neighborhood of 30 meters per second. This frictional contact causes wear on both the ring and the traveler at the areas where the surfaces are together. This frictional drag is necessary to induce enough tension in the yarn to control the balloon diameter, avoid balloon collapse, and produce a firm and stable yarn package. Tension in the yarn can be increased by the use of a heavier traveler with an upper limit of traveler weight determined by end-breakage rate, yarn mass evenness, and yarn elongation (high tension reduces elongation). Light travelers yield lower yarn tension which causes large balloon diameters, collapsed balloons, and soft (easy to damage) yarn packages.

There have been many attempts to reduce wear and the resulting high temperatures generated by the sliding friction. It was learned early that it is best for most of the wear caused by the rubbing zone to take place on the traveler rather than on the ring. Consequently, the rings are treated to be much harder than the travelers, and the standard procedure is to change travelers (remove worn and install new) at a regular frequency depending on the ratio of wear on the travelers. Both the rings and travelers are made of steel. Most travelers are plated with nickel which helps conduct the heat away from the contact zone. It is interesting to note that temperature as high as 649° Celsius (1200 °F) has been measured at the contact zone of the traveler and the ring. Because of wear rate and temperature generation, there have been traveler speed limits established as guides for the spinner; these limits are listed in Table 8.8. The traveler speed is usually calculated by the following formula:[8.11]

$$\text{Traveler Speed, m/sec} = \frac{\text{Spindle Speed (rpm)} \times \pi \times \text{Ring Diam. (mm)}}{60 \text{ sec/min} \times 1000 \text{ mm/m}} \quad (8.11)$$

From this discussion it can be concluded that the limiting factor on ring spinning production is the limit on traveler speed. One way to allow higher spindle speeds without exceeding traveler speed limits is to use smaller ring diameters. And this trend to smaller rings is being used universally to push production up on ring frames.

Travelers are made in several different shapes, two of which are shown in Figure 8.31. The C traveler has a higher "bow" than the elliptical traveler which has a lower center of gravity, allowing it to run at a slightly higher speed than the older C traveler. It is important that the traveler contour match that of the ring flange at the running contact zone so that the wear takes place over as large an area as possible. This larger contact area allows the heat to be conducted away more rapidly. The cross-sectional dimensions and shape of the steel wire that is formed into the travelers are of utmost importance for the best performance. All traveler suppliers provide any of their traveler shapes and circles (height and width of shape) in incremental weights.

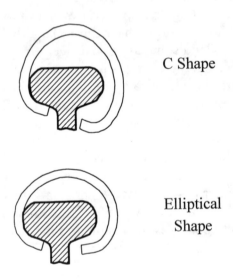

C Shape

Elliptical
Shape

Figure 8.31. Illustration of traveler shapes.

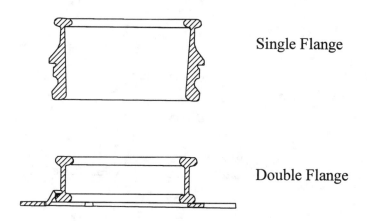

Single Flange

Double Flange

Figure 8.32. Cross-sectional view of single and double flange rings.

Rings are made as single flange and double flange as depicted by the cross-sectional sketches in Figure 8.32. Single flange rings are designed to be mounted directly into the ring rail, whereas the double flange ring must have a separate mounting for attachment to the ring rail. The advantage of the single flange ring is better accuracy for flatness and circle. The double flange ring can be turned when the first flange has become worn and needs to be replaced.

An important element associated with the ring-traveler system is the traveler "cleaner" as illustrated in Figure 8.33. As the yarn passes through the traveler on its route to the yarn package, some of the fibers, especially short fibers, get rubbed loose from the yarn strand to fly out into the surrounding space or become caught around the outside curve of the traveler. Without a blade (traveler cleaner) to knock them off the traveler, this accumulation of fibers hanging on the traveler builds up to the point that the extra weight and air-drag causes excessive tension in the yarn. The traveler cleaner setting is adjustable to accommodate the various sizes of travelers. It is normally set with a clearance with the traveler in running position of 0.5 mm.

Spindle Drives

Spindle drives can be separated into four basic types: tape drive, tangential drive, sectional tangential drive, and individual motor drive. Two methods of belt drive arrangements for the typical tape drives are illustrated in the drawings in Figure 8.34. In one case,

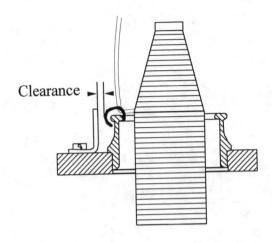

Figure 8.33. Illustration of a traveler cleaner showing the clearance setting to the traveler.

only one tensioning (jockey) pulley is needed; two jockey pulleys are utilized in the other system. Usually, the tape drives four spindles, two on each side of the spinning machine. When a tape becomes worn and breaks, only four spindles will stop, and the other spindles remain in operation; the broken tape can be replaced quickly.

The first design using a tangential belt drive incorporates only one belt which is threaded against the spindle whorls along both sides, driving all spindles. At every other spindle position, an idler pulley is utilized to maintain sufficient contact to the spindle. Even

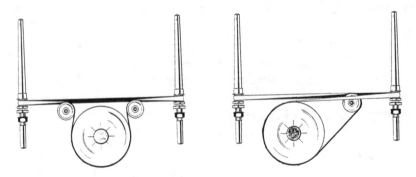

Figure 8.34. Illustration of two methods of driving spindle tapes. (Courtesy Spindelfabrik Suessen)

though the long continuous belt gives reasonably long service, it consumes more power and generates more noise than the tapes of the 4-spindle drive method. Because of these negative features and because so many ring frames now contain up to 1000 spindles on very long machines, the tangential belt system has been modified into a sectional tangential drive, as shown in Figure 8.35.

Spindles driven by individual motors (each spindle has its own motor) is the newest method for driving the spindle. Such an in-

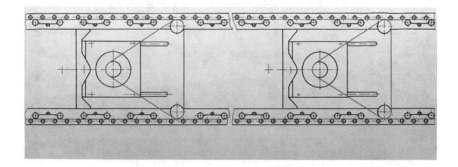

Figure 8.35. Picture showing the drive belt (tape) pulleys and the spindles of the sectional tangential spindle drive (top); and a view of a plan drawing for the drive belts (lower). (Courtesy Zinser GMBH)

stallation is shown in Figure 8.36. This drive method is said to offer savings in energy, less vibration and noise, infinitely-variable speeds up to very high speed, ease of speed control, and extremely uniform speeds spindle to spindle. This drive system is supplied by SKF GMBH. It has not as yet been strongly adopted, primarily because of its relatively high capital cost.

Automatic Doffing and Link Winding

Hand doffing (removing full packages from spindles) and donning (installing empty bobbins onto spindles) is a difficult and exhausting job for a person so assigned. In addition, it is costly from a labor and efficiency standpoint as the machine has to be stopped during the doffing operation. Over the years many systems for automating doffing have been proposed and commercialized, and the

Figure 8.36. View of spindles being driven by individual motors. (Courtesy SKF GMBH)

1. The ADP°system has stopped the frame and automatically prepared it for doffing. The tubes are in position on the transport tapes and the grasper bars are starting their outward motion to pick up the full bobbins.

2. The grasper bars have picked up the full bobbins and tilted out. All ends have been broken and are being retained by the travelers and underwind wraps of yarn placed on the spindles at wind-down. It is not necessary to re-position the node control rings in order to remove the full bobbins.

3. The grasper bars have deposited the full bobbins into the transport tapes which in turn have shifted position one-half the gauge of the frame; allowing the grasper bars to pick up the tubes. As shown, the grasper bars have lifted the tubes to the top of the spindles prior to dropping them on the spindles. After the tubes are positioned on the spindles, the frame automatically restarts.

4. While the frame is operating, the full bobbins are automatically delivered into doff boxes at the foot of the frame.

Figure 8.37. Sequential views of the built-in automatic doffer in operation. (Courtesy John D. Hollingsworth on Wheels, Inc.)

one system that has been adopted almost universally is the "built-in" doffer that was initially called "Co-We-Mat" by the inventor, Zinser GMBH. Shown in Figure 8.37 are four views of this type automatic doffer in operation on a Saco Lowell (Hollingsworth) Spinomatic ring frame. This system is very efficient regardless of the length of the frame as all bobbins are handled simultaneously. Normal doffing time is approximately 3 minutes.

This automatic doffer is ideally suited for feeding full packages to an automatic winder attached directly to the ring frame. In many modern plants the "linked" winder solution to automation of transport is being exploited. In Figure 8.38 is a view of a link winding operation.

Automated Ring Spinning

Soon after the advent of link winding, the machinery manufactures developed what is now referred to as an automated ring spinning system. One such system is shown in the diagram of Figure 8.39. As shown in the drawing, the roving frames are doffed automatically and the roving packages are installed on an overhead rail transport that allows them to be positioned at the creels of the spinning machines. These systems which are now in commercial use result in a substantial lower labor requirement, even though the initial capital investment is greater than a manual type operation.

Figure 8.38. View of a link winding operation with winder in the background. (Courtesy W. Schlafhorst AG & Co.)

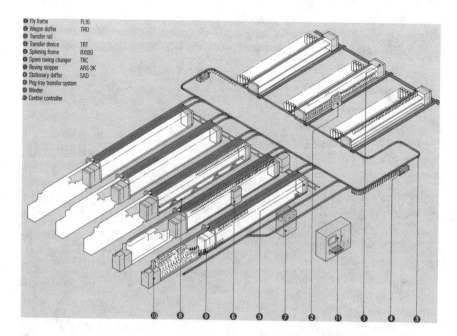

Legend:

1. Fly frame — FL16
2. Wagon doffer — TRD
3. Transfer rail
4. Transfer device — TRT
5. Spinning frame — RX100
6. Spare roving changer — TRC
7. Roving stripper — ARS-3K
8. Stationary doffer — SAD
9. Peg-tray transfer system
10. Winder
11. Central controller

Figure 8.39. Diagram showing one system of automated ring spinning. (Courtesy Toyota Automatic Loom Works, Ltd.)

Fly Lint Control at Ring Spinning and Winding

One aggravating problem associated with ring spinning is fly lint and lint accumulations on vital elements of the machinery. Fibers become separated from the roving strands, from the fiber ribbon exiting the front roll nip, and from the yarn strand in the twisting/balloon/traveler zones. And with link winders, fibers are dislodged profusely from the yarn package and yarn track during winding. Much of the flying lint that is generated has a tendency to be deposited on critical elements of the process, such as roving packages in the creels, guide rods, drafting rolls, and frame under parts. One traditional method that has been in use for many decades is the "traveling cleaner", a type of which is shown in Figure 8.40. The early traveling cleaners did not have a suction trunk which is useful in picking up lint "balls" which do accumulate on the floor. The basic principle of the unit is to prevent the lint from settling and becoming attached to the various machine and material surfaces.

The traveling cleaner normally moves along rails that are mounted over the creel structures of the spinning and winding ma-

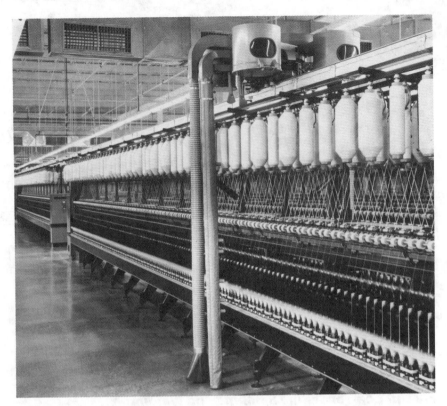

Figure 8.40. View of a traveling cleaner in use on a ring spinning machine.

chines in such a manner that one cleaner can travel over several machines. The blowing cleaner goes by each frame in its loop on a frequency of about 7 or 8 minutes, depending on the severity of the problem. Because many of the fibers in cotton are relatively short, the lint accumulation problem usually is more severe with cotton than with synthetic fibers.

The traveling cleaner has a pressure trunk and a suction trunk that hang down on each side very close to the frame. The pressure trunks have nozzles located to direct the air being blown to critical places on the frame elements. The suction trunks have a fan shaped mouth at their bottom end that is positioned very near the floor. There are compartments for collecting lint that is picked up by the suction trunks; these compartments are normally emptied automatically.

An important lint collecting system on the ring spinning machine has vacuum slots just under the front roll exit nips. For best operation these slots should be properly positioned and equipped with at least 50 mm (about 2 inches) of water suction at the frame end furthest away from the suction fan.

One of the latest methods now being employed to reduce the quantity of flying lint is the use of more air changes per hour in the spinning and winding areas. The typical number of air changes required for good conditioning control is in the range of 6 to 8 air changes per hour. For better control of lint in the room, some newer plants are designed for 25 to 30 air changes per hour.

Another consideration for lint control is the direction of air flow within the processing room. After many years of distributing conditioned air from the air washer units (conditioners) through ceiling grills and returning the air through wall openings at the conditioner units, it was found that under-floor return ducts gave more even control of flying lint, especially with 25 or so air changes per hour.

From all aspects it is absolutely critical to prevent lint accumulations from vital zones on the machinery, or when prevention is impractical, to remove the lint deposits before they impede good quality and efficiency.

Guidelines for temperature and relative humidity control are 28° Celsius (82.4 °F) temperature, 40 percent relative humidity for synthetic fibers other than rayon, 44 percent relative humidity for cotton, and 41 percent relative humidity for synthetic/cotton blends.

Summary of Critical Factors in Ring Spinning

As in all processes involved in yarn manufacturing, there are several factors of critical importance that must be optimized to give a good ring spinning operation for producing quality yarns. Listed in Table 8.9 are critical factors that can be used as a checklist for the process engineer responsible for the ring spinning machinery.

Checklist for Ring Frame Overhaul

It is usual practice to thoroughly overhaul ring spinning machines at a frequency of 9 to 12 months. Table 8.10 is a useful checklist to be used as a guide for the overhaul technicians.

Table 8.9. Checklist of Critical Factors for Best Ring Spinning Performance

1. No vibration in creels.
2. Roving guide rod location correct.
3. Roving hangers turn smoothly.
4. All roving paths, package to guide trumpet, the same.
5. No lint accumulation on roving packages.
6. No lint tags in creel zone.
7. Roving trumpets smooth and unchoked.
8. Roving traverse set correctly.
9. Steel drafting rolls clean and free of lint collars.
10. Top roll cots clean and free of defects.
11. Top roll weighting correct and consistent delivery to delivery.
12. Roll spacing correct and uniform.
13. Top roll positions correct and uniform.
14. Draft distribution consistent with roving critical draft.
15. Apron surfaces clean and free of defects.
16. Apron discharge openings correct.
17. Apron discharge plane just above front steel roll.
18. Roll stand alignment correct.
19. Top front drafting roll position is critical and should be set correctly.
20. Suction flute tube position correct.
21. Pigtail guide position correct with spindle axis.
22. Thread board (pigtail) movement smooth.
23. Spindle axis vertical and concentric with ring.
24. Ring horizontal and flat.
25. Ring wear not excessive.
26. Traveler cleaners set correctly.
27. Ring rail movement smooth without hesitation.
28. Coils per inch on package correct.
29. Yarn package 3 mm less than ring diameter.
30. Spindle speed range on frame less than 3%.
31. Spindle drive tape threading consistent.
32. Spindle drive tapes free of lint.
33. Bobbins in good condition; top ferrule not rough.
34. Temperature and relative humidity correct and uniform at all frames.

Ring Spun Yarn Quality

Yarns are tested everyday in the testing laboratory at the Institute of Textile Technology. The results of these tests are filed into a database classified according to spinning system, yarn style (fiber system), and yarn count. The sources of the yarns tested are many with the majority being furnished by the yarn plants in the Institute membership. These yarns are ring, rotor, and air jet spun, but in this chapter only ring spun yarns are considered. Naturally,

Table 8.10. Checklist for the Overhaul of Ring Spinning Frames

1. Replace defective bobbin holders and brakes.
2. Check level of roller beams, spindle rails, ring rails, and roll stand bearing alignment.
3. Check ring rail balance.
4. Replace bad spindle tapes.
5. Clean spindle tapes.
6. Lubricate idler pulley bearings and main shaft bearings.
7. Check main drive belt; lubricate motor bearings and pulley.
8. Clean and relubricate all gears; replace worn gears.
9. Check backlash in builder motion; lubricate all bearings.
10. Wash all steel drafting rolls with zero residue cleaner.
11. Install new aprons and freshly buffed top rolls.
12. Polish roller beam and thread board.
13. Replace all worn pigtail guides.
14. Plum spindles; check spindle bearings.
15. Check a 5% sample of rings for wear.
16. Reset traveler cleaners.
17. Replace worn lift rod bushings.
18. Clean roving guide rods and trumpets.
19. Replace broken or cracked separators.
20. Check apron discharge position and top front roll overhang.
21. Check top roll weight setting.
22. Make certain that all parts are identical in regards to make and style.
23. Wash vacuum box screen; check all flute tube boots; check all flute tubes for rough places.
24. Correct vibration.

there is a range in each yarn quality parameter in the database; these ranges are used to determine achievable levels of yarn quality. From these quality levels certain benchmarks, which are guides for superior yarn quality, can be determined. Examples of the 150 ITT Quality Index (superior quality average) for three counts of ring spun carded cotton yarns are given in Table 8.11.

Machine Suppliers

Popular ring spinning machine suppliers are as follows:

1. John D. Hollingsworth on Wheels, Inc., USA;
2. Rieter Machine Works, Ltd., Switzerland;
3. Fratelli Marzoli & C. spa, Italy;
4. Howa Machinery, Ltd., Japan;
5. Toyoda Automatic Loom Works, Ltd., Japan;
6. Zinser GMBH, Germany;

Table 8.11. The ITT Quality Index of 150 for 100% Carded Cotton Ring Spun Yarns (from Institute Quality Database)

Yarn Characteristic	18's Ne	26's Ne	34's Ne
Count Variability, %Vb	1.0	1.1	1.2
Skein Strength, lb	154	112	90
Skein Strength, %Vb	3.6	3.8	4.0
Single-End Strength, g	548	396	315
S.E.S., %Vb	2.3	2.6	3.0
Elongation, %	7.7	7.3	7.1
Elongation, %Vb	2.5	3.0	3.5
Yarn Appearance Index	128	125	122
Uster:			
Evenness (8 mm), %CVm	14.6	16.8	19.0
Variability, %Vb	1.7	1.8	1.8
Evenness (1-m), %CV	3.9	3.8	3.8
Evenness (3-m), %CV	2.8	2.9	3.0
Evenness (10-m), %CV	1.8	2.0	2.2
Evenness (50-m), %CV	0.7	0.9	1.0
Thin Places (-50%)	6	23	57
Thick Places (+50%)	247	445	683
Neps (+200%)	57	131	242
Classimat:			
Minors	123	254	431
Majors	0.9	1.0	1.0
Long Thick Places	0.4	0.4	0.5
Long Thin Places	31	50	69

7. Whitin Roberts Co., USA;

8. Spindelfabrik Suessen, Germany; and

9. Savio spa, Italy.

The Winding Process

As pointed out earlier in this chapter, the quantity of yarn on spinning bobbins is small compared to the package size needed for efficient processing at knitting and weaving. Thus, the primary purpose of the winding process is to transfer yarn from small spinning packages to larger packages, which yield more efficient downstream operations. In addition, defects in the yarn are removed by "clearing" during winding, and the completed wound packages should be of uniform weight, length, and density.

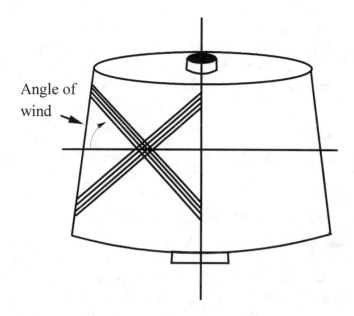

Angle of
wind

Figure 8.41. Diagram showing angle of wind on a cone of yarn produced at any winding operation.

The three leading vendors who traditionally compete for winding spindles in the USA are Murata (Muratec) of Japan, Savio of Italy, and Schlafhorst of Germany.

There are two types of winders, namely Drum Driven Winders (Random Winders) and Spindle Driven Winders (Precision Winders). The vendors listed above supply drum driven winders; spindle driven winders are supplied by SMM Scharer Schweiter Mettler Corporation. Drum winders are used in winding spun yarns for producing filling packages for weaving, supply packages for warping, and waxed yarn packages for knitting. Packages on the drum winder are driven by frictional contact of the package with the drum. Grooves in the drum cause the yarn to traverse onto the package. The angle of wind does not change as the package is filled. The angle of wind, as illustrated in Figure 8.41, is the angle formed by the intersection between the wrap of yarn on the package surface and a line drawn perpendicular to the vertical axis of the package.

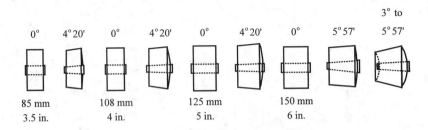

Figure 8.42. Illustrations of various wound package profiles.

Packages Produced at Winding

Winder packages are either "cheeses" or "cones". Cheeses have parallel ends and parallel sides, whereas cones have parallel ends and tapered sides. The angle of taper depends on the difference in the diameters of the two ends; it can be set over a range up to nearly 6 degrees. Some package profiles are shown in Figure 8.42. Generally, increased package taper gives less peak tensions when the yarn is withdrawn and, consequently, fewer package related breaks occur. The yarn is drawn off more freely with less contact with the package surface. Too much taper can lead to yarn sloughing.

Package weight is determined by the height and diameter of the "full" package. Thus, weight can range up to a maximum of approximately 3 kg (6.5 lb). If the spinning bobbins being wound contained 135 grams, the large wound package would contain yarn from approximately 22 full spinning bobbins. This means that the full wound package contains 21 joinings (knots or splices) if no breaks occurred.

Components of the Winder

For the purpose of describing a modern high speed winder, it is helpful to divide the winder into five zones as listed below:

1. Bobbin unwinding zone,
2. Tension zone,
3. Clearing zone,
4. Joining zone, and
5. Winding zone.

It will be helpful to refer to the diagram and legend in Figure 8.43 when studying the descriptions of the zones that follow.

Bobbin Unwinding Zone at the Winder

The bobbin unwinding zone is designed to supply a continuous feed of yarn to the winder and to regulate the tension in the yarn as it unwinds from the spinning bobbin at high speed, up to 1500 meters per minute. Each winder builder incorporates some type of balloon controller to give a reasonably steady unwinding tension. In Figure 8.43 this is called the balloon controller; another name that is often used is "accelerator". Another feature of the unwind-

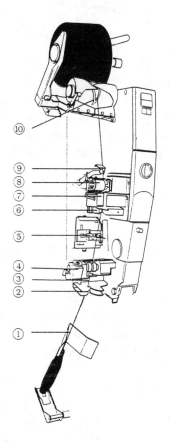

The elements in the yarn path are:

1. Balloon controller
2. Yarn shears
3. Preclearer
4. Yarn tensioner
5. Splicer
6. Electronic yarn clearer
7. Cut-clamp device
8. Waxing unit
9. Yarn trap
10. Yarn guide drum

Figure 8.43. Diagram of the winding unit for the Autoconer 238 winder. (Courtesy W. Schlafhorst AG & Co.)

ing zone is its ability to change bobbins quickly when all the yarn has been withdrawn from the bobbin. Critical settings in the unwinding zone are the alignment of the peg on which the supply bobbin is positioned, and the position of the balloon controller with respect to the supply bobbin. When the unwinding zone is not aligned correctly, certain product quality parameters can be adversely affected such as missed transfers, tension variations, increased hairiness, increased yarn breaks, and increases in the number of joinings.

Tension Zone at the Winder

In the tension zone, tensioning device "4" in Figure 8.43 is adjustable so that the tension can be set to provide uniform packages, break weak places in the yarn, and straighten out kinks in the yarn. In so doing, the number of thin weak places are reduced, and package densities from spindle to spindle are made more uniform. The preclearer, labeled "3" in Figure 8.43, prevents snarls or sloughs from transferring into the wound package. The traditional method of applying tension to the yarn is a pair of rotating spring loaded polished discs between which the yarn travels. Another method is the tension "gate". The following are critical factors applicable to the tension zone:

1. Preclearer set to yarn size being wound,
2. All surfaces free of lint buildup,
3. Tension discs/shoes in good condition,
4. Rotating disc clean and lint-free, and
5. All settings consistent.

For setting to the proper tension, the hand held tension tester should be positioned between the tensioning device and cone at the start of a full spinning bobbin. This tension should be set very low at approximately 11 percent of the yarn single-end strength.

Clearing Zone at the Winder

The purpose of the clearing zone is to eliminate yarn defects such as slubs, thick places, some thin places, and double yarn. The clearing devices used must be reliable and durable. There are two types of clearers; optical and capacitive. The optical type sees the yarn diameter profile electronically. They are supplied by Loepfe and by Zellweger Uster. The capacitive type supplied by Zellweger Uster measures yarn mass electronically as it passes between condenser plates. Optical clearers can be influenced by color and bulkiness, while the capacitive type can be influenced by moisture content of the fibers. The clearers must be set according to the yarn count being processed; they must be kept clean; the guide eyelets must be in good condition; and the cutter blades must be sharp and true, as they cut the yarn when the sensor gives a defect signal.

Joining Zone at the Winder

The joining zone ("5" in Figure 8.43) incorporates the mechanism that fastens together the leading and following ends of yarn that have become separated by a bobbin run-out, a defect cut, or a break. Joining methods used are knotting, mechanical splicing, and air splicing which is now the most commonly used. Splicing which was introduced to the industry in the late 1970s eliminates knots as a joining method. Knot types in use are weaver's knots and fisherman's knots, both of which tend to create problems in knitting and weaving, as well as being objectionable in finished fabric. Splice joints are nearly invisible and eliminate most of the difficulties caused by knots. The advent of splicing has allowed ring spinning bobbins to become smaller for increased production, even though there are considerably more joinings in an equivalent wound package weight. In order to guard against weak or bulky splices, careful and frequent attention must be directed to the splicing mechanism.

The state of the art splicing system involves the pneumatic preparation of the ends to be spliced by opening them up into a "paintbrush" configuration, followed by the interlacing of those brushed ends together to form a yarn joint that is barely visible and almost as strong as the parent yarn.

Winding Zone on A Drum Winder

The winding zone on drum driven winders consists of the drive drum, the cradle which holds the package and regulates its pressure against the drive drum, and the drive motor and controls. This is the area marked "10" in Figure 8.43. The purpose of the winding zone is to build a package of predetermined yarn length and proper density. The cradle is lifted the instant the yarn breaks or the supply bobbin is emptied. The package lifts off the drum, and the drum break is activated; the package brake is also activated. Each of these operations is independent of the other. The cradle lifter (package lifter) reduces joining failures and improves yarn quality. The drum is stepped at both ends for providing a small clearance between the drum and the take-up package to prevent damage to the yarn caused by slipping at the start of winding. This also ensures that synthetic fibers such as polyester do not melt. Each spindle is controlled by its own motor with its speed control to produce speeds up to a maximum of 1500 meters per minute, and to allow slow starts and ribbon breaking control. The drive drum is typically steel, light, and features excellent wear and corrosion resistance. It is critical for the drum to have a smooth, uniform surface with smooth reversal and crossover points. It must be balanced and properly mounted.

All leading manufacturers of winding machines are capable of offering an automatic doffing device which serves the function of removing full packages and restarting the wind spindle by placing an empty tube into the cradle and piecing up. The doffed packages can be automatically transferred to automated quality checking stations.

On a drum winder when the number of winds across the package is exactly a whole number or a whole number plus a half, each successive wrap of yarn will be laid exactly on top of the preceding wrap. This condition will exist for several successive traverses. This action is referred to as patterning, or ribboning, which increases the possibility of sloughing and breaks at the next process. It also creates difficulties in package dyeing. Each of the three major manufacturers has its particular method to eliminate patterning. Murata utilizes very quick changes in speed of the motor driving the winding drum. The method employed by Schlafhorst is frequent removal of the friction roller from the drum. On the Savio winders the cradle is rocked slightly to drive the package alternately from both ends of the package.

Murata can supply a pneumatic hairiness suppressor developed by the Institute of Textile Technology. This device uses an air vortex to reduce hairiness of spun yarns as they travel through the winder. Tests in the past have shown that winding operations can increase yarn hairiness as much as 200 percent compared to the hairiness of yarn on the spinning bobbins. The use of the ITT hairiness suppressor enables a winder to maintain hairiness levels equivalent to those on the spinning bobbin, and sometimes even reduce those levels. Other means are being researched for accomplishing this same benefit at the winder.

Winding Benchmarks and Critical Factors

Packages must be uniformly formed from each delivery on the winder. The following are suitable winding benchmarks for the package and the process:

1. Wound package of uniform density with Shore Hardness of 50 ± 2 A units;
2. Zero overthrows on package ends;
3. Zero spider webs on package;
4. Uniform color on package ends;
5. No dents or cuts visible on package;
6. Clean package;
7. No wild yarn or lint caught in package;
8. Correct transfer tails created;
9. Wind tension set preferably at 11 percent of single-end breaking strength;
10. Tension not to exceed 15 percent of yarn strength;
11. Splices with no protruding tails;
12. Splices at 15-20 mm (0.6-0.8 inch) in length;
13. Splice diameter at 120 percent of parent yarn diameter;
14. Splice strength at least 80 percent of parent yarn strength;
15. Packages lift from wind drum immediately when winder stops; and
16. Winders well maintained.

Link Winders

Before the advent of automation, the most expensive labor operation at winders was the manual feeding of spinning packages into the compartments of a carousel magazine at each winding unit, as shown in Figure 8.44. There have been several approaches to automating this feeding operation. One was the hopper feeding system which entailed the dumping of large boxes (wagons) filled with spinning packages into the special hopper which feeds the

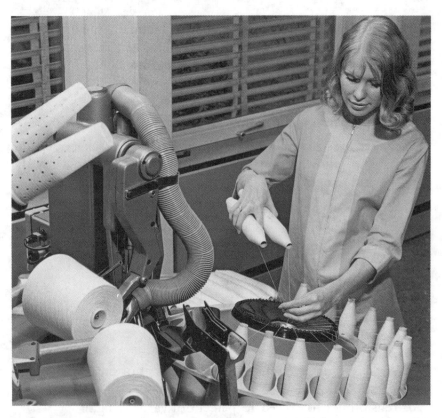

Figure 8.44. View of manual feeding of ring spinning bobbins into the winder feed unit.

bobbins, properly oriented into the winding heads that are fixed into parallel positions along the winding machine. With regard to the winding of spinning bobbins, the latest and most labor efficient system is the "linking" of the winding machine to the spinning machine. A picture of this linking is shown in Figure 8.38. In this system some type of mechanism physically transfers the automatically doffed spinning bobbins from the spinning frame and transports them, untouched by human hands, to the winding spindles. The spinning bobbins are never mixed in a random fashion, but are always maintained in an organized configuration that requires no special preparation prior to entering the unwinding zones of the winder positions. Each vendor of winding equipment is capable of supplying a winding system that links directly to a ring spinning machine.

Chapter 9

Rotor Spinning

Since the introduction of rotor (open-end) spinning in 1967, the number of rotors processing cotton and manufactured fibers in the United States has increased steadily, gaining an increasing share of the short staple spinning market. Because of productivity advantages over ring spinning, it is predicted that rotor spinning will represent between 45 percent and 50 percent of the USA spinning market by the year 2000. In the further discussions of this technology, rotor and open-end are used interchangeably.

In the early years of rotor spinning, the system was geared toward low quality and coarse count yarns. Large rotors were run at rotor speeds as low as 40,000 rpm. However, through developments of new bearings, smaller rotors, and better fiber control, rotor speeds of up to 150,000 rpm are now possible. With these changes to higher speeds (and to smaller rotors), high quality fibers and careful attention to sliver preparation are now required for efficient processing. What once was a low quality technology is now a technology capable of producing high performance yarns at productivity rates up to ten times higher than those for ring spinning.

Spinning Mechanism

Rotor spinning involves the separation of fibers by vigorous drafting, and then recollection and twisting of the fibers in a rotor as illustrated in Figure 9.1. In the actual rotor spinning machine, draw frame sliver is presented to a spring-loaded feed plate and feed roller. Fibers within the sliver are then individualized by a combing roller covered with saw-tooth wire clothing. Once opened, the fibers pass through a transport tube in which they are separated further

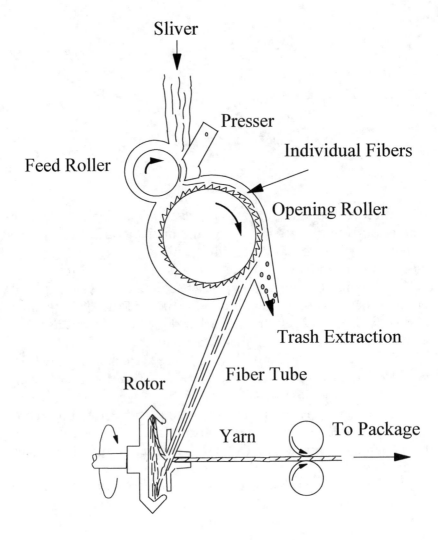

Figure 9.1. Schematic representation of the rotor spinning process.

and parallelized before being deposited on the inside wall of the rotor. A typical rotor spinning arrangement is shown schematically in Figure 9.2.

Centrifugal forces, generated by the rotor turning at high speeds, cause the fibers to collect along the wall of the rotor, forming a ring. The fiber ring is then swept from the rotor by a newly formed end of yarn which contains untwisted fibers. With each ro-

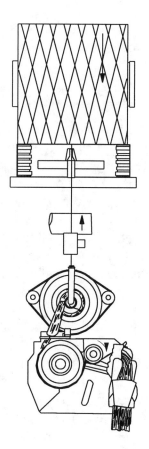

Figure 9.2. Illustration of a typical rotor spinning arrangement. (Courtesy W. Schlafhorst AG & Co.)

tation of the rotor, twist is inserted, converting the fiber bundle into yarn as it is pulled out of the rotor through a navel. The yarn is then taken up onto a cross-wound package, eliminating the need for a separate winding process as in ring spinning. As the yarn is drawn from the rotor, some fibers lying in the peeling point may wrap around the yarn, resulting in the formation of undesirable, random wrapper fibers which are characteristic of the open-end yarn structure.

For a more detailed description of the processing sequence and the handling of the fibers during the yarn formation process, rotor spinning can be divided into four major areas: fiber separation, fiber transport, fiber reassembly, and twist insertion.

Fiber Separation

Fiber separation is critical in rotor spinning for effective orientation of the fibers before yarn formation within the rotor. The sliver must be separated into individual fibers for effective delivery to the rotor. If the fibers are not separated effectively, a quality yarn with the best possible fiber orientation cannot be formed.

The most common method for fiber separation incorporates the use of combing rolls covered with saw-tooth wire. Sliver is fed into the rotational action of the combing roll by action of a feed roll/feed plate mechanism. As the sliver is fed into the wires of the combing roll, individual fibers are caught by the teeth on the roll and pulled from the sliver. At this point the centrifugal forces and aerodynamics of the system transport the fibers from the teeth on the surface of the combing roll to an air stream where the fibers are separated further and eventually deposited into the rotor in small layers over many revolutions.

Another critical function of the combing roll is the removal of trash from the sliver. Well-cleaned sliver should be presented to the system. Some dust and dirt particles, however, will still be present in the cleanest sliver, especially if cotton is being processed. The trash extraction unit of the combing roll is designed to allow lighter fibers to be carried by air to the transport duct while the heavy trash particles, because of their mass, will deflect through an opening below the combing roll and out of the system. If the fibers are not clean on delivery to the open-end system, excessive fine particles and dust will deposit in the rotor, preventing uniform fiber alignment. As a result of particle buildup, yarn of poor quality (with poor fiber orientation, lower strength, and increased imperfections) is produced.

It should also be realized that certain fibers, synthetic fiber finishes, and the amount of trash in the sliver can also contribute to the effectiveness of combing roll action. Certain fiber finishes have been found to accelerate wear of the combing roll teeth, which ultimately affects the treatment of the fibers during processing. In addition, excessive trash (or dust) in the sliver can abrasively wear down the teeth prematurely, leading to ineffective fiber separation.

Fiber Transport

Once removed from the combing roll, the fibers must be transported to the rotor without becoming excessively disoriented. The fiber transport tube is responsible for moving individualized fibers from the combing roll teeth and transporting them via air currents to the rotor. The transport tube is generally tapered to accelerate the air and fibers during movement through the tube. This fiber acceleration helps to straighten out some fiber hooks existing from the fibers leaving the combing roll.

Fiber Reassembly

Upon exiting the transport tube, the fibers are accumulated in the rotor which is the heart of the open-end spinning process. Within the rotor, fibers are collected into an untwisted strand against the rotor wall via centrifugal forces, and then the strand is drawn off as yarn.

As the fibers are delivered to the rotor wall, the centrifugal forces cause them to slide down the wall into a groove. It takes many layers of fiber to make up a strand of sufficient density for yarn; therefore, the yarn is built over a period of many revolutions. As a result, numerous doublings occur within the groove (approximately 100) wherein further blending takes place and short-term unevenness that occurs at drawing is reduced. Consequently, the rotor yarns are extremely even with few thick and thin defects.

For short staple spinning, rotor diameters range from 31 to 56 mm and may be constructed with a variety of shallow "groove shapes". The rotor design has a significant effect on the yarn structure and physical properties, resulting from the fiber orientation and the twist imparted on the yarn while it lies within the rotor groove. The rotor typically has a conical shape, and the

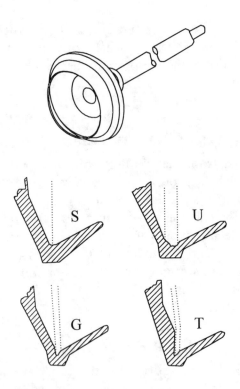

Figure 9.3. Illustrations of different rotor profiles available for rotor spinning. (Courtesy W. Schlafhorst AG & Co.)

inner surface along the wall is known as the collecting groove, the diameter of which is the specified rotor diameter. The rotor diameter depends on the machine speed, as well as on fiber properties, such as fiber length. As a rule of thumb, the rotor diameter should be no less than 1.2 times the staple length of the fiber; ends down at spinning otherwise increase.

The shape of the rotor groove should be considered because of the effects on twisting forces that occur in the groove to form the yarn. A variety of different rotor groove shapes exist to allow for different final yarn properties. Each one of the rotor profiles shown in Figure 9.3 is designed to yield different yarn strength, bulk, torque, and uniformity characteristics.

For instance, the T-rotor, because of its narrow groove diameter, produces yarns with a tight configuration more nearly like ring spun yarns than does the G-rotor. However, the bulk of the yarns produced from a G-rotor provides for better knit fabric hand and cover. As a result, specific rotors must be chosen to generate the appropriate yarn appearance and physical properties desired in the end product. S-rotors and U-rotors are generally used for sock, blanket, and towel yarns. G-rotors are normally used for apparel knitting yarns, and T-rotors are most often used for weaving yarns.

Twist Insertion

Twist occurs in the open-end spinning process as a result of the action of the rotor, navel, and take-up rolls. Once a sufficient number of fibers has collected in the rotor, twisting action from the rotation of the rotor propagates from the navel back to the peeling point at the rotor (the point at which the fibers leave the rotor).

At the peeling point, the fiber strand is slightly twisted and peeled off the collecting surface at which time full twist is imparted. The strand is then carried perpendicularly out through a navel along the axis of the rotor.

Figure 9.4 schematically diagrams the yarn formation process within the rotor during open-end spinning. The rotor rotates in direction "a" at a fixed rate. At point "B" the newly formed yarn moves through the yarn withdrawal tube (or navel) where it is removed from the rotor and wound onto a package. The actual yarn formation occurs in area "c", wherein the individual fibers begin to collect twist. Once slightly twisted, the fibers reach point "P", the peeling point, and the bundle is directed out of the rotor groove where it is fully twisted.

Each revolution of the rotor theoretically introduces about one turn of twist into the yarn; however, slippage occurring during actual twist insertion is believed to cause lower actual twist than the number of rotor rotations. Because the fibers are not held firmly by the nip of a pair of rollers, as in ring spinning, the fibers can migrate independently during twisting. In fact when twist is measured in rotor yarns, the measured twist is usually 15 percent to 40 percent lower than the machine twist. Machine twist is determined by the following formula:[9.1]

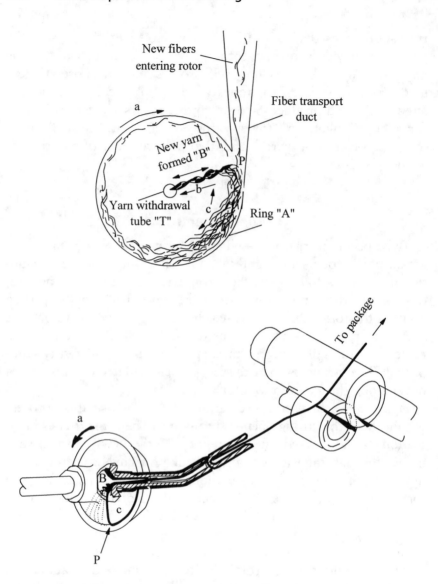

Figure 9.4. Schematic representation of the yarn formation process within a rotor.

$$\text{Twist (turns / m)} = \frac{\text{Rotor Speed (rpm)}}{\text{Delivery Speed (m / min)}} \quad (9.1)$$

Not all twist imparted to the yarn is directly caused by rotor rotation. As the yarn travels through the navel and doffing tube, a significant amount of contact occurs. This rolling action on the navel surface produces a false twist that is trapped in a section of the yarn inside the rotor. In addition, a proportion of the real twist arising from the rotation of the rotor projects backward into the rotor. Therefore, the total twist is the sum (or difference) of the two kinds of twist. Overall, the false twist provides for more stability of the yarn between the navel and the rotor groove than does the genuine preset twist.

The final yarn at the package contains only real twist, yet the false twist has a definite effect on final yarn characteristics. With increases in rotor speeds, false twist is increased correspondingly due to higher yarn tension and more centrifugal forces in the rotor. This increase in false twist tends to increase the amount of wrapper fibers in the yarn.

Wrapper Fiber Formation

The inner core structure of rotor yarn resembles that of ring spun yarn structure; however, rotor yarn has a unique structural buildup of outside yarn layers that affects the aesthetic as well as the physical characteristics of the yarn. Once each revolution some fibers entering the rotor from the transport tube interfere with the yarn peeling from the collecting surface. Portions of the fibers entering the rotor are captured inadvertently into the yarn. Instead of being twisted into the inner yarn structure, these fibers wrap around the outside of the yarn. The formation of these fibers, called "wrapper fibers" or "bridging fibers", is illustrated in Figure 9.5.

The fewer wrapper fibers that are present, the more that rotor yarns resemble ring spun yarns. However, methods to reduce wrapper fiber formation in rotor spinning cause reductions in productivity, as the minimum twist required for spinning increases. In general, wrapper fibers should be minimized to achieve an aesthetically appealing yarn while maintaining productivity.

Also, the presence of wrapper fibers in rotor yarn has been shown to contribute to increased needle wear in knitting. It is theorized that wrapper fibers move across the knitting needles like "speed bumps on a highway," sending waves of vibration through the needle and contributing to accelerated wear.

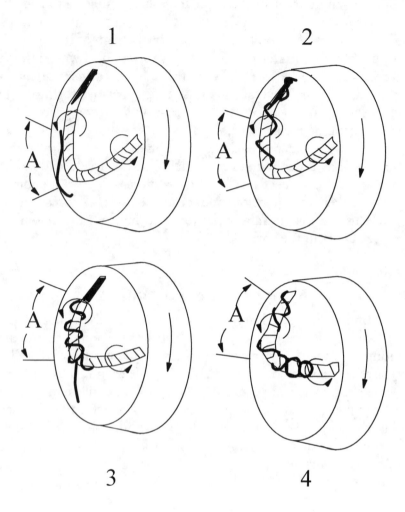

Figure 9.5. Sequence of illustrations showing one mechanism of bridging fiber formation on the surface of a rotor yarn with: (A) the fiber peeling point which moves slightly clockwise during the above sequence, (1) a fiber entering the rotor, (2) this fiber beginning to wrap around the body of the yarn rather than being twisted into the tail of the yarn, (3) the fiber continuing to wrap, and (4) the final view of such a bridging fiber. (Deussen, 1993)

The formation of wrapper fibers is largely affected by several machine-related and fiber-related factors including the following:

1. Rotor speed,
2. Rotor diameter,

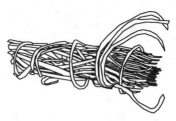

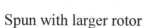

Spun with larger rotor   Spun with smaller rotor

Figure 9.6. Illustration of the comparative structures of rotor yarns spun with large and small diameter rotors.

3. Fiber length,
4. Friction between the fiber and rotor groove, and
5. Aggressiveness of the navel.

With increasing rotor speed, the levels of both false twist and yarn rotation become higher; hence, wrapper fibers are wrapped around the core more often. At higher speeds the rotor diameter or the navel should be changed to reduce false twist; otherwise, the yarn qualities will deteriorate.

With smaller rotors the presence of wrapper fibers is less pronounced than with larger rotors. Even though more wrapper fibers exist owing to the fact that more fibers are delivered to the peeling point of the rotor, the wrapper fibers are wound fewer times around the yarn core than with large rotors. Therefore, yarns produced on smaller rotors tend to be more hairy, but less bulky than similar yarns produced with larger rotors as illustrated in Figure 9.6.

Overall, the factors relating to wrapper fiber formation must be adjusted so that the minimum number of wrapper fibers are produced for a given speed. Wrapper fibers cannot be entirely removed, or productivity would be restricted; however, excessive wrapper fibers will result in low quality, aesthetically displeasing yarn.

Typical End Uses of Rotor Yarn

Rotor yarn is used in numerous products because of its versatility. Cotton and synthetic fibers can be spun on the rotor system in a typical count range from 3's to 40's Ne. About two thirds of the rotor yarns in the USA are used for knitting and the other one third for weaving. Some of the main products produced from rotor yarn are as follows:

1. Blankets,
2. Towels,
3. Sheeting,
4. Underwear,
5. T-Shirts,
6. Socks,
7. Denim,
8. Upholstery,
9. Knit apparel (shirts and dresses),
10. Woven apparel (shirting and pants),
11. Fleece fabrics,
12. Napery,
13. Gloves, and
14. Sweaters.

Advantages of Rotor Spun Yarn

The primary attraction of rotor yarn is its production cost advantage over ring spun yarn. Because of its high degree of automation and higher productivity, a pound of rotor yarn can be produced with approximately one third the labor needed to produce ring spun yarn. Part of the labor reduction is caused by the elimination of roving and winding, and the rest is attributed to the high automation of the system. Some other primary advantages include the following:

1. Lower defect levels compared to the other spinning systems, particularly fewer yarn long thick and thin places;
2. Superior knit fabric appearance;
3. Lower fiber shedding at knitting or weaving than ring spun yarn;
4. Less torque than ring spun yarn;
5. Less energy per unit produced required than for ring spinning;
6. Less floor space required compared to ring and air jet spinning;

7. Sophisticated real time quality and production monitoring on each yarn position; and
8. Superior dyeability compared to ring spun yarn.

Disadvantages of Rotor Yarn

As with any spinning system, some disadvantages exist with rotor yarns. From the initial development of rotor yarn, concerns have existed regarding the harshness of the yarn compared to ring spun yarn. Some developments have been made to offset the difference through special spinning setups or fabric finishing; however, fabrics produced from rotor and ring spun yarn are still readily distinguishable. These are other disadvantages of rotor yarn:

1. Low strength (approximately only 70 percent of ring spun yarn);
2. High pilling propensity compared to air jet yarn;
3. Accelerated needle wear at knitting compared to ring spun yarn; and
4. High maintenance costs compared to ring and air jet spinning.

Rotor Yarn Structure

As mentioned previously, during formation of rotor yarns some fibers entering the rotor come in contact with the yarn as it is exiting the rotor. These "bridging" fibers wrap around the outside of the yarn surface, and wrapper fibers are characteristic of rotor yarn. In addition, since the yarn is formed in the groove of the rotor, some fiber disorientation exists, and the yarn tends to be slightly bulkier than ring spun yarns. Figure 9.7 contains an illustration of the comparative structure of ring and rotor yarns.

Predominant Rotor Spinning Machines in the USA Industry

At the time of this writing, the large majority of rotor spinning machines in the USA textile industry (approximately 80 percent)

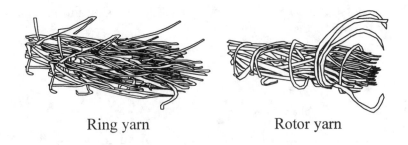

Ring yarn   Rotor yarn

Figure 9.7. Illustration of the comparative structures of ring and rotor yarns.

were produced by Schlafhorst. Several Schlafhorst models exist, including the SE-7, SE-8, SE-9 (ACO-240), SE-9 (ACO-288), and the newly introduced SE-10. The main differences between the models are rotor speed and electronic monitoring capability. Some adaptations to fiber transport channels have allowed for greater yarn count flexibility, and adaptations to bearings and belts have reduced energy consumption. A view of the SE-9 (ACO-288) is provided in Figure 9.8.

The SE-9 is capable of rotor speeds up to 130,000 revolutions per minute or a delivery rate of 200 meters per minute, generally on a 30-mm or 31-mm rotor. The new SE-10 utilizes as small as a

Figure 9.8. View of the SE-9 (ACO 288) rotor spinning machine. (Courtesy W. Schlafhorst AG & Co.)

28-mm rotor to achieve rotor speeds up to 150,000 revolutions per minute.

Another popular rotor spinning machine currently in the USA is the Rieter R1 (newest model R20). Some of the unique features provided on the R1 that have allowed for its acceptance in the industry include the following:

1. Elimination of the requirement for a starter package with a small amount of yarn to begin the spinning process so that each package has one less piecing;
2. An air bearing behind the rotor to eliminate rotor stem wear; and
3. A larger combing roll which helps to allow for heavier sliver weights to be fed into the machine without a quality loss.

A picture of the Rieter spinning machine is shown in Figure 9.9.

Figure 9.9. View of the R1 rotor spinning machine. (Courtesy Rieter Machine Works, Ltd.)

Table 9.1. Important Fiber Properties for Rotor Spinning

In Cotton:	In Synthetics:
Strength	Denier
Micronaire Fineness	Cohesion
Length	Strength
Short Fiber Content	Length

Raw Material Requirements

Over the years several authors have written about the importance of fiber properties and their effect on rotor spun yarn quality and spinning performance, especially at high speeds. Though few have agreed on which is the most important fiber characteristic, virtually all agree that the fiber properties as shown in Table 9.1 are critical.

Cotton Properties

Fiber Strength

A study was recently conducted at ITT for which three cottons were acquired with similar properties except for strength. When these cottons were spun into a 50/50 polyester/cotton blend, an increase in cotton strength of 1.5 g/tex was found to contribute to significant improvements as follows:

1. Yarn strength (+4 percent),
2. Elongation (+6 percent),
3. IPI thins (-40 percent), and
4. Uster Evenness (-2.5 percentage points).

In a study by graduate student Tutterow in 1992 on high speed spinning of fine count yarns, a 1 g/tex increase in fiber strength was found to contribute to the following:

1. A 69-point increase in yarn strength as measured by break factor,
2. A 0.4-point increase in yarn strength as measured in grams/tex, and
3. Significant improvements in Uster %CVm and IPI thin and thick places.

Micronaire Fineness

In general, the rotor spinning system performs better than all other spinning systems with low micronaire (finer) cotton. As

Yarn Single-End Strength, grams

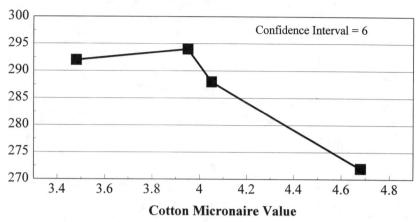

Cotton Micronaire Value

Figure 9.10. Graphical representation of the influence of cotton micronaire on the strength of 28's (Ne) 50/50 polyester/cotton rotor yarns.

shown in Figure 9.10, lower micronaire cotton provides for better yarn strength. In addition, IPI thins and thicks and Uster %CVm are typically improved with lower micronaire because of the higher number of fibers in a cross section of yarn. The relationship between micronaire and yarn evenness is shown in Figure 9.11.

The number of fibers per cross section of yarn can be estimated by these equations: (9.2 and 9.3)

$$\text{Fibers per cross section} = \frac{15{,}000}{(\text{Micronaire})(\text{Yarn Count, Ne})} \quad (9.2)$$

or

$$\text{Fibers per cross section} = \frac{5315}{(\text{Denier})(\text{Yarn Count, Ne})} \quad (9.3)$$

Thus, if 4.5 micronaire cotton was being used to make a 30's Ne yarn, the estimated fibers per cross section would be as follows:

$$\text{Fibers per cross section} = \frac{15000}{(4.5)(30)} = 111$$

For rotor spinning, a minimum of 110 fibers per cross section is desired for stable spinnability and good yarn quality. In this ex-

Yarn Uster Evenness, %CVm

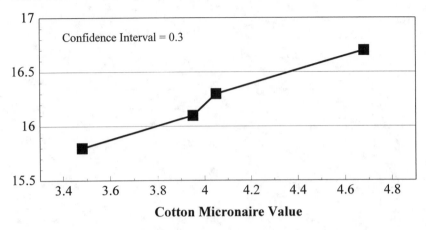

Cotton Micronaire Value

Figure 9.11. Graphical representation of the influence of cotton micronaire on the evenness of 28's (Ne) 50/50 polyester/cotton rotor yarns.

ample, the micronaire level is a bit high for fine count spinning; however, if the micronaire was 4.0 instead, fibers per cross section would be 125, which would be acceptable.

The main concern with using low micronaire cotton is poor dye uptake of the fabric, and a higher probability for undyed neps and/or pills.

Fiber Length

Though longer cotton fibers do provide for increased yarn strength, the contribution of cotton fiber length to yarn quality is less than that of fiber strength and micronaire. As shown in Figure 9.12, a 5-mm (about ³⁄₁₆ inch) increase in fiber length is expected to increase yarn strength by 2 percent on a 10's and 7 percent on a 30's Ne yarn.

In a study conducted at ITT on the effect of fiber properties on rotor yarn properties, fiber length was found to be highly correlated with Classimat minor defects as shown in Figure 9.13.

Short Fiber Content

At one time the level of short fiber content, or fibers less than 12.7 mm (½ inch) in length, was considered to have little effect on

Yarn Break Factor (CSP)

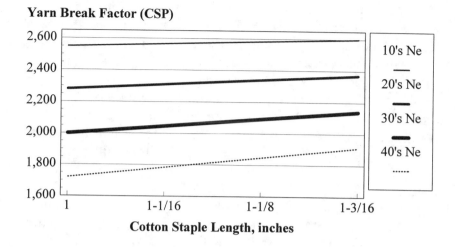

Figure 9.12. Graphical representation of the influence of fiber length on the Count Strength Product (CSP) of yarns with 4.8 twist multiple, spun from 100 percent, 4.0 micronaire, 28 grams/tex cotton. (Deussen, 1993)

Yarn Classimat Minor Defects/100M Meters

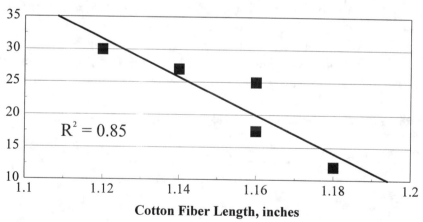

Figure 9.13. Graphical representation of the correlation between fiber length and the number of Classimat Minors in 26's (Ne) 100 percent cotton rotor test yarns.

rotor spinning. In fact many plants today have cotton laydowns with a high level of comber noils and reclaimed waste fiber being fed into rotor spinning. However, these fibers only tend to perform well when the yarn count is coarse (less than 12's Ne) and when rotor speeds are fairly conservative.

When yarn counts become finer and/or rotor speeds begin to exceed 100,000 rpm, short fiber content becomes critical. In the study done by Tutterow, it was shown that for every one percentage point that short fiber content increased, the spinning limit of the yarn (finest yarn capable of being spun) became 1.3 Ne heavier (Tutterow, 1992).

In another study at ITT, high short fiber content was found to negatively influence Uster evenness, yarn strength, and elongation in 26's Ne cotton yarn as shown in Table 9.2.

In summary, a presentation of the influence of cotton properties on yarn properties was developed by Deussen and is shown in Table 9.3.

Synthetic Fiber Properties

Fiber Denier

Currently, the denier of synthetic fibers used in the USA industry is typically between 0.9 and 2.25 denier per filament (dpf). A movement toward finer deniers (<1.2 dpf) has been made recently, particularly for yarn counts finer than 24's Ne. The fine denier has allowed for improvements in strength, evenness, and thin places by providing more fibers per cross section. A comparison of 28's

Table 9.2. The Influence of Short Fiber Content on 26's Ne Carded Cotton Rotor Yarn

Yarn Property	13% Short Fiber Mix	10% Short Fiber Mix
Yarn Count	26.0	26.0
Uster Evenness, %CVm	14.9	14.7
IPI Thins	28	26
Thicks	63	60
Single-End Strength, g	241	**251**
%Vb	3.3	3.9
Single-End Elongation, %	6.4	**6.6**
%Vb	7.2	**4.1**
Yarn Appearance Grade	B	B/B+

NOTE: Boldface type indicates significance difference @ 95 percent confidence level.

Table 9.3. The Influence of Cotton Properties on Rotor Yarn Properties (Deussen, 1993)

Yarn Property	Strength	Elongation	Length	Length Uniformity	Short Fiber Content	Fineness	Maturity	Trash Content	Dust Content	Color	Stickiness
Strength	++	0	++	++	++	++	0	+	+	0	+
Elongation	0	++	++	++	++	0	+	0	0	0	0
Non-Uniformity	+	0	++	++	++	++	++	++	+	0	+
Imperfections	+	0	++	++	++	++	++	++	+	0	+
Hairiness	0	0	++	++	++	++	0	+	+	0	0
Structure	0	+	++	++	++	++	0	+	+	0	0
Twist Level	++	0	++	++	++	++	0	0	0	0	0
Yarn Breaks (Spin Limit)	++	+	++	++	++	++	+	++	+	0	++
Dyeability	0	0	0	0	0	0	++	0	0	++	0

(++) Direct Influence, (+) Indirect Influence, (0) No Influence

Table 9.4. The Influence of Denier on 28's Ne 50/50 Polyester/Cotton Yarn Properties

Yarn Property	Denier Per Filament	
	0.9 dpf	1.2 dpf
Yarn Count	27.9	27.9
%Vb	1.7	1.1
Single-End Strength, g	**296**	281
%Vo	9.6	10.1
Single-End Elongation, %	8.4	8.3
%Vo	3.3	3.3
Uster Evenness, %CVm	**16.5**	17.0
%Vb	1.0	1.5
IPI Thins	**75**	132
Thick	**220**	264
Neps	42	64
Hairs/m (>3 mm)	**8.5**	12.7
%Vb	10.1	5.9

Note: Bold typeface indicates significant difference @ 95% probability.

Ne 50/50 polyester/cottons produced from 0.9 and 1.2 dpf polyester is shown in Table 9.4.

Fiber Length

As rotor speeds have increased, smaller rotors have been used to reduce tension on the yarn during spinning. However, the rotor can become too small for the fiber length being used. As a rule of thumb, the following relationship [9.4] is used to analyze the fiber length and rotor diameter dependency:

$$\frac{\text{Fiber Length (mm)}}{\text{Rotor Diameter (mm)}} < 1.2 \quad (9.4)$$

If the ratio of fiber length to rotor diameter exceeds 1.2, poor spinnability can be expected. Consequently, for high speed spinning of polyester/cotton blends, it is common for 32-mm (1.26-inch) fiber to be used, especially for knits. However, for woven products, for which higher yarn strength is generally required, 38-mm (1.50-inch) fibers are often used. In these cases the minimum rotor diameter is 33 mm, so spinning speeds are limited.

Fiber Tenacity

As would be expected, higher tenacity synthetic fibers provide for higher yarn strength; consequently, these fibers are used for weaving. Knitters, on the other hand, desire yarns with low pilling and improved dye uptake, causing lower tenacity fibers to be preferred.

Fiber Finish

As shown in Figure 9.14, the maximum rotor speed for 100 percent cotton is higher than it is for man-made fibers, because the natural fats and waxes on the cotton provide good lubrication, and the convoluted ribbon-like cross section provides sufficient fiber-to-fiber cohesion to hold the fiber bundle together (Deussen, 1993). Fiber producers are constantly experimenting with finishes to alter the fiber-to-metal friction in order to increase spinning speeds, reduce shedding, and improve yarn quality.

Sliver Preparation

To achieve acceptable levels of yarn evenness, count variability, strength variation, defect levels, and ends down, the sliver must be prepared appropriately for spinning. The critical sliver properties include the following:

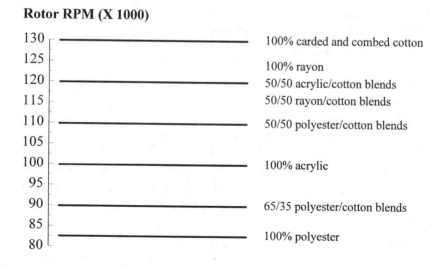

Figure 9.14. Maximum rotor speed levels in 1996 for various 30's (Ne) count yarns. (Deussen, 1993)

1. Sliver evenness (short-term and mid-term),
2. Fiber alignment,
3. Trash level in sliver,
4. Nep level in sliver, and
5. Slub level in sliver.

Sliver Evenness

To achieve satisfactory sliver evenness, two passes of drawing are typically used, preferably with the second passage incorporating the use of an autoleveler. Some plants use only one process of autoleveled drawing. For coarse count yarns, one drawing process is acceptable; however, the sliver evenness and fiber alignment are usually not good enough after one drawing process to achieve sufficient yarn strength for high speed weaving or yarn mass uniformity for superior knit fabric appearance.

Fiber Alignment

Improvements in fiber alignment, as measured by the ITT bulk tester, result in improved yarn strength, IPI thin places, and neps. A 5-point improvement in alignment index resulted in the following property improvements in 14's Ne cotton rotor yarns (Barnes, 1996):

1. Yarn strength improved 10 percent.
2. IPI thick places decreased 40 percent.
3. IPI neps decreased 20 percent.

Trash in Sliver

Trash level in the sliver has a strong influence on both the quality of the yarn produced and the stop level at spinning. Some cleaning machines and cards can combine to remove as much as 95 percent or more of the initial trash level in the cotton, leaving little residual trash. However, any trash or dust that remains in the sliver has the potential for accumulating in the rotor groove and preventing the yarn bundle from being formed appropriately, as schematically illustrated in Figure 9.15. Contaminated rotors cause a repeating pattern of thick and thin places in the yarn called moiré.

The accumulation of dust and trash in the rotor groove is accelerated with smaller grooved rotors. Schlafhorst has developed some boundary values for the level of sliver trash, dust, and fiber

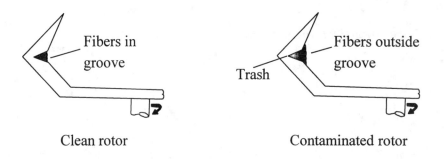

Figure 9.15. Schematic comparison of a clean rotor to a rotor contaminated with trash buildup.

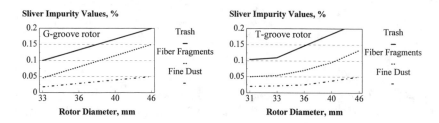

Figure 9.16. Boundary levels for trash, fiber fragments (>500 microns), and dust (>15 microns) in sliver for G-groove and T-groove rotors of various diameters. (Deussen, 1993).

fragments as measured on the Zellweger Uster MDTA3. These boundary values, which are shown in Figure 9.16 for G-groove and T-groove rotors of various diameters, are the values below which acceptable spinning performance should result.

In a study conducted by Artzt, a strong relationship was found between trash level in the card sliver and ends down at spinning. In this study a reduction of trash in sliver of 0.05 percentage points resulted in approximately 30 percent fewer ends down when spinning 20's Ne cotton rotor yarns as illustrated in Figure 9.17 (Artzt, 1990).

Neps in Sliver

The nep level in the sliver is indicative of cleaning and carding performance. One false perception in the industry is that sliver nep level is of secondary importance because rotor yarns "hide neps within the yarn structure." It is true that the nep level of rotor yarns

Ends Down/1000 Rotor Hours

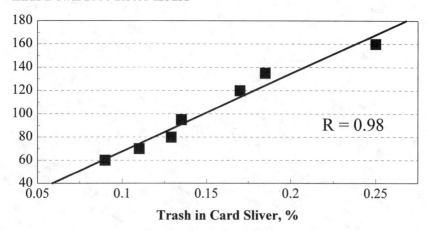

Figure 9.17. Graphical representation of the relationship between trash percentage in card sliver and ends down per 1000 rotor hours for 20's (Ne) cotton yarn spinning. (Artzt, 1990)

Yarn IPI Neps/1000 Meters (+200%)

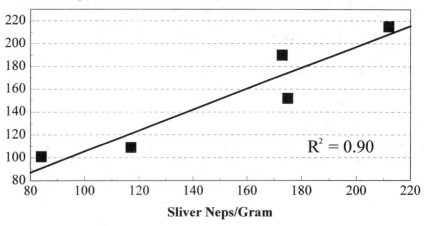

Figure 9.18. Graphical representation of the relationship between card sliver neps and IPI Neps in 18's (Ne) cotton rotor yarns.

is lower than that of ring spun yarns when the same sliver is used (because some trash and neps are removed by the combing roll), but a strong correlation still exists between nep level in the sliver and nep level in rotor yarn as illustrated in Figure 9.18. In fact the level of AFIS neps per gram in the sliver is approximately equal to the number of neps (+200 percent) per 1000 meters of yarn.

Slubs in Sliver

Thick places or slubs in the sliver are not measured often in plants unless on-line sliver monitoring is incorporated at the drawing process. These slubs, which typically are caused by fiber buildup on the steel rolls at drawing, improperly set draw frame roll clearers, damaged card wire, or by unclean cards, will cause excessive quality stops at spinning due to long thick places recognized by the monitoring system. For best results, slub level should be targeted below 0.5 per pound as measured on a Sliver Analyzer. Read about the Sliver Analyzer in Chapter 5 in the section titled "Slub Generation."

Machinery Description

Unlike ring spinning, the rotor spinning machine has the capability to form a package that is suitable to be taken directly to knitting, weaving, or warping. In addition, modern machines are automated with the ability to doff full packages and replace them with new starter packages (or empty cones). Full packages are moved by conveyor to the end of the machine after they are doffed, where they can be packed for shipment or for transfer to the next process.

When a yarn breaks at an individual position, a traveling piecer will attempt to automatically rejoin the yarn, thus resuming the spinning process without any human intervention in most cases. Theoretically, a human operator is only needed to replace sliver that has run out, keep wax on the machine (for knitting yarns), and attend to rotors that the piecer fails to restart.

To fulfill the automatic functions, these accessories are needed:

1. *Starter Winder* to wind approximately 500 yards of yarn on the new cones, providing the seed yarn needed to begin the spinning process (Schlafhorst machines);
2. *Piecer* to rejoin broken ends;

A. Drive unit
B. Intermediate machine sections
C. End unit
E. Starter winder
F. Automatic piecer
G. Automatic package doffer
K. Spinboxes
L. Winding units
M. Doffer shuttle

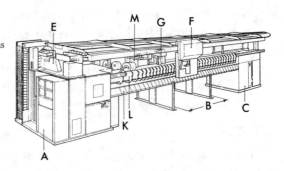

Figure 9.19. General location of the major components of an automated rotor spinning machine. (Courtesy W. Schlafhorst AG & Co.)

3. *Doffer* to remove full packages and insert starter packages (Schlafhorst machines); and

4. *Doffer Shuttle* to carry replacement starter cones to the doffer.

A schematic showing the location of the major machine accessories is provided in Figure 9.19.

It should be noted that the Rieter and Savio open-end spinning machines do not use starter packages. A seed yarn is instead carried by the piecer to initiate spinning on a new package. In addition, the doffer and piecer are combined into a single unit.

Since the actual spinning of the yarn is accomplished in the spinbox, the majority of the remainder of this chapter is devoted to the discussion of spinbox features, with some discussion on the accessory features at the end of the chapter.

Critical Spinbox Factors for Spinning Performance and Quality

Draft

One of the first decisions that must be made when beginning to produce a yarn at rotor spinning is the weight of the sliver that should be fed into the machine. The relationship between the sliver weight and the yarn weight is the *draft* required by the machine.

The machine draft can be calculated with this equation:[9.5]

$$\text{Draft} = \text{Sliver Weight (gr / yd)} \times \frac{\text{Yarn Count (Ne)}}{8.33} \quad (9.5)$$

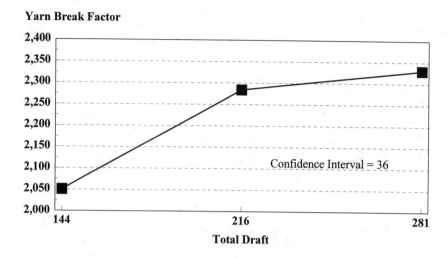

Figure 9.20. Graphical representation of the influence of Rieter R1 machine draft on the strength of 30's (Ne) cotton yarns.

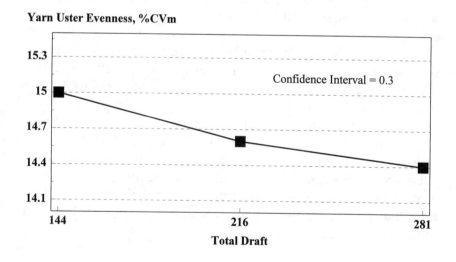

Figure 9.21. Graphical representation of the influence of Rieter R1 machine draft on the evenness of 30's (Ne) cotton yarns.

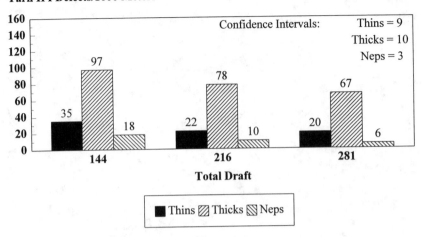

Yarn IPI Defects/1000 Meters

Figure 9.22. Graphical illustration of the influence of Rieter R1 machine draft on the number of IPI defects in 30's (Ne) cotton yarns.

The preferred draft is different for the various machines available. For the Rieter R1, draft levels above 200 generally help yarn strength, evenness, and IPI defects as shown in Figures 9.20 through 9.22.

Because the Schlafhorst machines have a smaller combing roll, the preferred draft level is lower, usually less than 200. As shown in Figure 9.23, for a 30's Ne yarn count the preferred draft is approximately 195. As a result, a sliver weight of 54 grains per yard (gr/yd) would be ideal. Unfortunately, few plants have the luxury of spinning only one yarn count, but production of only one sliver weight is preferred. Thus, a sliver weight should be selected that prevents the draft from being too high on the finest count and too low on the coarsest count. As a rule of thumb, if draft is too high, long thin places, shedding, and ends down deteriorate. If draft is too low, thick places increase.

Rotor Speed

Rotor speed has a strong correlation with yarn strength, elongation, evenness, shedding, and yarn breaks if all else is held con-

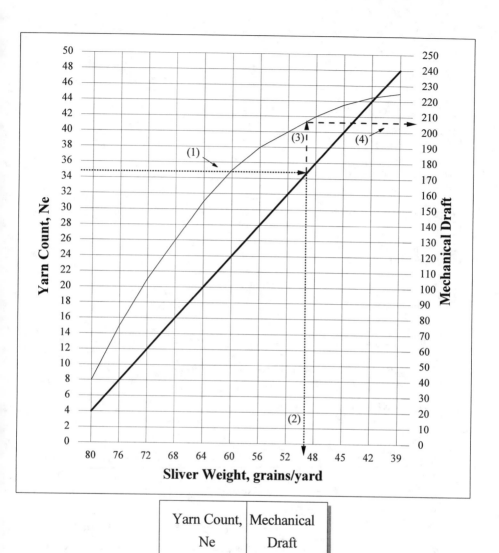

Yarn Count, Ne	Mechanical Draft
———	———

Figure 9.23. Chart for determining the preferred draft on a Schlafhorst rotor spinning machine with an example 35's yarn used as follows: (1) tracing from the Y1 axis to the Ne line and (2) directly down to the X axis, a sliver weight of 49 grains/yard (± 4) is preferred; then, (3) tracing from the Ne line to the mechanical draft line and (4) directly over to the Y2 axis, a mechanical draft of 206 is indicated. (Deussen, 1993)

Table 9.5. The Influence of Rotor Speed on the Quality of 20's Ne 50/50 Polyester/Cotton Test Yarns

Yarn Quality	% of Quality Variance Attributed to Rotor Speed Change (R² x 100)
Lint Shedding	95.4
Uster Evenness, %CVm	98.1
Uster IPI Thins (-50%)	89.2
Uster IPI Thick (+50%)	96.1
Uster IPI Neps (+200%)	77.2
Single-End Strength, g	88.9
Single-End Strength, %Vo	60.6
Elongation, %	96.1
Elongation, %Vo	75.7
Uster Hairiness Index (UHI)	79.4
Shirley Hairiness, >3 mm	73.4
Zweigle Hairiness:	
>1 mm	70.6
>2 mm	8.1
>3 mm	56.3
>4 mm	92.6
>6 mm	96.2

stant. Increases in rotor speed cause increases in spinning tension, which disrupt fiber formation in the rotor. However, if a rotor speed increase is made in conjunction with a rotor diameter decrease, it is possible to avoid a spinning tension increase and preserve yarn quality and ends down levels.

The effects of rotor speed on yarn quality are illustrated in Table 9.5 and in Figures 9.24 through 9.27. In this study numerous 20's Ne 50/50 polyester/cotton yarns were produced in 31-mm rotors at speeds from 87,000 to 110,000 rpm. The majority of yarn properties were highly correlated with rotor speed (Butenhoff, 1995). These data were generated on a Schlafhorst machine, but similar results have been documented on Rieter equipment.

Suction

A vacuum is generated at the end of the rotor spinning machine to provide suction at each spinning position. The suction helps to remove the fibers from the combing roll and to move them through the fiber transport channel shown in Figure 9.28. The removal of fibers occurs before the fibers make a full turn on the combing roll.

Yarn Single-End Strength, cN/tex

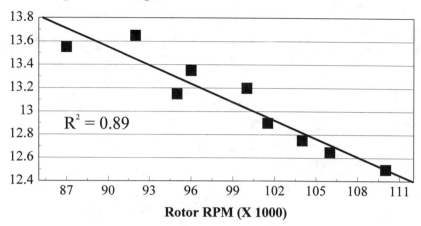

Figure 9.24. Graphical representation of the effect of rotor speed on the strength of 20's (Ne) 50/50 polyester/cotton rotor yarns. (Butenhoff, 1995)

Yarn Single-End Elongation, %

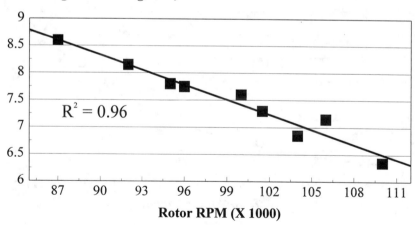

Figure 9.25. Graphical representation of the effect of rotor speed on the elongation of 20's (Ne) 50/50 polyester/cotton rotor yarns. (Butenhoff, 1995)

Yarn Uster Evenness, %CVm

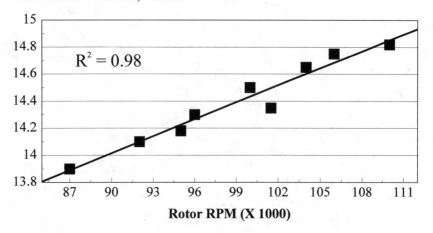

Figure 9.26. Graphical representation of the effect of rotor speed on the evenness of 20's (Ne) 50/50 polyester/cotton rotor yarns. (Butenhoff, 1995)

Lint Shedding, milligrams/1000 meters

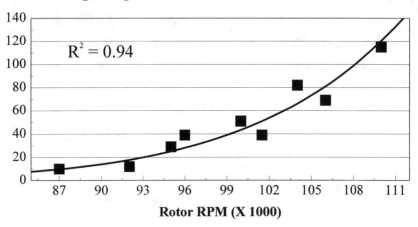

Figure 9.27. Graphical representation of the effect of rotor speed on the lint shedding propensity of 20's (Ne) 50/50 polyester/cotton rotor yarns. (Butenhoff, 1995)

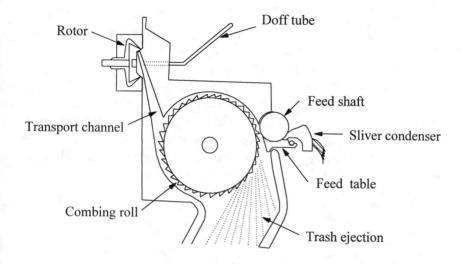

Figure 9.28. Schematic of a rotor spinning "spinbox." (Deussen, 1993)

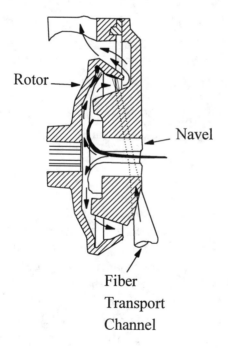

Figure 9.29. Schematic diagram of the path of air flow through the rotor chamber. (Courtesy W. Schlafhorst AG & Co.)

The air that travels with the fibers through the transport channel is accelerated by approximately 50 percent as it moves into the rotor because of the taper of the channel and the extra air generated by the rotor and vacuum. The air then exits around the edge of the rotor as shown in Figure 9.29. If any turbulence exists (because of improper setting of the rotor), if air leaks occur due to worn seals, or if the vacuum level is insufficient, yarn formation in the rotor will be adversely affected, and quality and efficiency will deteriorate. It is desired that suction be no lower than 65 mbar on SE-7 and SE-8 Schlafhorst machines. On high speed machines (SE-9), suction levels above 80 mbar are normal.

Combing Roll Zone

The combing roll zone is schematically illustrated in Figure 9.28. The sliver is delivered to the combing roll by the feed shaft which turns at a speed that is based on both the draft and the yarn delivery speed set on the machine. When the yarn breaks, an electromagnetic clutch stops the individual spinning position.

The critical factors in the combing roll zone for optimal quality and running performance include the following:

1. Feed clutch wear,
2. Feed tray-to-combing roll spacing,
3. Combing roll speed,
4. Combing roll wire selection, and
5. Combing roll wire condition.

Clutch Wear

When the feed clutch becomes worn on the rotor spinning machine, the sliver is no longer fed consistently into the machine. Consequently, long thick and thin defects occur in the yarn. On the Schlafhorst machine, the condition of the clutch can be evaluated by measuring the torque on the feed roll. A new clutch will generally provide for a torque above 15 inch-pounds. When the torque on the feed roll becomes less than 10 inch-pounds, replacement is most likely necessary. Table 9.6 contains data on yarn produced from positions that had worn versus satisfactory clutches. In this case long thin places were most adversely affected by the worn clutches.

Table 9.6. The Influence of Clutch Wear on 35's Ne 50/50 Polyester/Cotton Rotor Yarns

Yarn Property	Feed Roll Torque (inch-pounds)	
	<10	16
Minor Defects	129	121
Long Thick Places	1	0
Long Thin Places	37	2

Note: Boldface type indicates significant difference @ 95% probability.

Feed Tray-to-Combing Roll Spacing

On most Schlafhorst machines, the feed tray-to-combing roll distance is adjustable. This setting can have a significant influence on quality. Also, if the feed tray gets out of alignment with the combing roll, problems can occur.

On the Rieter R1, the feed tray-to-combing roll distance is not adjustable, but different feed trays are available to change the distance, particularly when a change is made from cotton to 38-mm (1.50-inch) synthetic fibers.

If the feed tray is set at the wrong distance or is misaligned with the combing roll, there is a strong probability that thick places will exist in the yarn. The distance between the feed tray and the combing roll is checked on the Schlafhorst machine with a plug that is put into the machine in place of the combing roll as shown in Figure 9.30. The feed tray is loosened and then set flush with the edge of the plug. An improperly set feed tray can cause yarn evenness to deteriorate by over 10 percent and short thick places to increase by over 100 percent.

Combing Roll Speed

Selection of the appropriate combing roll speed is usually a compromise between yarn strength, defect levels, and shedding. As shown in Table 9.7 from a study by Dailey at ITT, increases in combing roll speed will usually contribute to increased long thin defects in cotton yarns and to elongation and uniformity losses in 50/50 polyester/cotton yarns. Also, more short fiber is generated at higher combing roll speeds, causing yarn shedding to increase. In this study short fiber content of the fibers in the rotor were determined to be higher than the input sliver regardless of the combing roll speed, but higher combing roll speeds caused greater short fiber increases in the spinbox. Most plants set combing roll speed between 7600 and 8600 rpm.

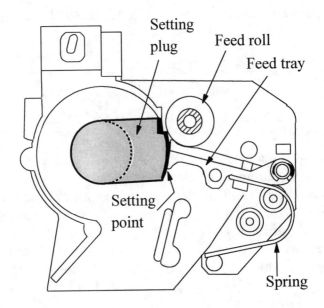

Figure 9.30. Illustration of the plug used to set the feed tray-to-combing roll distance.

Table 9.7. The Influence of Combing Roll Speed on the Quality of Various Rotor Yarns

Yarn Type and Yarn Property	Combing Roll RPM			Significant Difference
	7600 (a)	8100 (b)	8600 (c)	
100% Cotton:				
Single-End Strength, g/tex	10.14	10.01	9.94	None
Elongation	6.06	6.00	5.95	None
Uster %CVm	14.60	14.41	14.38	None
Thicks	66	55	56	None
Thins	65	61	65	None
Neps (+280%)	4.1	3.2	2.3	None
Long Thins	2.4	8.7	17.0	a<c
Lint Shedding, μg/m	12.4	13.1	13.4	a<b, a<c
50/50 Polyester/Cotton:				
Single-End Strength, g/tex	14.85	14.72	14.57	None
Elongation, %	8.43	8.19	8.08	a>b, a>c
Uster %CVm	14.34	15.08	15.05	a<b, a<c
Thicks	87	86	92	None
Thins	21	22	23	None
Neps (+280%)	13.1	13.8	14.0	None
Long Thins	2.3	3.1	5.6	None
Lint Shedding, μg/m	8.0	8.3	9.3	a<b, a<c

General guidelines that apply regarding combing roll speed are as follows:

1. Trashy fibers require higher combing roll speed.
2. Synthetic fibers require higher combing roll speed.
3. Lower micronaire cotton requires higher combing roll speeds.
4. Coarser yarn counts require higher speed.

Combing Roll Wire

Several choices exist for combing roll wire. Most wire used in the industry is "saw-tooth" shaped and is hardened with diamond and nickel. Depending on the fiber being used, the angle of the wire is selected to allow for the most effective combing and fiber removal. Several common wire profiles are shown in Figure 9.31. Synthetic fibers usually require less of an angle of the wire to allow for easier removal of these highly cohesive fibers from the teeth. The cotton wire has a more aggressive angle for trash removal and fiber alignment. In some cases S21 (synthetic) wire is used on cotton for the flexibility of using the same wire for cotton and polyester/cotton blends. In these applications significant deterioration in thin places and evenness of the cotton yarns can be expected. Recent research has shown that lower density cotton wire is a better wire for applications involving cotton and blend yarns spun from the same combing roll.

Combing Roll Wear

Over time, the movement of fibers across the teeth of the combing roll will cut a groove in the leading edge of the teeth and dull the tips. Coatings of diamond, nickel, and boron are available to prolong the life of the teeth at a significant expense. When the grooves become severe, the fibers will release in bunches, causing thick places in the yarn. Combing roll wear can be detected by an increase in quality monitor stops for short thick places, an increase in the amplitude of the spectrogram between 50 cm and 91 cm (20 inches and 1 yd), or a deterioration in yarn evenness. Combing roll wire usually lasts for 6800 to 9100 kg (15,000 to 20,000 lb) of production, depending on the trash level in the raw material and/or additives to the synthetic fibers.

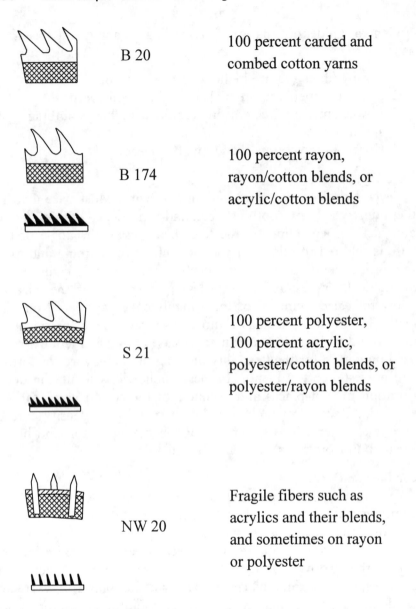

Figure 9.31. Various rotor spinning combing roll wire profiles and potential fiber use applications. (Deussen, 1993)

Fiber Transport Channel

As shown in Figures 9.1, 9.4, 9.28, and 9.29, the fiber transport channel is the transition piece from the combing roll to the rotor. The air flow through the tapered channel accelerates the fiber flow

Rotor

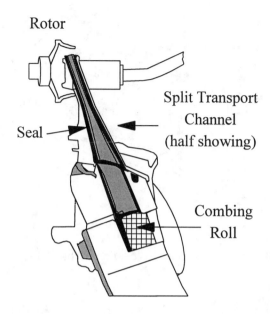

Split Transport
Channel
(half showing)

Seal

Combing
Roll

Figure 9.32. Illustration showing a cross section of the split fiber transport channel as used on the Rieter and Schlafhorst rotor spinning machines.

to help straighten hooks generated by the combing roll. In the Schlafhorst and Rieter machines, the transport channel is split, as shown in Figure 9.32, to allow the spinbox to open for easy access to the rotor (Deussen, 1993). These are key control factors in the transport channel section:

1. Seal condition,
2. Upper and lower channel alignment, and
3. Channel wear.

In between the upper and lower channel on machines with split channels is a seal to prevent air leaks. Opening and closing of the spinbox causes the seal to wear. As the seal deteriorates, air will leak out of the channel and fibers will become caught between the two channel sections, causing thick places in the yarn. The rotor also has a seal as shown in Figure 9.33. Opening and closing of the spinbox can cause the alignment of the transport channel to become skewed, causing thick places in the yarn.

Finally, after several years the abrasion of the fibers against the transport channel wall will cause grooves to form. This wear is ex-

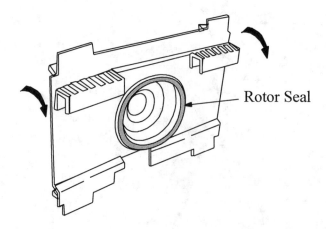

Figure 9.33. Illustration showing the location of the main rotor seal.

tremely difficult to detect, because increases in defects and ends down that are caused would typically be blamed on the combing roll.

Rotor Zone

There are different drive mechanisms for the rotor depending on the type machine being used. The direct rotor drive turns the rotor on ball bearings and is usually limited in speed to under 100,000 rpm; thus, it is not a popular choice for modern spinning, except for coarse counts and 100 percent synthetic end uses. The Schlafhorst and modern Rieter machines (R1, R14, and R20) incorporate the indirect rotor drive. This system, shown in Figure 9.34, uses a pair of twin discs to drive the rotor stem. The discs are driven by a tangential belt. Because the twin discs are ten times larger than the rotor, they are driven at one tenth the speed. Thus, their bearings are subjected to less load and are able to drive the rotor up to 150,000 rpm.

In order to keep the rotor position constant, the twin disc drive exerts a force on the rotor toward the back of the machine, causing the tip of the rotor to rotate against a lubricated steel ball. This system is schematically illustrated in Figures 9.35 and 9.38.

The rotor zone can be defined as the combination of the rotor, twin disc assembly, and rotor drive belt. Critical factors influenc-

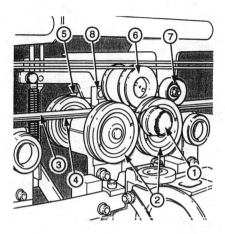

1. Rotor shaft
2. Twin discs
3. Tangential driving belt
4. Tangential driving belt
5. Air bearing, absorbing axial force
6. Tension roller
7. Tension roller
8. Setting sleeve

Figure 9.34. Schematic illustration of the twin disc drive system for the spinning rotor. (Courtesy Rieter Machine Works, Ltd.)

ing quality and/or machine performance in this zone include the following:

1. Rotor speed/diameter,
2. Rotor groove,
3. Rotor stem wear,
4. Rotor cup wear,
5. Twin disc wear, and
6. Rotor belt wear and alignment.

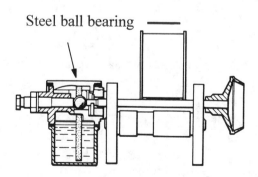

Steel ball bearing

Figure 9.35. Illustration showing the lubricated steel ball that the rotor stem rotates against. (Deussen, 1993)

Rotor Speed/Diameter

As mentioned previously, rotor speed has a significant influence on wrapper fiber formation and yarn characteristics. In general, increasing rotor speed deteriorates yarn quality when all else is held constant. However, high rotor speeds (up to 130,000 rpm) can be run with the appropriate rotor, transport channel design, and raw material selections. To run maximum rotor speeds, a rotor must be used that is small enough in diameter to allow centrifugal forces to be kept under control. With a 30-mm rotor, the centrifugal force at a 120,000 rpm rotor speed is approximately equal to the centrifugal force exerted by a 36-mm rotor at 80,000 rpm. Smaller rotors are being selected to maximize productivity as a result.

A restriction on the size of the rotor that can be used exists based on yarn count and fiber length. For yarns coarser than 16's Ne, 30-mm and 31-mm rotors cannot be effectively used because too much fiber mass exists. In addition, when the fiber length is 38 mm (1.50 inches) or greater, 31-mm rotors are too small for stable spinning. In these cases maximum rotor speed must be sacrificed. Figure 9.36 contains practical rotor speed ranges and rotor sizes for various yarn counts (Deussen, 1993). As illustrated in this figure, the rotor speed range for a G-240 (40-mm) rotor is 65,000 to 80,000 rpm as compared to 90,000 to 115,000 rpm for a G-231 (31-mm) rotor. Consequently, smaller rotors can be more economical provided the yarn is fine enough and appropriate properties can be achieved. Use of the smaller rotors, however, will typically cause fabric hand and cover to deteriorate, so compromises must be made.

Rotor Groove

Various rotor groove sizes are available to provide flexibility in yarn bulk. For high bulk yarns that would be used in sweaters, hosiery, or napped fabrics, an S-groove or U-groove rotor is often used. For most apparel knitting applications, a G-groove rotor is used, which provides better strength than the S- and U-rotors, but still sufficient bulk for knit yarns. The T-rotor has the narrowest groove, thereby providing the leanest, least hairy, strongest open-end yarn. Yarns produced with the T-rotor tend to have more torque than the other yarns. In recent years a K-groove rotor has been used in some of the smaller diameter rotors as a compromise between G- and T-rotors. The primary drawbacks of the small groove rotors tend to be loss of yarn bulk, higher yarn torque, and

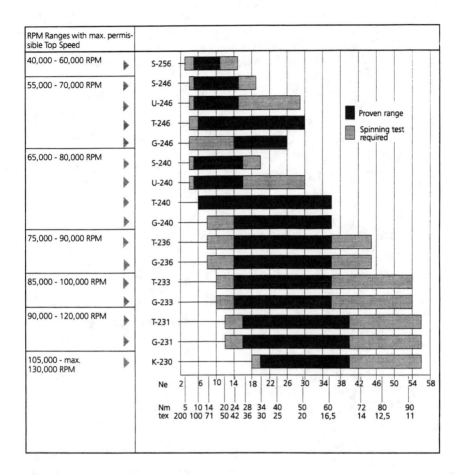

Figure 9.36. Guide for selecting the appropriate rotor size and type for a given yarn count. (Courtesy W. Schlafhorst AG & Co.)

higher trash buildup in the rotor groove. Each of the aforementioned rotors are shown in Figure 9.37.

Rotor Stem Wear

The contact between the rotor stem and the thrust bearing on the Schlafhorst machine can cause wear on the stem of the rotor if the thrust bearing is not properly lubricated. When the rotor rubs against the thrust bearing, shown in Figure 9.38, a concave wear pattern occurs on the stem. When this wear occurs, it causes the axial positioning of the rotor cup to change. Consequently, the fibers do not enter the rotor at the correct location, and yarn de-

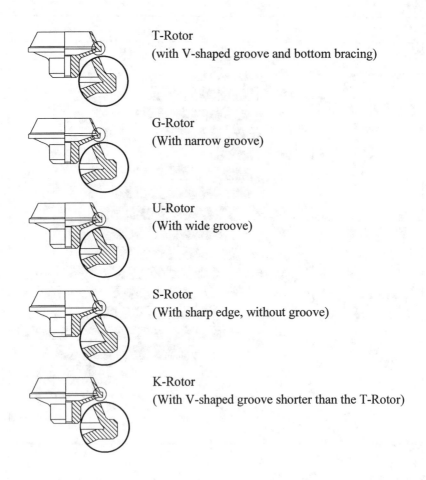

T-Rotor
(with V-shaped groove and bottom bracing)

G-Rotor
(With narrow groove)

U-Rotor
(With wide groove)

S-Rotor
(With sharp edge, without groove)

K-Rotor
(With V-shaped groove shorter than the T-Rotor)

Figure 9.37. Illustrations of five rotor types available for rotor spinning (Courtesy W. Schlafhorst AG & Co.)

fects increase. As a general rule, a concave wear pattern on the rotor stem greater than 3 mm in diameter necessitates replacement of the rotor. Table 9.8 provides data documenting the influence of stem wear on yarn evenness and IPI defects.

To minimize stem wear, whenever rotors are removed from a machine for maintenance, the rotors should be replaced into different spinboxes. The Rieter R1 machine does not have a rotor stem wear problem because an air bearing is used to support the rotor, thus no metal-to-metal contact exists as is illustrated in Figure 9.38.

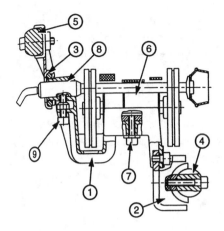

1. Rotor thrust bearing
2. Dampening element
3. Ball bearing dampening element
4. Thrust bearing mounting rod
5. Thrust bearing mounting rod
6. Twin disc bearings
7. Twin disc fastener screw
8. Setting sleeve
9. Setting sleeve fastener screw

Figure 9.38. Illustration showing the mechanical elements of the rotor twin disc drive assembly. (Courtesy Rieter Machine Works, Ltd.)

Rotor Cup Wear

Rotor cups generally begin to show significant wear after 20,000 hours of operation. As a rule of thumb, rotors require replacement when the yarn evenness is 1.0 point higher than it is with new rotors. Another test for worn rotors involves evaluating the moiré (repeating defect) channel on the on-line quality monitors. As rotors become worn, moiré defects will increase.

Twin Disc Wear

The twin discs are covered with rubber tires that contact the rotor stem. If the tires become damaged, a smooth drive of the rotor is not possible and spinning performance will suffer. Any visual damage to the tires warrants replacement. Trash or wax buildup on the tires can also hurt quality and performance. These tires can and should be cleaned periodically (every six months) with detergent.

Table 9.8. The Influence of Rotor Stem Wear on Rotor Yarn Properties (24's 100 Percent Cotton)

Yarn Property	Worn Rotors	New Rotors	95% Confidence Limit
Evenness, %CVm	16.7	15.9	0.3
IPI Thins	64	40	14
IPI Thicks	274	191	40
IPI Neps	50	40	3

Twin disc bearing wear can also cause quality problems. When twin discs are removed for the 6-month maintenance, they should be checked for how freely they spin. Any twin discs with restricted movement should be removed.

Rotor Belt Wear and Alignment

The rotor belt drives every rotor on each side of the machine. When a belt gets a wear spot, it causes the rotors to be driven abnormally, causing yarn breaks to increase. Similarly, if the belt is misaligned, yarn breaks will be higher than normal. Often a bad belt can be recognized by a high pitch noise. Worn belts can cause up to a 50 percent increase in yarn breaks.

Yarn Withdrawal Zone (Navel and Doff Tube)

The fiber bundle formed in the rotor is withdrawn through a navel and doff tube as shown in Figure 9.39. These components not only guide the newly formed yarn out of the spinbox, but contribute significantly to spinning performance and to yarn characteristics. As mentioned previously, movement of yarn against the navel introduces a false twist that strengthens the yarn between the rotor and delivery roll. Similarly, inserts can be added into the doff tube to increase friction on the yarn and therefore to increase false twist. However, some measures taken to increase false twist cause evenness of the yarn to deteriorate. Critical factors in the

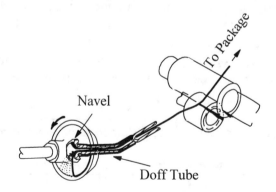

Figure 9.39. Illustration highlighting the navel and doff tube as used for rotor spinning. (Courtesy W. Schlafhorst AG & Co.)

yarn withdrawal zone include navel selection, navel spacing, and doff tube selection.

Navel Selection

To provide different levels of friction on the yarn, different navel surfaces are available. Most navels are made out of ceramic to provide maximum resistance to wear; however, steel navels are used in applications involving 100 percent synthetic fibers, such as acrylic or polyester, because of the ability of the steel to dissipate heat better and prevent thermal damage which can occur to these fibers at high spinning speeds.

Other modifications to navels involve grooves and ridges which are cut or molded into the navel. As shown in Figure 9.40, navels typically are available with 0, 3, 4, or 8 grooves around the orifice. Also called knurls, extra grooves around the rim are available for applications requiring low twist and high hairiness. A spiral ridge is available for low bulk applications, and a cross-shaped and fluted insert can be used for extra yarn bulk.

As the number of grooves increases, in general the hairiness of the yarn will increase, the shedding propensity of the yarn at knitting or weaving will increase yarn evenness will deteriorate, and IPI defects will increase, as shown in Table 9.9 (Copeland, 1996).

Because of the increased hairiness when the knurled rimmed or whorl insert navels are used, fabrics produced from these yarns exhibit significantly more cover. In a test conducted on fabrics

Table 9.9. The Effect of Navel Type on Rotor Yarn Quality (18's Ne 100 Percent Cotton)

Yarn Property	Navel Type						95% Confidence Limit
	KN	KN4	KN8	KN8R	KN4R4	KSR4	
Uster %CVm	13.2	13.4	13.7	14.0	13.9	13.3	0.3
IPI Thins (-50%)	4	6	5	10	6	3	4
IPI Thicks (+50%)	13	18	26	35	35	15	15
Lint (lb/10,000 lb of yarn)	3.0	2.1	2.3	3.3	3.9	3.6	1.2
Roughness Index	36.9	39.5	42.2	45.6	55.4	48.3	5
Hairiness (Uster Index)	4.0	4.3	4.3	5.1	6.3	5.3	0.2

Smooth navel without grooves to spin non-hairy yarns of cotton and acrylics.

3-Groove navel for low yarn hairiness with all raw materials.

4-Groove navel for low yarn bulk with all raw materials.

8-Groove navel for use with smaller diameter rotors for low twist yarns.

8-Groove navel with knurled rim for high hairiness and high bulk in yarns with all raw materials.

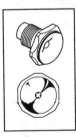

Smooth navel with fluted insert in navel throat for greater yarn hairiness and bulk with materials with low short fiber content.

Ceramic navel with a spirally arranged ridge for low hairiness and low bulk in 100 percent cotton yarns.

Figure 9.40. Illustrations and general descriptions of various navels available for use in rotor spinning. (Deussen, 1993)

produced from 18's cotton rotor yarns, fabrics produced with KN8R and KSR4 navels had 20 to 25 percent more cover as measured by fabric permeability tests.

Navel Spacing

The distance from the navel surface to the peeling point of the fibers from the rotor can be changed by adding or removing a washer under the navel collar. Adding washers causes the friction between the yarn and navel to increase, thereby causing more false twist. Due to the increased false twist, some improvement in yarn strength is possible; however, higher ends down are possible with the washer in place because of the increased spinning tension (Deussen, 1993). In a study by Copeland, no significant influence was found on evenness or IPI defects with or without a 1-mm washer behind the navel (Copeland, 1996).

Doff Tube Selection

The doff tube, shown in Figure 9.41, is located behind the navel and guides the yarn out of the spinbox. Early rotor spinning machines contained doff tubes with a 90 degree deflection of the yarn. As speeds have increased, the deflection of the doff tube has been reduced to a range of 30 to 37 degrees in order to reduce spinning tension. Another modification to the doff tube in modern machines has been the addition of a twist trap within the doff

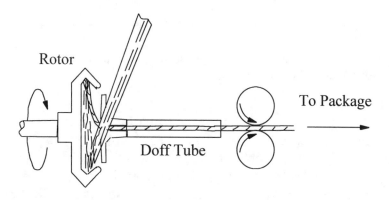

Figure 9.41. Simplified schematic illustration of the doff tube as used in rotor spinning.

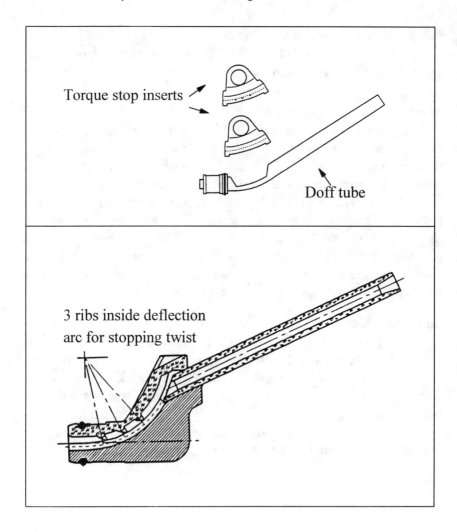

Figure 9.42. Two schematic views of the torque stop or twist trap device. (Courtesy W. Schlafhorst AG & Co.)

tube. The twist trap, called a torque stop on Schlafhorst machines, is similar to "speed bumps" on a roadway. The torque stop, shown in Figure 9.42, consists of three ridges inside the lower section of the doff tube. The extra friction provided by these ridges helps to increase false twist in the rotor groove. As a result, higher speed or lower twist yarns can be spun without as much of an increase in ends down at spinning.

Table 9.10. The Effect of Twist Traps on Rotor Yarn Evenness (18's 100 Percent Cotton)

Yarn Property	Smooth Doff Tube	Twist Trap	95% Confidence Limit
Uster Evenness, % CVm	13.8	14.4	0.3
IPI Thins (-50%)	19	54	8
IPI Thicks (+50%)	39	70	20
IPI Neps (+200%)	202	308	32

The drawback of a twist trap is the deterioration caused to yarn evenness and IPI defects as shown in Table 9.10. Consequently, if a twist trap is not needed for ends down reduction, it should not be used (i.e., weaving yarns, coarse count yarns).

On the newer model Schlafhorst machines (ACO 288), doff tubes are available that can be converted easily back and forth from smooth doff tubes to torque stops of various aggressiveness simply by switching interchangeable clips. This feature makes optimization of the doff tube much more convenient.

Critical Winding and Piecing Factors for Spinning Performance and Quality

Winding Zone

The winding zone consists of the area from which the yarn exits the spinbox to the drum that turns the package, as illustrated in Figure 9.43. The optimal setup of the winding zone is strongly dependent on the end use of the yarn. For instance, if the yarn is to be dyed, the desired package density would be low to allow dye to pass through the package. Thus, tension would be set lower than for yarns for weaving or knitting applications. Weaving packages are usually wound with relatively high tension to allow for a high density, heavy package, so it has to be changed less frequently at the next process, thereby helping processing efficiency. If a knitting yarn is being produced, wax must be applied at the winding zone to help to lubricate the yarn in order to reduce yarn-to-metal friction at the knitting machine.

Critical factors to control in the winding zone to produce a high quality package include the following:

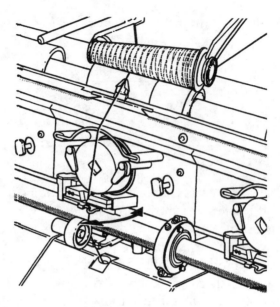

Winding at start of package

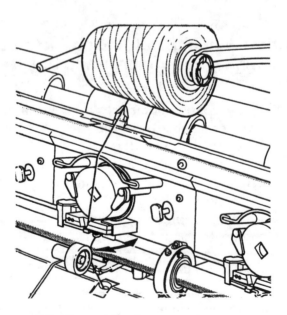

Winding with a larger package

Figure 9.43. Illustrations of the winding zone of a rotor spinning position depicting the start-up and partially filled yarn packages. (Courtesy Rieter Machine Works, Ltd.)

1. Yarn tension setting,
2. Angle of wind,
3. Cradle pressure,
4. Cradle alignment,
5. Yarn traverse displacement,
6. Delivery roll wear,
7. Wax application, and
8. Drive tire condition.

Yarn Tension Setting

The winding tension is the ratio in speed of the winding drum and the delivery roll that pulls the yarn out of the spinbox. This ratio is adjusted using gears or a dial, depending on the type machine. As a rule of thumb, the higher the tension, the lower the yarn elongation, and the harder the package. It is expected for normal applications that the winding tension be 1/3 to 1/2 of the spinning tension. It is easy to measure the spinning tension (between spinbox and delivery roll) and the winding tension (between delivery roll and the winding drum) with a hand held tensiometer as shown in Figure 9.44.

Angle of Wind

The angle of wind is influenced by speed of the traverse guide relative to the speed of the winding drum. A schematic illustrating how the angle is measured is shown in Figure 9.45. In most cases a wind angle of 33 degrees is standard, but 30-degree wind angles have been used for finer counts because of the lower winding tension provided as illustrated in Figure 9.46.

Cradle Pressure

Besides tension gears and wind angle, the density of a package can be affected by cradle pressure. A central setting can be made that controls the amount of spring pressure applied to the package when it is small. This pressure is reduced as the package builds. Less loss of yarn elongation occurs when package density is increased using cradle pressure instead of with a change in the tension gear.

Cradle Alignment

All of the settings mentioned to this point are centrally controlled on modern spinning frames, yet winding tension still varies

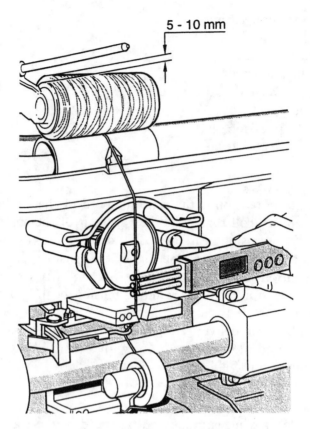

5 - 10 mm

Mean tension is measured with a wound
yarn body between 5 and 10 mm thickness
on the Rieter R1 machine.

Figure 9.44. Illustration of the use of a hand held tensiometer for checking
the tension on the yarn during winding at the rotor spinning position. (Courtesy Rieter Machine Works, Ltd.)

considerably in some plants from position to position. A winding
tension %CV under 10 should be targeted for a machine, and
when this value is not achieved, it is highly probable that inconsistent cradle alignment is at least partly to blame. Each cradle needs
to be aligned such that the package is not being driven too much
from either side of the package. This is particularly critical on conical take-ups for which driving of the small end of the cone will in-

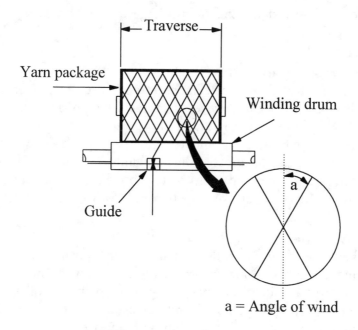

Figure 9.45. Illustration showing the angle of wind on a rotor spinning yarn package or any similarly wound yarn package.

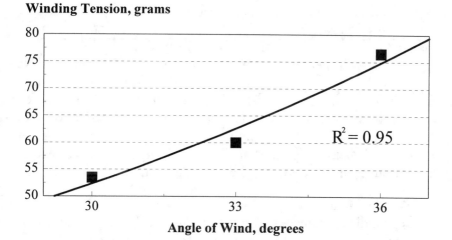

Figure 9.46. Graphical representation of the influence of angle of wind on winding tension.

crease tension and driving of the large end of the cone will decrease tension.

Yarn Traverse Displacement

The yarn traverse displacement is governed by a mechanism built into the yarn guide drive. It causes the yarn traverse to shift periodically from zero to four millimeters in order to adjust the hardness of the package shoulders. If the package shoulders are too hard, the yarn can be stressed during winding. If the shoulders are too soft, package density could be too low for optimal unwinding. Also, it is critical to control edge hardness for dye packages.

Delivery Roll Wear

When the delivery rolls become worn, position-to-position winding tension becomes more inconsistent. Furthermore, if the alignment of the delivery roll cot with the delivery roll becomes skewed by repeated removal and replacement of worn delivery cots over time, the winding tension will increase.

Wax Application

Wax is applied to knitting yarn in most cases to reduce the friction of the yarn against the knitting needles. The amount of wax applied can be influenced by the softness of the wax and/or the yarn deflection against the wax. The yarn deflection can be changed by using washers that adjust the position of the wax.

Unwaxed 100 percent cotton yarn between 12's Ne and 20's Ne will typically have a coefficient of friction of approximately 0.20 as measured by the Lawson-Hemphill Friction Tester. When waxed, the yarn friction will be reduced to approximately 0.11 to 0.14, depending on the amount of wax applied. It is important when wax is being applied that the wax disc turns and that excessive amounts of lint do not build up around the wax disc.

Drive Tire Condition

The winding drums on rotor spinning machines are equipped with rubber rings called drive tires that provide extra friction against the package in order to prevent slippage. For cylindrical tubes, two drive tires exist on the drum with one at either end. For conical packages, a single drive tire exists in the center of the drum.

When the drive tires wear, friction on the package becomes lower and more inconsistent, having the same effect on winding tension.

Automatic Piecer Consideration

The modern rotor spinning machines are equipped with a piecer that travels along the machine, automatically rejoining the yarns that have stopped spinning. When the piecer reaches a stopped position, the spinbox is opened and cleaned automatically using a plastic scraper that cleans the groove and a brush that removes loose residue.

The actual piecing cycle occurs as follows:

1. The rotor is cleaned.
2. The feed shaft delivers approximately 1.0 to 1.5 inches of sliver.
3. The suction arm finds the broken end on the package.
4. The yarn is moved from the package to the exit of the doff tube.
5. The yarn is fed into the doff tube and stops at the navel.
6. The feed shaft feeds enough sliver to form a ring in the rotor.
7. The yarn is moved from the navel into the rotor to begin collecting the fiber ring.
8. The rotor begins to turn and a piece-up occurs.
9. The feed shaft begins turning.
10. The piece-up is checked for thick or thin places.
11. The yarn is transferred from the piecer to the winder head.

Depending on the machine, the piece-up will occur between half speed and full speed of the rotor. For the machines that piece up yarn at less than full speed, some additional twist must be added to the piece-up to make it consistent with the twist in the normal yarn (Remember: twist is a function of rotor speed and delivery rate). However, if additional twist is too high, the piece-ups will be kinky and unstable.

It must be recognized that piece-ups are a "weak link" in the yarn, and it can be a weak link from an efficiency standpoint. It is critical that the piecer be successful a high percentage of the time, but not to the point where it is successful at the expense of aesthetically satisfactory piece-ups. These are critical factors involving piece-ups:

1. Efficiency,
2. Strength, and
3. Appearance.

Piecer Efficiency

The success efficiency of piecing up should be no less than 75 percent. In many cases 90 percent efficiency has been achieved in fine-tuned piecers. If efficiency is low, settings influencing the amount of sliver and yarn fed into the rotor, additional twist, draft, timing, and transfer speed should be evaluated.

Piecing Strength

It is expected that a yarn piecing should be 70 to 80 percent of parent yarn strength. Particularly on weaving yarns and fine count knitting yarns, piece-up strengths below 70 percent of parent yarn strength can be detrimental to downstream performance. A portable single-end tester is available on the market for convenient testing of the strength of piece-ups at the spinning machine.

Piecing Appearance

The appearance of a piece-up depends on the end use. For weaving yarns the strength is more critical than the appearance, whereas for knitting yarns some strength can be sacrificed to achieve a piece-up that is undetectable in the fabric. As a process control procedure, piece-ups should be removed from each machine weekly and placed on a blackboard (especially for knitting end uses) to assess piece-up appearance. If piece-up appearance is objectionable, the yarn and/or sliver feed into the rotor likely needs to be adjusted.

Rotor Spinning Benchmarks

From a production standpoint, it is expected that spinning efficiency for 18's cotton rotor yarn be 94 to 97 percent depending on the speed, raw material, and number of piecers. Yarn breaks should be between 100 and 200 per 1000 rotor hours, and quality cuts should be less than 100 total per 1000 rotor hours.

Table 9.11. World Class Quality Targets for Cotton and Polyester/Cotton Rotor Yarns

Yarn Property	8's Ne Cotton	18's Ne Cotton	18's Ne 50/50 Polyester/Cotton
Yarn Count, %Vb	0.7	0.8	1.0
Single-End Strength, g	900	400	475
Single-End Strength, %Vo	5.5	7.5	7.5
Elongation, %	9.0	7.0	9.5
Elongation Variation, %Vo	3.0	3.0	2.0
Evenness, %CVm	13.5	14.0	13.5
Evenness Variation, %Vb	1.5	1.5	1.0
Periodicities	0	0	0
IPI Thins	1	4	5
IPI Thicks	20	35	50
Neps	20	35	45
Classimat Minors	15	15	20
Classimat Majors/Long Thicks	0	0	0
Classimat Long Thins	0	5	3

Table 9.11 contains data on world class targets for 8's and 18's 100 percent cotton yarns and for 18's 50/50 polyester/cotton yarns.

Chapter 10

Air Jet Spinning

In the early 1980s air jet spinning was introduced to the textile industry as an alternative to ring spinning. At that time the 160 meters per minute (m/min) delivery rate of the air jet system, which was made by Murata Machinery Company, was ten times faster than that of ring spinning. Furthermore, it was able to produce finer yarns than the rotor system (up to 40's).

The first Murata machine was called the MJS 801. This machine contained a 3-roll drafting system as shown in Figure 10.1. In the late 1980s Murata introduced the MJS 802 machine. With the modification to a 4-roll drafting system and a change to a nozzle that provided better control of the yarn, an increase in speed up to a maximum of 210 m/min was achieved. The cross section of the new nozzle known as the cotton nozzle is shown in Figure 10.2.

At the 1991 International Textile Exhibition in Hanover, Germany, Murata introduced the 802H model, which had another new nozzle design and a modification to the delivery roll. This new model could run at speeds in excess of 300 m/min. Another version was introduced in 1995 incorporating a friction drive to work in conjunction with an air jet to produce yarn. This machine is called the Murata RJS (roller jet spinner) and has produced yarn in production facilities at 350 m/minute. It is predicted that speeds of air jet spinning machines will continue to increase to above 500 m/min as the nozzles become more efficient, the winding systems are further adapted, the yarn joining method is improved, and as quality monitoring is enhanced.

The high productivity of this spinning system, excellent yarn uniformity, and exceptional yarn and fabric pilling performance

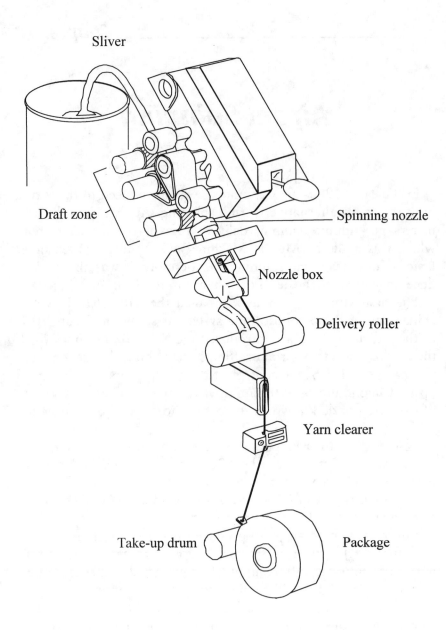

Figure 10.1. Schematic illustration of the MJS 801 machine. (Courtesy Murata Machinery, Ltd.)

Standard Nozzle

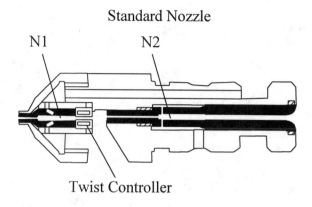

Cotton Nozzle

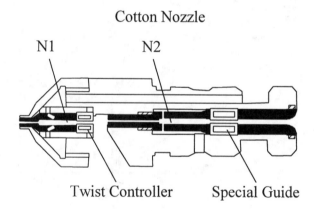

Figure 10.2. Cross-sectional views of the standard and cotton nozzles for MJS 802 spinning. (Courtesy Murata Machinery, Ltd.)

allowed air jet spinning to account for approximately 10 percent of the short staple yarn pounds produced in the USA in 1996.

Air Jet Spinning Principle

A schematic of the Murata MJS 802 machine is shown in Figure 10.3. Four main zones exist which include the creel zone, draft zone, nozzle zone, and winding zone. In the creel zone, sliver is fed into the machine such that a total draft is maintained within the

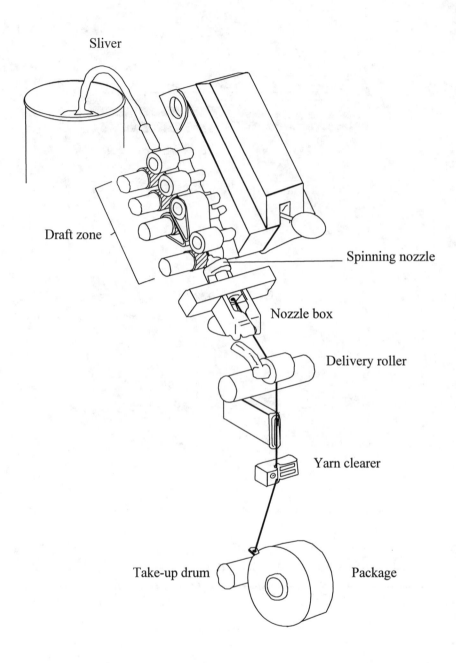

Figure 10.3. Schematic representation of the MJS 802 machine. (Courtesy Murata Machinery, Ltd.)

appropriate range. The sliver is usually placed on plastic guides that rest on stationary creel rods and is condensed by a trumpet behind the back drafting roll as shown in Figure 10.4.

In the draft zone, as shown in Figure 10.5, four sets of drafting rolls are typically used to reduce the mass per unit length of the fiber assembly. Conventionally, these rolls are numbered from front to back. Main components in the draft zone are rubber coated top cots, fluted steel bottom rolls, a condenser between the second and third roll, and a set of aprons between the first and second rolls. Roll spacing, draft distribution, apron pressure, apron height, condenser size, apron material, cot material, and front roll speed can all be adjusted within the draft zone to change the characteristics of the yarn.

In the nozzle zone, shown in Figure 10.6, two nozzles are used to form the yarn. The N2 nozzle imparts a false twist to the yarn bundle that migrates back to the front roll of the draft zone. The N1 nozzle provides a lower air flow in the opposite direction of the N2 nozzle, causing fibers on the edge of the spinning triangle to break free. These separated fiber ends twist around the bundle in the direction of the N1 air flow. As the yarn exits the second nozzle, the false twist is removed, so that the core fibers become aligned with the yarn axis, but the fibers that were separated remain twisted around the bundle. The untwisting of the core causes the bundle to expand in size, thus tightening the binder fibers. Factors that can be altered in the nozzle zone to affect yarn characteristics include: N1 nozzle pressure, N2 nozzle pressure, type of nozzles and/or guides, spinning tension (feed ratio), and distance between the nozzle and the front drafting roll.

The winding zone is depicted in Figure 10.7. In the winding zone, the yarn is taken up on a straight or tapered package. The package build is influenced by take-up tension, traverse speed, distance of the guide to the drum, wear on the delivery roll, wear on the drum, and cradle pressure.

Typical End Uses of Air Jet Yarn

In the initial stages of air jet spinning, the yarn was predominantly used for woven sheeting and bed products (comforters, pillow shams, etc.). In the early 1990s MJS yarns were also used in

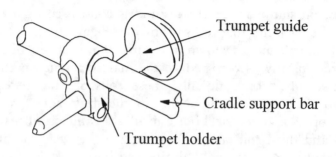

Figure 10.4. Schematic of the sliver condensing trumpet guide.

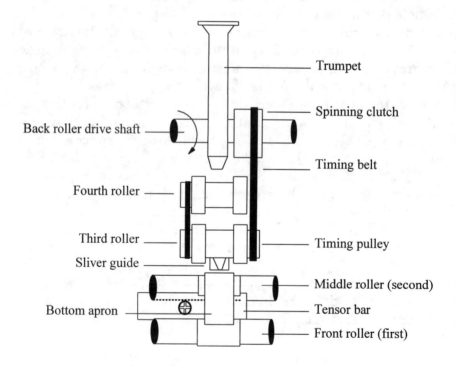

Figure 10.5. Schematic of the MJS four-roller draft zone. (Courtesy Murata Machinery, Ltd.)

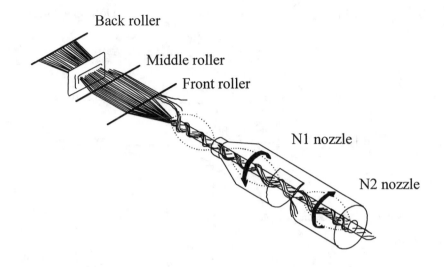

Figure 10.6. Illustration of the double-nozzle yarn assembly mechanism of MJS air jet spinning machines. (Courtesy Murata Machinery, Ltd.)

the knitting industry for apparel products, mostly lightweight shirting. Advantages were also found for MJS yarns in fleece, primarily because of the low pilling tendency of these yarns. MJS yarns to date have not been readily accepted in the jersey knit market. Air jet yarns in many cases do not have acceptable appearance in loose jersey knit structures, primarily because of long thin places.

The most common markets for MJS yarns currently include the following:

1. Sheeting, comforters, bedding accessories;
2. Napery (cloth napkins);
3. Pocketing material;
4. Men's pin point oxford shirting;
5. Fleece;
6. Gloves;
7. Bottom weight fabrics (trousers);
8. Knit apparel;
9. Automotive upholstery;
10. Industrial fabrics; and
11. Sewing thread.

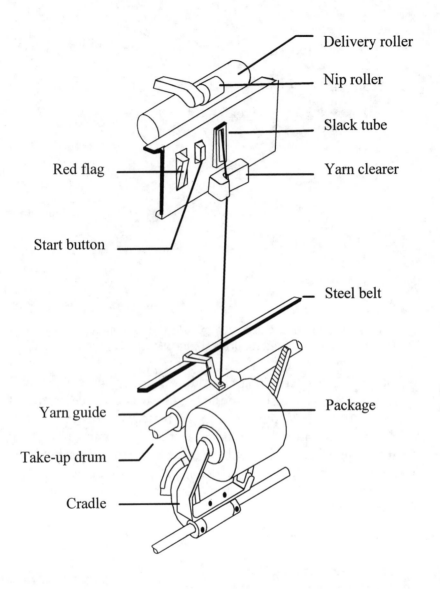

Figure 10.7. Schematic illustration of the MJS winding zone. (Courtesy Murata Machinery, Ltd.)

Advantages of MJS Yarns

The biggest advantage of MJS yarns is the productivity at which they are produced compared to ring spinning and rotor spinning, particularly in counts finer than 20's. Other advantages are as follows:

1. Low pilling propensity;
2. Few long thin defects (compared to ring yarns);
3. Excellent yarn evenness;
4. Better depth of shade and color fastness (compared to ring yarns);
5. Less yarn shedding (compared to ring yarns); and
6. Lower yarn torque (compared to ring yarns).

Another advantage of the MJS technology over ring spinning is the delivery of yarn to a large package. As a result, MJS yarn packages used at knitting or weaving have far fewer splices than packages wound from ring spun bobbins. Thus, fewer weak links are present. In fact a 2.7-kg (6-lb) MJS yarn package would have approximately 60 fewer splices per package compared to a ring spun yarn.

In a study conducted by Fite at the Institute of Textile Technology (Fite, 1994), MJS yarns were compared to ring and rotor yarns for pilling, strength, and color of dyed fabrics, before and after washings. Figures 10.8 and 10.9 contain the data, which confirm the lower pilling propensity and better dye uptake of the fabrics produced from MJS yarn versus ring yarn. In the dye uptake comparison, the MJS and rotor yarns were similar.

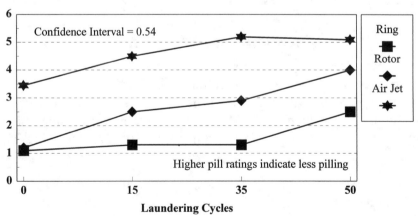

Figure 10.8. Comparison of pilling propensity before and after multiple washings among twill fabrics produced from ring, rotor, and air jet 14's (Ne) 65/35 polyester/cotton yarns. (Fite, 1994)

K/S Values

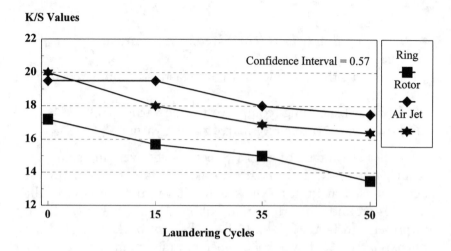

Figure 10.9. Comparison of depth of shade before and after multiple washings among twill fabrics produced from ring, rotor, and air jet 14's (Ne) 65/35 polyester/cotton yarns. (Fite, 1994)

With respect to yarn evenness and defects, Table 10.1 contains data on 26's Ne 50/50 polyester/cotton average yarn properties within the USA textile industry, according to the Institute of Textile Technology quality database. These data indicate that the tensile properties of air jet yarns are better than those of rotor yarns, and the defect counts (thicks and thins) are lower than those of ring spun yarns.

Disadvantages of MJS Yarns

As with any yarn manufacturing system, some disadvantages exist with the air jet spinning system. The biggest disadvantage is that 100 percent carded cotton yarns cannot yet be produced commercially. Too much short fiber exists for adequate binder fiber formation with 100 percent cotton. Other disadvantages include the following:

1. Low strength compared to ring spun yarn (approximately 25 percent less);
2. Harsher fabric hand compared to ring spun yarn;
3. Less resultant cover in fabric compared to ring spun yarn; and
4. Higher fabric shrinkage compared to rotor yarn.

Table 10.1. USA Industry Average Quality for Ring, Rotor, and MJS Yarns (26's Ne 50/50 Polyester/Cotton)

Yarn Property	Ring	Rotor	MJS
Yarn Count, %Vb	2.4	1.4	1.4
Skein Strength, lb	124.1	78.6	94.3
Skein Strength, %Vb	6.5	5.1	5.6
Single-End Strength, g	464.5	293.1	343.7
Single-End Strength, %Vo	10.1	9.1	10.1
Single-End Strength, %Vw	8.9	8.1	8.8
Single-End Strength, %Vb	4.6	4.0	4.7
Elongation, %	9.3	7.9	8.5
Elongation, %Vb	3.9	4.0	5.2
Yarn Appearance Index	105.5	114.0	111.3
Uster Evenness, %CVm	18.3	15.8	16.3
Uster Evenness, %Vb	3.4	1.8	2.4
1-yd %CV	6.1	4.0	5.5
3-yd %CV	4.6	3.3	3.6
10-yd %CV	3.2	2.4	2.2
50-yd %CV	1.7	1.2	1.0
IPI Thin Places (-50%)	48.8	44.1	41.5
IPI Thick Places (+50%)	791.9	181.8	259.6
IPI Neps (+200%)	634.0	162.7	404.7
Classimat Minors	1993.0	89.3	466.0
Classimat Majors	11.2	0.1	1.1
Long Thick Places	7.7	0.4	0.5
Long Thin Places	337.4	5.5	167.2

The difference in yarn strength among the systems is shown in Table 10.1. The differences in fabric hand, cover, and shrinkage are displayed in Figures 10.10 through 10.12 (Fite, 1994).

Air Jet Yarn Structure

Because of the separation and wrapping of fibers around the yarn bundle, the MJS yarn has a fasciated structure. The yarn has a central core of mostly parallel fibers wrapped with binder fibers as illustrated in Figure 10.13.

Surface fibers twisting tightly around the core cause the yarn to be slightly buckled. The wavy structure of the yarn in turn causes the yarn to be well suited for use as filling in air jet weaving machines, as it can be propelled across the shed more quickly. Ac-

Total Fabric Hand Values

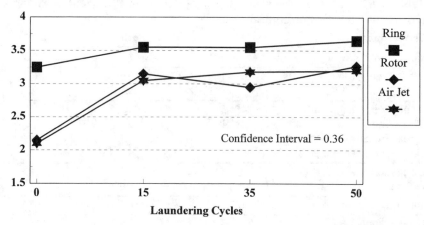

Figure 10.10. Comparison of Kawabata total hand values before and after multiple washings among twill fabrics produced from ring, rotor, and air jet 14's (Ne) 65/35 polyester/cotton yarns. (Fite, 1994)

Air Permeability, cu. ft./sq. ft./min.

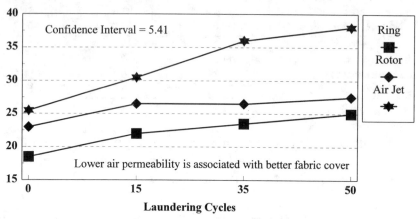

Figure 10.11. Comparison of air permeability before and after multiple washings among twill fabrics produced from ring, rotor, and air jet 14's (Ne) 65/35 polyester/cotton yarns. (Fite, 1994)

cording to Mackey, air jet yarns are propelled across a weaving machine 10 percent faster than ring spun yarns and 7 percent faster than rotor yarns (Mackey, 1995).

The lack of twist in the air jet yarn core is believed to contribute

Fabric Picks/Inch

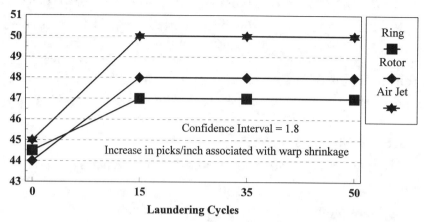

Figure 10.12. Comparison of warp shrinkage after multiple washings among twill fabrics produced from ring, rotor, and air jet 14's (Ne) 65/35 polyester/cotton yarns. (Fite, 1994)

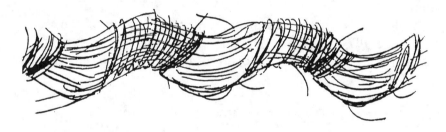

Figure 10.13. Illustration of the physical structure of Murata air jet yarn.

to the low pilling propensity of these yarns. The pills can still be created from abrasion, but they are not "locked" into the structure because of the absence of twist. Pills break away soon after they are formed as a result.

The yarn harshness and low cover factor are influenced by the tight binder fibers along the yarn surface. This harshness can be decreased by manipulating settings, but such a change usually results in loss of yarn strength. Unlike the rotor yarn wrapper fibers, the binder fibers that wrap around the MJS yarn are critical to yarn strength as they hold the internal bundle tightly together.

Raw Material Requirements for Air Jet Spinning

Because of the high productivity and requirement for consistent binder fiber formation in air jet spinning, raw material requirements are critical. In the industry most air jet yarns are produced from 50/50 or 65/35 polyester/cotton. Some yarns are being produced from 100 percent polyester, acrylic, and rayon as well.

Synthetic Fiber Properties

For the synthetic fibers the critical characteristics include the following:

1. Cohesion (as influenced by finish and crimp),
2. Cut length,
3. Denier, and
4. Tenacity.

Cohesion

Cohesion is perhaps the most influential property for yarn quality. Generally, the higher the cohesion, the higher the strength, as shown in Figure 10.14. However, increasing cohesion usually

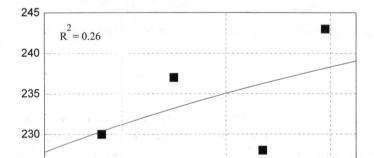

Figure 10.14. Graphical illustration of the influence of fiber-to-fiber cohesion on the strength of 35's (Ne) 50/50 polyester/cotton MJS yarns.

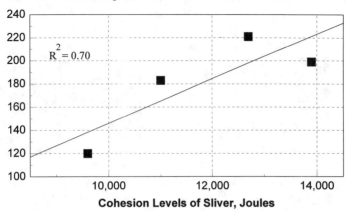

Figure 10.15. Graphical illustration of the influence of fiber-to-fiber cohesion on the number of Classimat Long Thin Places in 35's (Ne) 50/50 polyester/cotton MJS yarns.

causes long thin places to increase, as shown in Figure 10.15, because of difficulty in drafting. Phosphate based finishes are typically used in the industry on polyester; however, mineral oil finish provides higher cohesiveness. As a result, some polyester producers use a combination finish to achieve the benefits of both bases.

Cut Length

In a study conducted at ITT, it was proven that cut length has a significant impact on yarn evenness, 1-yard %CV, and long thin places. When fiber length was reduced from 38 mm (1.50 inches) to 32 mm (1.26 inches), tighter roll spacings could be used on the Murata machine, which helped reduce defects and improve knit fabric appearance. Figures 10.16 through 10.18 represent this study and illustrate the influence of polyester fiber denier and cut length on yarn evenness and defect levels when spun at normal and tighter MJS roll settings. The drawbacks are a loss of yarn strength with shorter fibers and difficulty in changing from the wide roll spacings to the tight spacings. As a result, few manufacturing plants have taken advantage of this opportunity for lower defects. Changeover cost, lower yarn strength, and limited flexibility with the tight roll spacings and shorter fibers have been the deterring factors.

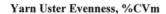

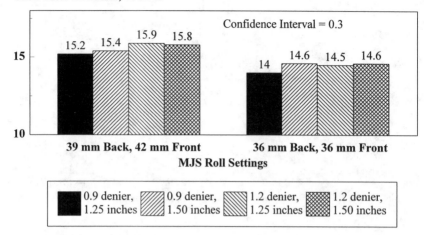

Figure 10.16. Graphical illustration of the combined influence of fiber denier, fiber length, and MJS roll settings on the evenness of 35's (Ne) 50/50 polyester/cotton MJS yarns.

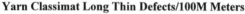

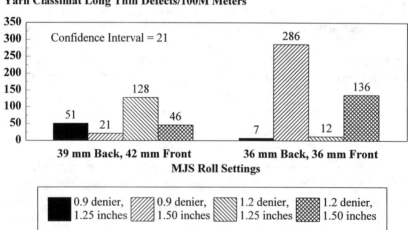

Figure 10.17. Graphical illustration of the combined influence of fiber denier, fiber length, and MJS roll settings on the number of long thins in 35's (Ne) 50/50 polyester/cotton MJS yarns.

Yarn IPI Thick Places/1000 Meters

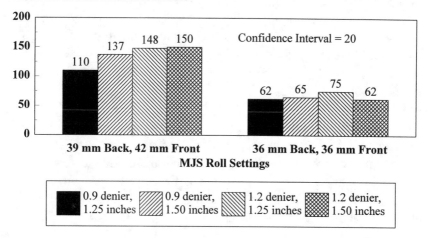

Figure 10.18. Graphical illustration of the combined influence of fiber denier, fiber length, and MJS roll settings on the number of IPI thicks in 35's (Ne) 50/50 polyester/cotton MJS yarns.

In another study conducted at ITT, different polyester cut lengths were evaluated with the conventional roll spacings of 39 mm (1.54 inches). In this study 35-mm (1.38-inch) and 38-mm (1.50-inch) fibers were compared. The yarns produced from the shorter fibers had lower defect levels as shown in Table 10.2. Several fiber producers have recognized that cutting fibers shorter than 38 mm offers better performance of the fibers at spinning. Whereas the standard cut length for a nominal 1.5-inch fiber was 38 mm, current 1.5-inch fibers are being cut between 35 and 37

Table 10.2. The Effect of Polyester Fiber Lengths on MJS Yarn Properties (35's Ne 50/50 Polyester/Cotton)

Yarn Property	Fiber Cut Length		95% Confidence Limit
	35 mm	38 mm	
Uster %CVm	16.1	16.8	0.2
IPI Thicks	203	279	38
1-Yard %CV	5.4	5.8	0.1
Classimat Minor Defects	728	879	166
Classimat Long Thins	120	183	63
Single-End Strength, g	230	237	5

mm for air jet spinning, depending on the producer. If a problem exists when polyester fibers are being cut, and the longest fibers in the distribution are greater than 41 mm (1.61 inches), excessive stops can be expected at spinning.

Denier

Some researchers suggest denier has a significant impact on MJS yarn strength, evenness, thin places, elongation, neps, and Classimat defects; however, in recent plant comparisons, little difference has been recognized in yarns produced from 1.2 versus 1.0 dpf fibers. The largest difference in yarns produced from different deniers is that the yarn produced with the finer denier will tend to be characterized by the following:

1. Lower elongation,
2. Slightly higher strength,
3. Fewer long thin places,
4. More neps, and
5. Slightly higher number of Classimat minor defects.

Fiber Tenacity

Fiber tenacity is selected based on the end use of the yarn. With knitting yarns, for which appearance is critical and some strength can be sacrificed, low and mid-tenacity polyesters are used. The low and mid-tenacity fibers allow for reductions in pilling severity and improvements in dye uptake. The direct and/or indirect influence of fiber tenacity on certain yarn properties is shown in Table 10.3.

Table 10.3 Influence of Fiber Tenacity on MJS Yarn Properties (38's Ne 50/50 Polyester/Cotton)

Yarn Property	Mid-Tenacity (5.2 g/denier)	High Tenacity (6.8 g/denier)	95% Confidence Limit
Single-End Strength, g	178	209	12
Single-End Strength, g/tex	11.48	13.48	0.7
Elongation, %	8.0	8.7	0.5
Evenness, %CVm	17.4	17.7	0.4
IPI Thins	141	146	20
IPI Thicks	355	378	29
Classimat Minors	1404	1709	47
Classimat Long Thins	400	458	20

Cotton Properties

The critical cotton properties for air jet spinning include the following:

1. Short fiber content,
2. Fiber strength,
3. Fiber length, and
4. Micronaire fineness.

Short Fiber Content

As shown in Table 10.4, short fiber content was found to have a significant impact on nearly every measured yarn property in a study done in an ITT member plant. Single-end strength, Uster evenness, thin places, and Classimat minor defects were affected the most. In addition, several plants currently ensure that laydowns average less than 10 percent short fiber content to avoid excessive efficiency losses at spinning. Short fiber content is the *most critical* cotton property for control of air jet spinning quality and ends down level.

Table 10.4. The Influence of Short Fiber Content on 30's Ne 65/35 Polyester/ Cotton MJS Yarns

Yarn Property	Short Fiber Content in Laydown		Significant Difference
	(9.9%)	**(12.1%)**	
Count %Vb	0.9	1.3	No
Break Factor	2541	2423	No
Single-End Strength, g	314.2	301.3	Yes
%Vo	8.9	9.8	No
Single-End Elongation,%	7.6	7.7	No
%Vb	3.1	5.0	Yes
Yarn Appearance Grade	B/B-	B	—
Uster Evenness, %CVm	16.7	17.7	Yes
%Vb	1.4	1.1	No
IPI Thins	65	136	Yes
Thicks	207	311	Yes
Neps	308	348	Yes
Classimat Minors	701	991	Yes
Majors	3	3	No
Long Thins	230	495	Yes
Hairs/Meter (>3 mm)	1.6	2.4	No

Table 10.5. Expected Yarn Quality Improvements Associated with a Cotton Strength Increase

Yarn Property	Improvement with 1 g/tex Strength Increase
Single-End Strength, g	5%
Single-End Elongation, %	3%
Uster Evenness, %CVm	0.25 Points

Fiber Strength

Higher fiber strength generally results in improved MJS yarn strength, elongation, and evenness. Based on ITT research, the expected improvement in yarn for a given cotton strength improvement is listed in Table 10.5.

Fiber Upper Quartile Length

As cotton fiber length increases, improvements are expected in MJS yarn strength, elongation, and hairiness as shown in Table 10.6. Uster evenness and defects are not significantly affected by these changes in cotton upper quartile length.

Micronaire

Increasing cotton micronaire causes a reduction in MJS yarn strength and elongation as shown in Table 10.7, but allows for improvements in Uster evenness, thick places, and neps. Hairiness also increases with the higher micronaire, or coarser cotton fibers.

For best results raw cotton characteristics should be chosen with the following targets:

1. Fiber strength, > 28 g/tex;
2. Fiber length, > 1.12 inches;

Table 10.6. The Influence of Cotton Upper Quartile Length On 35's Ne 50/50 Polyester/Cotton MJS Yarn Properties

Yarn Property	Cotton Length (inches)			95% Confidence Limit
	1.09	1.13	1.18	
Skein Strength, lb	58.6	61.3	64.9	3.7
Break Factor	2035	2188	2269	130
Single-End Strength, g/tex	12.4	12.9	13.5	1.0
Elongation, %	5.7	6.6	6.9	0.4
Yarn Appearance Grade	B-/B	B-/B	B/B+	—
Hairs/Meter (>3 mm)	10	7	7	2.5

Table 10.7. The Influence of Micronaire on 35's Ne 50/50 Polyester/Cotton MJS Yarn Quality

Yarn Property	Micronaire				95% Significant Difference
	3.48	3.95	4.05	4.68	
Count, %Vb	0.7	0.4	0.4	0.5	No
Break Factor	2480	2422	2308	2259	158
Single-End Strength, g	252	244	241	238	9.8
%Vo	9.9	10.6	9.8	10.4	No
Single-End Elongation,%	7.0	6.9	6.7	6.3	0.4
%Vo	8.2	10.9	11.0	8.5	1.1
Yarn Appearance Grade	B-/B	B-/B	B	B/B+	—
Uster Evenness, %CVm	17.4	17.6	17.0	16.6	0.4
IPI Thin Places	98	101	69	80	20
Thick Places	382	404	312	230	55
Neps	494	615	503	375	114
Hairs/Meter (>3 mm)	9.6	10.1	10.3	12.6	2.1

3. Short fiber content, < 10.0 percent; and
4. Micronaire, 4.1 to 4.3.

Sliver Requirements for Air Jet Spinning

Special emphasis must be placed on sliver preparation because any sliver defects will cause a stop at spinning. It is all too true what was said by Mr. Jeff Alverson of Russell Corporation at a recent spinning conference:

"The best tester of yarn quality is an air jet weaving machine, but the best tester of *sliver quality* is an air jet spinning machine."

Unlike the rotor spinning machine which has the combing roll to help remove impurities and the rotor to provide extra blending, the quality of the sliver fed into the air jet spinning machine completely influences yarn quality. In addition, with the requirement for consistent formation of binder fibers to provide consistent yarn characteristics, the sliver must be extremely uniform.

For top quality and high efficiency spinning, the following sliver properties are desired:

1. Short-term evenness, < 3.2 %CVm;
2. High fiber alignment, < 40 on ITT Bulk Density Tester;
3. Minimal slubs, < 0.5 per lb;

4. Neps per gram, < 70;
5. Short fiber content, < 4.5 percent; and
6. No periodicities.

Sliver Evenness

The Uster evenness of finisher sliver is correlated with long thins, 10-meter %CV, and 50-meter %CV of the yarn. In a study done at ITT with 50/50 polyester/cotton MJS yarns produced from several slivers, the slivers with evenness below 3.2 %CVm resulted in under 60 long thin places. The slivers with evenness above 4.0 %CVm resulted in over 200 long thin places. In the same study it was shown that 96.5 percent of the changes in 10-meter %CV of a yarn can be attributed to changes in Uster %CVm of the sliver used to make the yarn as shown in Figure 10.19.

Fiber Alignment

Almost every air jet spinner claims that alignment of fibers in the sliver is most critical for air jet spinning, yet few people check fiber alignment regularly. Using the ITT bulk tester, an alignment value below 40 is preferred. Expected yarn quality levels from various sliver fiber alignments are listed in Table 10.8.

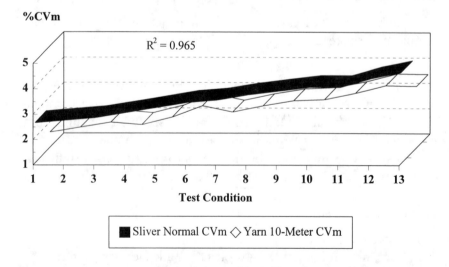

Figure 10.19. Graphical illustration of the influence of sliver normal Uster evenness on the mass uniformity of 10-meter lengths of MJS yarns.

Table 10.8. Influence of Fiber Alignment in Sliver on Yarn Properties (30's Ne 50/50 Polyester/Cotton)

Yarn Property	Sliver Fiber Alignment Index			95% Confidence Limit
	35	40	45	
Single-End Strength, g	270	280	290	12
Uster %CVm	15.6	15.8	16.0	0.3
Classimat Minors	525	650	775	121
Long Thins	41	66	97	18

Slubs

Slubs in the sliver have a detrimental effect on stops at spinning. As shown in Table 10.9, a strong correlation exists between slubs in sliver and stops called slub cuts at spinning. In addition, an increase in slubs also causes thick places, neps, and Classimat minor defects to increase in the yarn.

Neps

Neps in the sliver are a strong indicator of the quality of carding in a plant. For MJS yarns, sliver nep level correlates with several yarn characteristics. As shown in Figure 10.20, sliver neps correlate with yarn neps, thick places, Classimat minor defects, and evenness. The sliver neps apparently have a direct influence on small defects in the yarn, and seem to indirectly be associated with the number of larger defects and with yarn short-term unevenness.

Table 10.9. The Influence of Slubs in Sliver on Slub Cuts on the Air Jet Spinning Machine

Sample	Sliver Slubs/lb	Slub Cuts (per 60 hours)
1	0	0.0
2	3	11.5
3	3	14.4
4	4	14.2
5	11	18.1
6	12	36.7
7	29	24.8
8	29	39.2
9	61	178.5
10	79	77.0

R^2 (Coefficient of determination) = 0.77

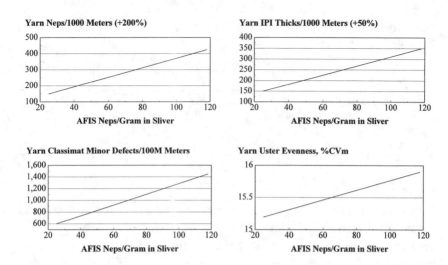

Figure 10.20. Graphical representations of the influence of neps in sliver on MJS yarn neps, thick places, Classimat minor defects, and evenness. (Sensitivity analyses from ITT Neurospin database.)

Sliver Periodicities

In a recent comparison of two slivers from the same plant, it was determined that one of the slivers had periodic defects at 50 and 125 mm (about 2 and 5 inches) as shown in Figure 10.21. All other sliver properties were the same. When these slivers were spun into 30's Ne 50/50 polyester/cotton yarns, a significant difference in quality resulted as shown in Table 10.10. In particular, Uster %CVm, 1-yard %CV, and long thin places were of much higher magnitude in the yarn spun from the poor quality sliver. The severity of the influences of sliver periodicities on MJS yarn properties causes on-line sliver monitoring to be preferred for sliver that is to be used for MJS spinning.

Critical Factors for Yarn Quality

Several factors are important for producing a top quality yarn that will spin efficiently. Unfortunately, when a setting on the machine is changed, it often helps some yarn properties and deteriorates others. That is why the end product must be kept in mind. For example, a weaving yarn often requires low hairiness and high strength, whereas a knitting yarn requires soft hand and as much bulk as possible. As a result, a weaving yarn would typically be

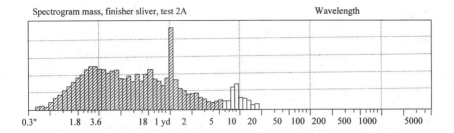

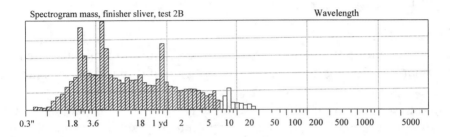

Figure 10.21. Spectrograms comparing two different test slivers from finisher drawing.

run at a lower speed and higher N2 pressure than would a knitting yarn to get higher strength and less hairiness, at the expense of yarn softness and bulk.

The influence of each critical factor is discussed in the section relating to the zone on the machine where each is applicable.

Table 10.10. The Effect of Sliver Periodicities on 30's Ne 50/50 Polyester/Cotton MJS Yarn Characteristics

Yarn Property	Normal Sliver	Sliver w/Periodicity	Significant Difference
Yarn Count, %Vb	1.2	2.1	yes
Uster %CVm	16.0	18.5	yes
1-yd %CV	5.4	8.2	yes
3-yd %CV	3.3	5.1	yes
10-yd %CV	2.0	2.7	no
50-yd %CV	1.1	1.4	no
IPI Thins	38	186	yes
Thicks	227	336	yes
Classimat Minors	564	830	yes
Long Thins	146	2304	yes

Draft Zone

A schematic for the draft zone is shown in Figure 10.22. The *trumpet* alignment is critical for providing consistent yarn strength from position to position. This trumpet is held in place by a set screw, and it can easily move. As a result, it is often set improperly. If the trumpet is set too far left or right, the sliver will not travel straight through the drafting zone and a yarn strength loss will exist. This trumpet should also be set at a distance of 3 mm (0.118 inch) from the back steel roll.

Total Draft

A dilemma that always must be solved in the spinning plant is what sliver weight to deliver to the spinning machine. In many cases several yarn counts are spun in the plant, and it would be desired to use one sliver weight for all counts. Also, the heavier the sliver weight, the greater the amount of stock that can be produced at drawing in a given amount of time. Unfortunately, the air jet spinning machine has a certain total draft range which must be adhered to for good results. In general, as draft gets too high (above 170), spinnability deteriorates and yarn thin defects increase. As draft gets too low, yarn strength and Classimat minor

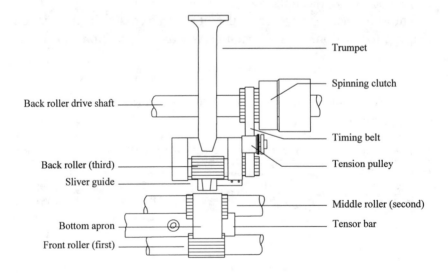

Figure 10.22. Schematic of the MJS three-roller draft system. (Courtesy Murata Machinery, Ltd.

Table 10.11. Recommended Sliver Weights for Air Jet Spinning

Ne Yarn Count Range	Sliver Weight	
	gr/yd	g/m
34's to 40's	35	2.48
28's to 33's	40	2.83
20's to 27's	45	3.19
15's to 19's	50	3.54
10's to 14's	55	3.90
9's and coarser	60	4.25

defects are usually worse. Recommended sliver weights for various counts are listed in Table 10.11.

Main Draft

Draft distribution is defined as the combination of draft in each zone of the drafting system. In the current MJS machines, the draft in the back zone (between the third and fourth rolls) is fixed at 2. The draft in the main (front) zone is chosen, and the middle zone gets the leftover draft.

The drafts in each zone multiplied together equal the total draft. Thus, if the main draft is set at 42 and the total draft is 150, the intermediate draft will be determined as follows:

$$42 \times \text{Intermediate Draft} \times 2 = 150$$

$$\text{Intermediate Draft} = \frac{150}{2 \times 49} = 1.79$$

The main draft should not be set so high that it forces the intermediate draft to be below 1.15, or spinning stability will be adversely affected. However, previous research has shown that a high main draft, which causes intermediate draft to be low (under 1.5), helps to improve yarn strength, elongation, and long thins. The influence of main draft on these yarn characteristics is shown in Table 10.12.

Roll Spacings

The top roll spacings on the Murata air jet spinning machine are set by a component called the side plate which is shown in Figure

Table 10.12. The Influence of Main Draft Ratio on 35's Ne 50/50 Polyester/Cotton Yarn Properties

Yarn Property	Main Draft			95% Confidence Limit
	35	42	50	
Single-End Strength, g	226	232	244	5
Single-End Elongation, %	6.3	6.5	6.9	0.25
Uster %CVm	16.6	16.9	17.1	0.35
IPI Thicks	240	270	350	39
IPI Neps	320	380	510	111
1-Yd %CV	5.5	5.2	5.0	0.19
Classimat Long Thins	220	120	50	120
Classimat Minors	750	900	1350	47
Uster Hairiness Index	4.80	4.65	4.55	0.10

10.23. There are three side plate sizes that are typically used for the 4-roll 802 machine. These are available as 39/42, 41/42, or 36/36 combinations. The first number designation on the side plate is the distance between the second and third rolls, and the second number denotes the distance in mm between the third and fourth rolls.

In most instances the 39/42 side plate is used with counts finer than 20's and with 38-mm (1.50-inch) polyester. The 41/42 side plate is used for coarse counts, and the 36/36 is used in conjunction with 32-mm (1.26-inch) polyester at 100 percent or in blends, or in conjunction with 100 percent cotton.

Seal

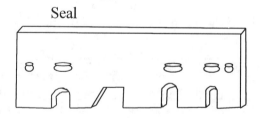

Figure 10.23. Schematic view of a Murata MJS side plate.

In a study done at ITT comparing 39/42 to 41/42 side plates on 35's 50/50 polyester/cotton yarn, the 39/42 side plates provided better Uster %CVm, IPI thins, and yarn strength. However, when counts coarser than 20's were spun, machine performance (nozzle chokes) tended to deteriorate with the tighter roll spacings.

The 36/36 side plates, when used with 32-mm polyester, can provide significant advantages over the 39/42 side plates. As shown in Table 10.13, the use of the tighter side plates improved the properties of yarn produced from 32-mm fiber, but caused defects to increase in the yarns produced from 38-mm fibers.

Cot and Apron Selection

Top cots for the Murata MJS machine can be selected with various Shore hardness levels. The typical hardness range for the industry is between 75° and 83° Shore hardness. The cots that are on the soft side of the range typically provide for better yarn evenness and fewer IPI thin places; however, the wear life is shorter.

Table 10.13. The Influence of Side Plate Selection on the Quality of 35's 50/50 Polyester/Cotton MJS Yarns (802H Machine)

Yarn Property	Polyester Fiber Length (mm), Side Plate Combination				95% Confidence Limit
	32 mm, 36/36	38 mm, 36/36	32 mm, 39/42	38 mm, 39/42	
Yarn Count, Ne	34.4	34.5	34.4	34.3	—
Single-End Strength, g	273*	241	260	257	5
% Vo	10.3	12.9	10.7	9.3	1.2
Minimum	176	162	166	197	—
Single-End Elongation, %	8.7*	8.0	8.0	8.1	0.2
%Vo	9.9	9.5	15.8	9.4	1.2
Uster %CV	14.4*	15.3	15.1	15.2	0.2
1-Yd %CV	4.2*	6.3	5.2	5.2	0.2
3-Yd %CV	2.7*	3.9	2.9*	3.2	0.2
IPI Thins (-50%)	14*	27	34	33	7
Thicks (+50%)	107	119	114	121	27
Neps (+200%)	122*	294	196	208	33
Classimat Minors	958	3507	688*	679*	67
Long Thins	62*	664	107	88	23
% Shrinkage	3.5	4.5	3.4	3.6	—
Pilling Rating	3.3	3.7	3.7	2.3	—
Hairs/m (>3 mm)	3.8*	4.2*	11.5	12.6	1.7

* Bold typeface and asterisks indicate best quality achieved.
Note: To convert from mm to inches, mm/25.4.

Consequently, the decision needs to be made at the individual plant on cost versus quality benefit.

It is more difficult to differentiate aprons with a quantitative measurement. Some feel more slick than others, and some collect more lint than others, but normally the aprons are simply differentiated by the manufacturer. In a recent survey of ITT members, it was determined that five different apron manufacturers are supplying aprons, and that apron life ranges from eight to thirteen weeks. In most cases the decision for apron use depends on wear life, but the apron material can have a significant effect on quality as well. As shown in Table 10.14, comparison of new aprons from two manufacturers showed that one apron type contributed to half the Classimat defects of another type. The interesting thing to mention here is that most in-plant studies exclude Classimat testing, and if the comparison were just made on evenness and strength, no difference would have been recognized.

Roll Pressure

The standard roll pressure in the Murata air jet spinning machine is 25 kg (55 lb) on the front rolls, and 22 kg (48.5 lb) on the other rolls. Some attempts have been made to increase the pressure on the third and fourth rolls to 25 kg, with slight improvements in Uster evenness and 1-yard %CV. However, most plants still run with the standard setup.

Table 10.14. Comparison of MJS Yarn Properties Using Two Apron Brands (30's Ne 50/50 Polyester/Cotton)

Yarn Property	Brand X	Hokushin	95% Confidence Limit
Single-End Strength, g	290	292	13
Single-End Elongation, %	8.7	8.7	0.8
Uster %CVm	16.7	16.5	0.5
1-Yd %CV	5.3	5.1	0.1
3-Yd %CV	3.7	3.5	0.1
10-Yd %CV	2.1	2.0	0.1
IPI Thins	84	82	31
IPI Thicks	282	276	75
IPI Neps	388	372	70
Classimat Minors	823	441	57
Long Thins	110	56	21

Apron Spring Pressure

On the Murata air jet spinning machine, the pressure exerted by the apron is controlled by a spring in the roll weighting arm. Consequently, the aprons are pressed together, unlike ring spinning where they are held apart. In most cases a spring weighting of 3 kg (6.6 lb) is used; however, for more or less spreading of the bundle, and in turn the presence of more or less binder fibers, a 2-kg (4.4-lb) or 4-kg (8.8-lb) spring can be used. A schematic illustration of the effect of apron spring pressure on yarn structure is shown in Figure 10.24.

Typically, when apron spring pressure is increased, strength and hairiness increase as shown in Table 10.15. However, higher apron spring pressures usually produce a harsher yarn with higher propensity for pilling because of the increase in both hairiness and binder fibers.

Tensor Bar Height

The tensor bar lies between the first and second steel rolls and serves as a guide for the aprons. The height at which it is set controls how the fibers are delivered to the front roll of the spinning system. If the tensor bar height is too low, the fibers will be delivered into the front steel roll instead of to the nip between the top cot and the bottom steel roll. Defects will be higher in number if

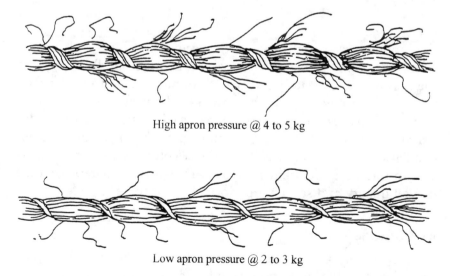

High apron pressure @ 4 to 5 kg

Low apron pressure @ 2 to 3 kg

Figure 10.24. Schematic illustration of the effect of apron spring pressure on MJS yarn structure. (Courtesy Murata Machinery, Ltd.)

Table 10.15. The Influence of Apron Spring Pressure on Yarn Properties (30's Ne 50/50 Polyester/Cotton)

Yarn Property	Apron Spring		95% Confidence Limit
	3.0 kg (6.6 lb)	4.0 kg (8.8 lb)	
Single-End Strength, g	16.1	16.8	1.0
Single-End Elongation, %	8.3	8.8	0.4
Uster %CVm	14.0	13.8	0.3
1-Yard %CV	3.9	3.9	0.3
IPI Thins	8	8	11
IPI Thicks	89	70	36
Classimat Minors	301	222	49
Long Thins	5	4	6

this is the case. To avoid such a situation, Murata recommends that tensor bar height be 2.88 mm (0.113 inch) above the bottom steel roll (the recommendation was originally 2.38 mm (0.094 inch), but this height does not leave enough tolerance at the lower end).

Setting the tensor bar too high also results in excessive defects. As tensor bar height is increased above 2.88 mm, deterioration can be expected in Classimat minors, strength, 1-yard %CV, and long thin places. In a study done at ITT, the fewest defects were actually found at a setting of 2.5 mm, (0.098 inch). On the newest high speed machines, tensor bar settings of 2.88 mm to 3.38 mm are preferred for quality reasons.

Condenser Size

A condenser is used between the second and third rolls on the Murata machine to control the size of the fiber bundle going into the apron as shown in Figure 10.25. In most cases a smaller condenser will provide for higher strength; however, an increase in yarn thick places and poor spinnability will result when the condenser is too tight. Table 10.16 illustrates the effect of different condensers on 36's Ne 50/50 polyester/cotton MJS yarn properties. In this study the tighter condensers significantly improved the overall quality and spinning performance for these yarns.

Nozzle Zone

A schematic of the nozzle zone is shown in Figure 10.26. The critical factors in this section include the following:

1. Nozzle-to-front roll distance,

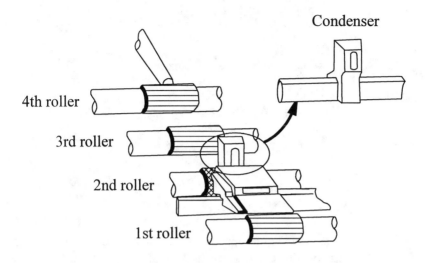

Figure 10.25. Schematic illustration showing the location of the condenser in the MJS four-roller draft zone. (Courtesy Murata Machinery, Ltd.)

Table 10.16. Comparison of 36's Ne 50/50 Polyester/Cotton MJS Yarns Produced Using 3-Millimeter and 4-Millimeter Condensers

Yarn Property	3-mm (0.118-inch) Condenser	4-mm (0.157-inch) Condenser	95% Confidence Limit
Uster %CVm	16.7	17.2	0.3
Uster Thins, (-50%)	86	133	23
Uster Thicks, (+50%)	303	368	55
Uster Neps, (+200%)	495	540	62
Uster Hairiness	4.80	5.12	0.10
Skein Strength, lb	62.2	56.3	3.9
Single-End Strength, g	209.6	194.9	5.6
Single-End Elongation, %	7.51	7.65	0.28
Machine Efficiency, %	92	84	N/A

2. Nozzle pressures,
3. Feed ratio (spinning tension),
4. Nozzle type,
5. Nozzle wear, and
6. Nozzle cleaning.

These factors are discussed in the following paragraphs.

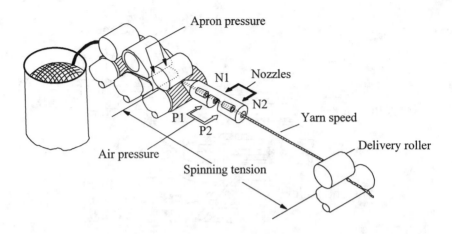

Figure 10.26. Schematic illustration of the nozzle zone of the Murata 801, 802, and 802H spinning machines. (Courtesy Murata Machinery, Ltd.)

Nozzle-to-Front Roll Distance

The nozzle-to-front roll distance has an effect on yarn evenness, IPI defects, and running performance. As the yarn count gets finer, tighter settings can be used. Usually, for yarns finer than 30's, the N1-to-front roll setting, which is actually based on the distance from the bar holding the nozzle to the first roll, is between 38.5 and 39.5 mm (1.52 and 1.56 inches). The closer the setting, the better the yarn evenness. The setting must be wider as the count gets coarser to avoid chokes in the nozzle. As a result for a 15's yarn, a setting of 41 mm (1.61 inches) is usually standard. The effect of nozzle-to-front roll distance on yarn evenness and defects is shown in Table 10.17.

Nozzle Pressures

Nozzle pressure is the setting that is most unique from plant to plant. As mentioned in the initial part of this chapter, the nozzle pressures are critical for formation of the binder fibers. The N2 pressure, which is highest of the two nozzle pressures, provides false twist back to the front roll. This pressure affects how tightly the binder fibers are wrapped around the yarn. As more air pressure is applied to the N2 nozzle, the strand becomes smaller because the increased false twist provides more twist contraction. The binder fibers that break away from the strand, as shown in

Table 10.17. The Influence of Nozzle-to-Front Roll Spacing on Characteristics of 35's Ne Polyester/Cotton Yarns

Yarn Properties	% Polyester in Blends and Nozzle-to-Front Roll Space, mm				95% Confidence Limit
	100% 39 mm	100% 41.5 mm	50% 39 mm	50% 41.5 mm	
Yarn Count, Ne	35.4	35.6	33.7	33.7	1.3>0.6
%Vb	0.5	1.3	0.6	0.9	0.9>0.5
Single-End Strength, g	412	395	259	256	16
%Vo	11.8	17.3	9.6	10.8	All
Single-End Elongation, %	11.8	11.4	8.0	8.0	0.4
Uster Evenness, %CVm	12.65	12.90	15.27	15.44	0.27
%Vb	1.7	2.5	1.7	1.9	No
IPI Thins	9	16	26	36	8
Thicks	13	14	160	172	12
Neps	16	16	300	351	110
Classimat Minors	177	208	1542	1898	115
Majors	1	0	5	2	No
Long Thins	30	17	93	115	29
Hairs/meter (>3 mm)	6.7	6.2	11.5	9.5	2.7

Note: To convert from mm to inches, mm/25.4.

Figure 10.24, wrap around the twisted core. As the core twist is removed (upon the exit of the yarn from the N2 nozzle), the binder fibers tighten. With higher initial twist in the strand (from high N2 pressure), the bundle will expand more, thus causing more tightening of the binder fibers.

The N1 nozzle pressure influences the amount of binder fibers that exist. As the N1 air pressure increases (with all else held constant), the percentage of binder fibers will increase. With this increase in binder fibers, yarn strength will often increase, but the yarn will become more harsh. As a result, knitting yarns are often produced with an N1 pressure of 2.5 kg/cm^2 or less, while weaving yarns are produced with at least 3.0 kg/cm^2.

A schematic illustration of the influence of N1 nozzle pressure on yarn structure is shown in Figure 10.27. It is worth noting here that increasing N1 pressure does not always increase strength because the tensile properties of MJS yarns are most largely influenced by fiber-to-fiber friction of the core fibers. Thus, if too many binder fibers are present (from high N1 pressure), not enough fibers are left in the core to carry the load. As a result, N1 pressure that is set too low or too high will be detri-

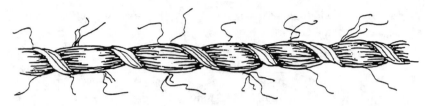

High N1 air pressure @ 3.5 kgf/cm^2

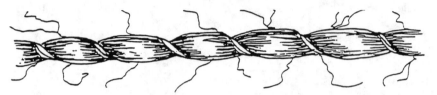

Low N1 air pressure @ 2.5 kgf/cm^2

Figure 10.27. Schematic illustration of the effect of N1 nozzle air pressure on MJS yarn structure. (Courtesy Murata Machinery, Ltd.)

mental to yarn strength. If nozzle pressures are used to improve strength, the N2 pressure is usually increased to tighten the binder fibers, instead of increasing the number of binder fibers with the N1 nozzle in an attempt to improve fiber-to-fiber friction in the bundle.

As shown in Figure 10.28, an increase in N1 pressure did not provide for strength improvement in 26's Ne 50/50 polyester/cotton yarns, likely because too many binder fibers were generated. However, with a 4.5 kg/cm^2 N2 pressure, higher N1 pressure resulted in increased yarn strength.

The optimum air pressures on the 802 machines, which tend to incorporate larger N2 nozzles (type H26), are usually different from those on the 802H machine, which utilizes high speed (type H3) nozzles. Typical nozzle pressures on these two models are listed in Table 10.18. On the 802H machine, the difference between the optimum N1 and N2 nozzle pressures tends to be greater than it is on the 802 machine, with a higher N2 pressure necessary for the higher speed spinning.

Regardless of the machine being used, an optimization trial is necessary when a new product is being produced on any MJS machine. In most cases the experimental setup would be similar to that shown for the 802H machine in Table 10.19.

Yarn Single-End Strength, grams

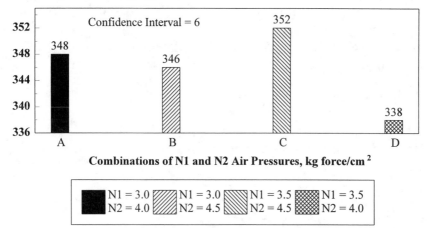

Figure 10.28. Graphical illustration of the influence of N1 and N2 air pressure on the strength of 26's (Ne) 50/50 polyester/cotton yarns produced on an MJS 802 machine.

Table 10.18. Typical Nozzle Pressures on 802 versus 802H Machines

Machine Type	N1 Pressure (kg/cm²)	N2 Pressure (kg/cm²)
802	3.0 - 4.0	3.5 - 4.5
802H	1.5 - 3.0	4.5 - 5.5

The nozzle combinations marked with the "x" would be spun and tested for strength, strength variation, and elongation. If the product is for knitting, hairiness would also be evaluated. Based on which section of the box provides the best tensile properties, a fine-tuning test would be done around those nozzle pressures to determine the optimal settings.

In a nozzle pressure analysis done by Anderson, Uster evenness and IPI defects deteriorated with higher N1 pressures and lower

Table 10.19. Experimental Approach for Determining Optimum Air Pressures for 802H MJS Spinning

N2 Pressure (kg/cm²)	N1 Pressure (kg/cm²)		
	1.5	2.0	2.5
4.0	x		x
4.5		x	
5.0	x		x

Note: "X" indicates initial yarn conditions to produce and test for quality.

N2 pressures as shown in Figure 10.29 (Anderson, 1995). However, the nozzle combination that resulted in the best evenness often contributed to lower strength. As shown in Figure 10.30, strength generally improved with higher N1 pressure on the 802H machine. In many cases compromises must be made in one

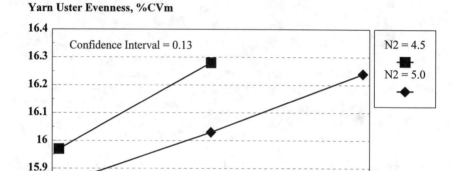

Figure 10.29. Graphical illustration of the influence of N1 and N2 air pressure on the evenness of 35's (Ne) 50/50 polyester/cotton yarns produced on an MJS 802H machine. (Anderson, 1995)

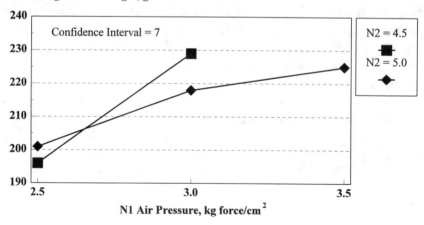

Figure 10.30. Graphical illustration of the influence of N1 and N2 air pressure on the strength of 35's (Ne) 50/50 polyester/cotton yarns produced on an MJS 802H machine. (Anderson, 1995)

High feed ratio @ 0.98 to 0.99

Low feed ratio @ 0.96 to 0.97

Figure 10.31. Schematic illustration of the effect of feed ratio on MJS yarn structure. (Courtesy Murata Machinery, Ltd.)

property to improve another property with the air jet spinning system.

Spinning Tension (Feed Ratio)

On the Murata machine, spinning tension is influenced by a setting called feed ratio that controls the speed ratio between the front roll and the delivery roll. The feed ratio influences how taut the yarn is in the nozzle, causing yarns produced from low feed ratios to have more binder fibers and a more wavy structure as shown in Figure 10.31.

Selecting the optimal feed ratio can be difficult because of trade-offs involved. In most cases feed ratio is set at 0.98 on the 802H machine. Decreasing the feed ratio is associated with the following:

1. Decreased yarn strength (Figure 10.32);
2. Increased yarn thick places (Figure 10.33);
3. Decreased yarn shrinkage (Figure 10.34);
4. Decreased fabric cover (Figure 10.35);
5. Decreased yarn long thin places; and
6. Increased yarn hairiness.

Nozzle Selection

The nozzle assembly in the Murata MJS spinning machine consists of four major components: The N1 and N2 nozzles, the N1 nozzle guide, and the N2 nozzle guide. There are also five O-rings and a cap for the N2 nozzle within the assembly that ensure a

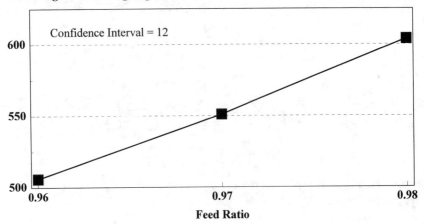

Figure 10.32. Graphical illustration of the influence of feed ratio on the strength of 17's (Ne) acrylic MJS yarns.

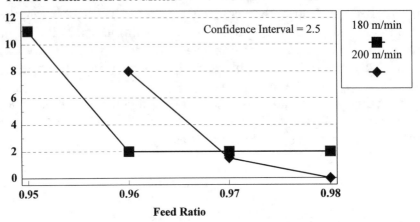

Figure 10.33. Graphical illustration of the influence of feed ratio on the number of Uster IPI Thicks in 17's (Ne) acrylic MJS yarns.

proper fit of components and resistance to air leakages. The nozzle assembly is shown in Figure 10.36.

Currently in the model 802 frame, the combination of the "coarse" N1 and the H26 N2 nozzle is used, and on the 802H machine, smaller nozzles are used which are referred to as H3

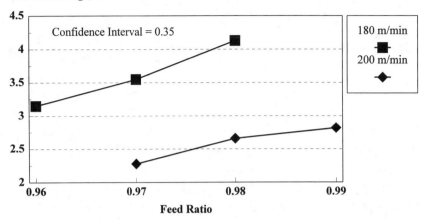

Figure 10.34. Graphical illustration of the influence of feed ratio on the shrinkage of 17's (Ne) acrylic MJS yarns.

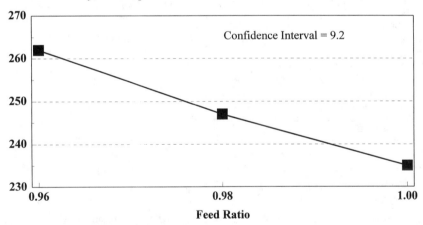

Figure 10.35. Graphical illustration of the influence of feed ratio on the air permeability of woven fabrics produced with 35's 65/35 polyester/combed cotton MJS yarns in the filling direction.

nozzles. However, in research conducted at ITT, a potential for improvement was shown on 802 machines when the H3 N2 nozzle was used instead of the H26 nozzle (Anderson, 1995). A comparison of resultant yarn properties with the standard H26/H26

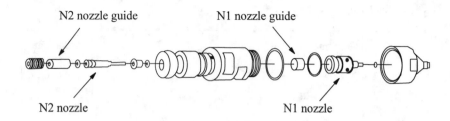

N2 nozzle guide   N1 nozzle guide

N2 nozzle   N1 nozzle

Figure 10.36. Schematic illustration of the nozzle components for the MJS machine. (Courtesy Murata Machinery, Ltd.)

Table 10.20. A Comparison of 35's Ne Polyester/Cotton Yarn Properties Produced with Different N2 Nozzles on an 802 Machine (Anderson, 1995)

Yarn Property	Nozzle Combination (N1/N2)		95% Confidence Limit
	H26/H26	H26/H3	
Single-End Strength, g	208	231	7
Uster Evenness, %CVm	15.79	15.62	0.15
1-Yard %CV	5.28	5.23	0.20
IPI Thins	58	45	8
IPI Thicks	144	129	20
IPI Neps	146	201	25
Air Permeability, ft^3/ft^2/min	235	223	10
Classimat Long Thins	294	101	42
Classimat Minors	352	215	48

* Note: Lower Air Permeability indicates more cover.

nozzle combination versus a H26/H3 nozzle combination is shown in Table 10.20.

The nozzle guides control the ballooning of the yarn as it passes through the nozzle assembly. By altering the shape of the nozzle guides, yarn characteristics such as hairiness and bulk can be influenced. Some of the common N1 and N2 nozzle guides are shown in Figure 10.37.

The six-pointed star N1 and N2 designs were specifically developed to provide a hairier yarn for knitting applications. It was confirmed by Anderson that the 2.0-mm star shaped N2 guide did improve the cover of yarns produced on the 802 machine by approximately 8 percent, but the cover of the yarns produced with this guide on the 802H machine was 10 percent lower. A similar

H26 N1 Guide   H3 N1 Guide   B-Type N1 Guide

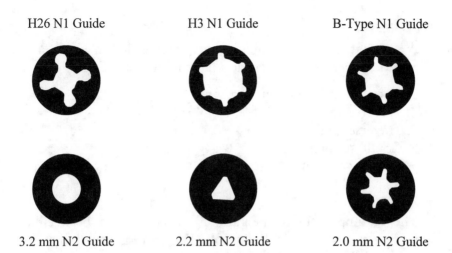

3.2 mm N2 Guide   2.2 mm N2 Guide   2.0 mm N2 Guide

Figure 10.37. Cross-sectional shapes of common N1 and N2 nozzle guides for MJS machines.

trend held true for the "B" type N1 guide; it improved cover of yarns spun on the 802 machine, but made yarn cover worse when spun on the 802H machine (Anderson, 1995).

Nozzle Wear

Within the nozzle assembly, O-rings deteriorate over time, and dirt clogs the holes in the nozzles. Consequently, the nozzles must be periodically cleaned. The recommended frequency for cleaning nozzles and replacing O-rings is every two years, but many plants service the nozzles annually. In a study done within an ITT member plant, changing the nozzle components (O-rings, nozzles, and nozzle guides) after 18 months contributed to a significant improvement in the following yarn properties:

1. Single-end strength,
2. Minimum strength,
3. Single-end elongation,
4. Elongation variation (position to position), and
5. Hairiness.

It was determined that worn nozzles provided less air flow than the new nozzles, which is likely the cause of yarn strength and elongation loss with the old nozzles.

Winding Zone

A schematic of the winding zone was provided in Figure 10.7. The critical factors in this section include the following:

1. Take-up ratio (winding tension),
2. Traverse speed,
3. Traverse guide setting,
4. Cradle settings, and
5. Delivery cot condition.

Take-Up Ratio

The take-up ratio is the ratio in surface speed between the winding drum and the delivery roll. This ratio is usually between 0.98 and 1.00. The winding drum must run at the same speed or slightly slower than the delivery roll to avoid overstretching of the yarn. An overfeed can physically occur because of the back and forth traverse of the package.

Take-up ratio has the strongest influence on package density and elongation. If the take-up ratio is too low, less yarn can be put on a package which increases the cost of spinning (more doffing) and shipping (less weight per case). In addition, soft packages tend to have more unwinding problems. However, as usual, a trade-off exists. Lowering take-up ratio improves yarn elongation significantly as shown in Figure 10.38. As a result, a compromise usually must be made between optimal package density and optimal elongation.

Traverse Speed

The traverse speed and spinning speed on the Murata MJS machine affect the angle of wind on the package. As spinning speed increases, traverse speed must be increased to maintain the same wind angle, and thus consistent package density. In most cases a wind angle between 10 and 12 degrees is used; however, the optimal angle often depends on the yarn count and material being processed. In a study done at ITT, traverse speed was found to have an effect on strength, elongation, and elongation variation. It was determined that setting traverse speed too low (on an 802H machine) was more detrimental to the tensile characteristics of the yarn than setting it too high.

Yarn Single-End Elongation, %

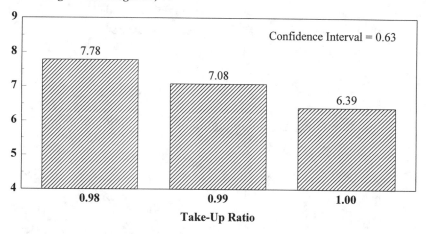

Figure 10.38. Graphical illustration of the influence of take-up ratio on the elongation of 35's (Ne) 50/50 polyester/cotton yarns produced at 240 m/min on an MJS 802H machine.

Traverse Guide

The traverse guide, as shown in Figure 10.39, moves the yarn from side to side on the package. The traverse guide should be set 4 mm (0.157 inch) from the winding drum. A poorly set traverse guide will produce packages with overthrows and/or ridges that often cause stops at the next process. In addition, improperly set traverse guides cause tension variation.

Cradle Settings

Critical settings of the cradle include the pressure of the cradle to the drum and the alignment of the cradle with the drum. The pressure of the cradle is checked with a spring gauge and should be 3 kg ± 0.3 (6.6 lb ± 0.7). Too much variation in cradle pressure can cause variation in yarn elongation as well as knotting problems.

The alignment of the cradle to the drum is checked with a new package. When the package is against the drum, it should not be cocked. If one side is not touching the drum, the potential for high winding tension exists. Such a condition will cause yarn elongation to deteriorate. In a recent study done by ITT in a member plant, positions that had misaligned cradles had winding tension twice as high as the norm for the machine, resulting in a 20 percent loss in yarn elongation.

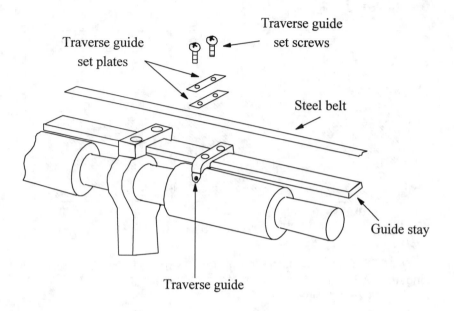

Figure 10.39. Schematic illustration of the traverse guide setting mechanism. (Courtesy Murata Machinery, Ltd.)

Delivery Cot Condition

A predominant cause of high position-to-position elongation variation and/or select positions that do not have success with knotter attempts is worn delivery cots. The delivery cots get grooved with wear and cause the yarn to track in one place instead of traversing across the cot. Such a situation causes the tension to be highly variable. Delivery cots can be checked with a Shore Durometer to determine how much hardness has increased since they were new. In addition, feeling the cots on a regular basis to ensure grooves do not exist is highly recommended.

Influence of Spinning Speed on Yarn Properties

Unlike the other spinning systems, increasing spinning speed on the air jet system does not guarantee loss of quality. In fact softness of the yarn is virtually always improved when spinning speed is increased.

As shown in Table 10.21 in a recent ITT study, most yarn prop-

Table 10.21. The Influence of MJS 802H Spinning Speed on Yarn Properties (35's Ne 50/50 Polyester/Cotton)

| Yarn Property | Spinning Speed, m/min | | Significant Difference |
	240	260	
Single-End Strength, g	241	247	Yes
%Vo	10.1	10.0	No
Single End Elongation, %	7.3	7.9	Yes
% Vo	14.0	12.5	Yes
Uster Evenness, %CVm	16.0	16.0	No
IPI Thins	52	52	No
IPI Thicks	195	213	Yes
IPI Neps	272	303	Yes
Zweigle Hairs/m (>2 mm)	2150	3730	Yes

erties were the same or better when the speed was increased from 240 to 260 m/min (262 to 284 yd/min) on an 802H machine. It is believed that improvements in yarn tensile properties occurred with higher spinning speed in this trial because of the reduction of spinning tension that took place.

In most cases when spinning speed is increased, the following yarn properties are expected to deteriorate:

1. IPI thicks and neps,
2. Classimat minor defects, and
3. Classimat long thin defects.

In production, spinning speeds of up to 300 m/min (328 yd/min) on 802H machines have been used. For weaving yarns, speeds between 240 and 260 m/min (262 and 284 yd/min) are used on 802H machines to control hairiness and long thin places, and knitting yarns are spun between 260 and 300 m/minute. On 802 model machines, most yarns are spun between 190 and 210 m/min, while some plants have increased speeds to 230 m/min (252 yd/min) with application of the H3 nozzles.

Automatic Knotting/Splicing

The Murata air jet spinning machines are equipped with either automatic knotters or splicers to initiate the spinning process after a yarn stop. The knotter offers a stronger joint which is preferred by the majority of weavers using air jet yarns, whereas a less visibly detectable joint exists when the yarns are air spliced. Some limitations have existed with the performance of the splicers (success efficiency and splice strength) as speeds have increased above 260 m/minute.

The Murata knotter or splicer, shown in Figure 10.40, utilizes a suction arm to catch yarn as it comes out of the nozzle. At the same time a reversing roll turns the package while a suction mouth catches the end of the yarn on the package. These components then bring the two ends of the yarn to the knotting or splicing unit, wherein they are joined. The yarn is then released, and spinning begins while excess slack in the yarn is taken up in the slack tube until enough tension exists to remove the yarn.

It is important that the timing is set appropriately on a knotter used for joining yarn ends. If too little slack is provided during knotting, the tails of the knot will be too short, and the knots will fail downstream. However, if too much slack is provided, the yarn will stay in the slack tube too long and kink. This yarn kinking causes unwinding problems at later processes, especially warping. One test to judge the timing effectiveness is to measure the time in the slack tube after knotting. On a new package it is expected that the yarn should be in the slack tube 5 to 8 seconds. On a full package the yarn should not be in the slack tube over 12 seconds. Positions with yarn in the slack tube only 2 to 3 seconds are potentially problem positions from a slip knot perspective, and those positions with yarn in the slack tube over 12 seconds will likely cause kinking in the yarn. Routine checks of time in the slack tube are most helpful for checking knotting performance.

Knotter Performance

The performance of the knotter (or splicer) is critical to efficiency of the machine. It is typically expected that the knotter success rate be over 90 percent, and if it is not, the operator's job becomes more difficult because of the increased number of positions to service. In a normal situation, positions stopped because of slubs (short thick places) will be restarted without operator intervention. However, if the knotter fails to restart the position, the position is flagged for the operator.

Unsuccessful knotting can be caused by position-to-position problems or knotting unit problems. The best method of trou-

Figure 10.40. Photograph of the MJS 802H splicer. (Courtesy Murata Machin

bleshooting is to compare the misknot percentage of the machine to the individual positions using the Intelligence Analyzer (computer) on the machine. If the knotter is the problem, the position-to-position misknot rate will be higher than that on other machines. However, if most positions have misknot rates comparable to those throughout the plant, and only a few positions are rogue, the problem is likely at those positions.

Some positional problems that cause a high misknot level include the following:

1. Poor cradle alignment with the take-up drum, causing tension to be very high;
2. Excessive wear on the take-up drum;
3. Excessive wear on the rubber friction disks on the take-up drum;
4. Worn delivery cots causing winding tension to be too low; or
5. Insufficient suction in the slack tube.

When the knotting performance loss appears to occur across the entire machine, it could be influenced by the following:

1. Worn cams or cam followers driving the suction arm, suction mouth, or knotting heads in the knotter;
2. Incorrect timing of the knotter;
3. Poor alignment of the suction arm with the N2 nozzle and/or knotter head;
4. Poor alignment of the suction mouth with the packages; or
5. Incorrect machine settings (traverse speed, take-up ratio, feed ratio, nozzle pressures, or total draft).

Benchmarks for Air Jet Yarn Performance and Quality

It is expected that MJS frames run with the following performance targets:

1. Above 95 percent spinning efficiency,
2. Less than 20 slub cuts per machine hour,
3. Less than 6 red flags per machine hour, and
4. Less than 10 percent misknots.

A slub cut is defined as a stop caused by the clearing head. This type of stop does not require operator intervention to get the end knotted together again. The majority of these stops can be traced back to the fiber characteristics or sliver quality. The level of slub cuts can also be influenced by the sensitivity set to cut. In most cases the sensitivity setting for a slub is 140 percent of the normal yarn thickness for yarn lengths of 2 cm (0.079 inch) or longer.

A red flag is a stop caused by missing sliver, a nozzle clog, a tension break, or an off-quality condition recognized by the quality monitor. These stops require operator attention to get the position running again.

A misknot is a situation where the knotter attempted to restart a position and failed. High misknots could be caused by a poorly set knotter or a poorly set winding drum at a spinning position. Misknots have a significant influence on machine efficiency.

From a yarn quality standpoint, Table 10.22 contains targets for several common MJS yarn counts.

Table 10.22. Yarn Quality Benchmarks for Various MJS Yarns

Yarn Property	18's 65/35 Polyester/Cotton	26's 50/50 Polyester/Cotton	35's 50/50 Polyester/Cotton
Yarn Count, %Vb	1.2	0.7	0.7
Skein Strength, lb	170	100	72
Single-End Strength, g	615	375	275
%Vo	8.5	9.0	9.5
Single-End Elongation, %	11.4	9.2	8.4
%Vb	3.5	3.5	4.5
Uster Evenness, %CVm	13.5	15.4	16.0
%Vb	2.8	1.6	1.3
1-yd %CV	4.7	4.9	5.1
3-yd %CV	3.3	3.0	3.0
10-yd %CV	2.2	1.8	1.8
IPI Thins (-50%)	3	20	50
IPI Thicks (+50%)	75	150	220
IPI Neps (+200%)	80	175	300
Classimat Minors	125	225	450
Majors	0	0	0
Long Thins	20	50	150

Chapter 11

Other Spinning Systems

During the twenty years of the 1970s and 1980s, there were a number of new and different short staple spinning systems developed, some which became commercially viable and some which did not prove to be economically or product acceptable to the industry. In this chapter those systems that were the most prominent are discussed briefly. They are as follows:

1. Wrap spinning,
2. Friction spinning,
3. Integrated composite system (ICS)-Bobtex, and
4. Ply spinning.

All of these systems offer a similar feature; that is, their productive capacity per each spinning delivery is substantially higher than the capacity of the ring spinning system. Nominal delivery rates for some of these machines and a nominal rate for ring spinning are listed in Table 11.1.

The machines (and yarns) as listed in Table 11.1 are described in the following sections of this chapter.

Coverspun (Wrap Yarn)

The Coverspun yarn is basically a bi-component yarn made of staple fibers as a core and a filament continuous yarn as a wrap binder on the outside of the parallel staple fibers. The standard Coverspun yarn configuration is represented by the diagrams in Figure 11.1.

The process of producing the Coverspun wrap yarn is illustrated by the drawing shown in Figure 11.2. The input fiber for the ma-

Table 11.1 Delivery Rates of Certain Short Staple Spinning Systems Compared to the Ring System

Spinning System Type	Nominal Delivery Rate, Meters/Minute (m/min)
Ring	16
Wrap Spinning:	
Leesona Coverspun (USA)	44
Suessen Parafil (Germany)	150
Friction Spinning:	
Hollingsworth Masterspinner (England)	300
Dref 3 (Austria)	300
ICS-Bobtex (Canada)	600
Ply Spinning:	
Murata Twin Spinner (Japan)	200
Suessen Plyfil (Germany)	250

chine is conventional short staple roving, similar to that fed into a ring spinning machine. The roving is drafted in a standard SKF drafting unit, again similar to that used on a standard ring spinning machine. The yarn forming mechanism following the drafting unit is very different from that on the ring spinning system.

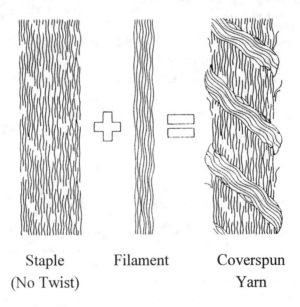

Staple   Filament   Coverspun
(No Twist)   Yarn

Figure 11.1. Diagram showing the configuration of Coverspun wrap yarn. (Courtesy Leesona Corp.)

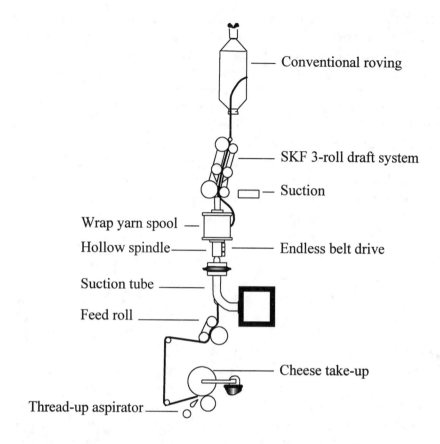

Figure 11.2. Illustration of the Coverspun process. (Courtesy Leesona Corp.)

The fibers flow from the nip of the front roll into a hollow spindle on which is mounted a spool containing the filament strand to be wrapped around the drafted parallel bundle of fibers, thus forming the final yarn structure. The hollow spindle is positioned so that the top of the spindle is very close to the front roll nip, allowing the filament strand from the rapidly rotating spool to wrap around the fiber bundle as it emerges from the drafting rolls. The wrapped yarn continues down through the hollow spindle to emerge from the suction tube to proceed to the feed roll and then to the package take-up.

Production per delivery is limited by the maximum allowable speed of the spindle-filament spool system that wraps the filament strand around the fiber strand. The maximum spindle speed on the Coverspun machine is near 30,000 revolutions per minute, which allows a delivery rate somewhat higher than that of the ring frame. Filament wraps per inch can be slightly lower than twist per inch on comparable ring yarn. One interesting feature of this yarn is its extremely low torque such that the yarn is free of liveliness. Such liveliness often causes fabrication difficulties with the use of ring yarn. Count range is from 5's to 100's Ne (118 to 6 tex).

The yarn has been used in both jersey and warp knits, double knits, light weight woven fabrics, and some hosiery products. This particular machinery had a measure of success for a few years, but now seems to have faded from the textile scene.

Parafil (Wrap Yarn)

The Parafil Model 1000 yarn wrap spinning machine, designed and built by Spindelfabrik Suessen of Suessen, Germany, is for use with short staple fibers up to 60 mm (2.36 inches) in length. The yarn produced by this machine is known by the term "parallel yarn." As stated in the previous section, the fibers are not twisted as in ring and open-end yarn, but remain parallel and are held compactly together by a continuous filament strand which is wrapped in the form of a helix around the parallel core fibers. The wrapping filament stabilizes the fiber bundle and imparts the necessary friction to the staple fibers by its radial pressure, thus giving the yarn its strength.

A photograph of the Parafil 1000 machine is shown in Figure 11.3, showing the sliver input at the top, the roller drafting units just above the spindle zone which is covered, and the delivery rolls and winding unit in the lower section of the photograph. Yarn packages produced are 250 mm (10 inches) in diameter and weigh near 1.7 kg (3.7 lb).

Maximum filament package speed is 35,000 revolutions per minute, allowing delivery rates up to 150 meters per minute for coarse yarn counts. The drafting system is a 4-roll unit with spring weighted synthetic rubber covered top rolls, allowing drafts up to 185. Each spindle of the machine can be stopped and started independently. When an end break occurs during spinning, the sliver is

Figure 11.3. The Suessen Parafil 1000 spinning machine. (Courtesy Spindelfabrik Suessen)

automatically clamped at the back roller pair so no fibers are lost in the suction system.

The spindle containing the filament yarn package is not mounted close to the discharge point of the drafting rollers, as is the case with the Coverspun machine. The exiting bundle of fibers is given a "false" twist by its unique gripping by a mechanism associated with the spindle. This false twist disappears as the yarn is nipped at the gripping point, but it prevents breaks in the region before the filament wrap occurs. The output of a spinning position is five times greater, or more, as compared to a conventional ring spindle.

Doffing can be done with the machine running by the use of a suction gun with integrated cutter. The tubes for the filament packages are wound on the package winder which is an integral part of the machine. Sewing thread and home textiles are markets now using the Parafil yarns.

Hollingsworth Masterspinner (Friction)

In the Hollingsworth Masterspinner machine, the bundle of fibers that is formed into a continuous yarn is twisted by frictional contact with two precision parallel cylindrical rolls that turn in the same direction, and that are almost touching. The fibers are laid by means of directional air flow into the crotch formed at the line where the rolls are very close to each other. One roll is perforated and the other is ceramic coated. Air flow into the perforated cylinder holds the fibers into position to be twisted. The ratio of the diameters of the twisting cylinders (they have equal diameters) and the fiber bundle (yarn) diameter produces a large mechanical advantage between the twisting elements and the yarn. This allows high twisting capability at relative low speed of the twisting mechanism, which on the Masterspinner has a range from 2,200 revolutions per minute to a maximum speed of 10,000 revolutions per

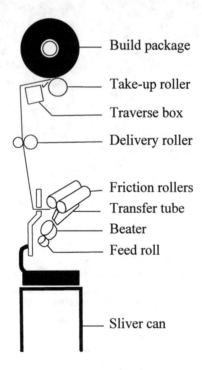

Build package

Take-up roller

Traverse box

Delivery roller

Friction rollers

Transfer tube

Beater

Feed roll

Sliver can

Figure 11.4. Schematic of a friction spinner threadpath. (Courtesy John D. Hollingsworth on Wheels, Inc.)

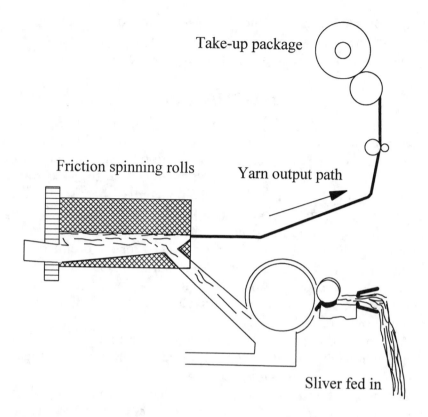

Take-up package

Friction spinning rolls

Yarn output path

Sliver fed in

Figure 11.5. Schematic drawing showing the general concept of the Master-spinner for friction spinning. (Courtesy John D. Hollingsworth on Wheels, Inc.)

minute. Schematic diagrams of the machine thread path are shown in Figures 11.4 and 11.5.

The machine spins 100 percent cotton, 100 percent man-made fibers, or blends of cotton with man-made fibers in the count range from 59 to 15 tex (10's to 40's Ne), with staple lengths up to 40 mm (1.57 inches) at delivery speeds up to 300 meters per minute.

A number of multi-machine installations were operated in different parts of the world with several in the USA. End uses for the yarns produced were primarily in knit fabrics. This experience revealed that twist in the yarns produced on these machines was inconsistent, which caused serious difficulty in fabrication. The machines have been set aside for further development until a solution to this problem can be found. However, many knowledgeable textile machine analysts believe that friction twisting similar to that employed in the Masterspinner has great potential.

DREF 3 (Friction)

In the early 1980's the friction spinning machine DREF 3 was developed and offered to the market by Dr. Ernst Fehrer, founder of Fehrer AG of Linz, Austria. The friction twisting principle in this machine was already proven commercially in the long staple spinner DREF 2. The DREF 3 machine handles short staple fiber in the range from 30 to 50 mm (1.18 to 1.97 inches) for the sheath fibers, and up to 60 mm (2.36 inches) for the fibers in the yarn core. The DREF 3 machine is arranged for the production of component yarns in the count range from 33 to 650 tex (18's to 0.9's Ne). A view of the machine with 12 deliveries is shown in Figure 11.6, and a schematic diagram of the spinning unit is shown in Figure 11.7.

As shown in Figure 11.7, there are two drafting units; drafting Unit I is for the core fibers, and drafting Unit II is for the sheath fibers (covering fibers) which cover the core fibers in the nip between two perforated spinning drums. Sliver fed to drafting Unit I produces the core yarn, which consists of parallel and aligned fibers. These fibers pass through the twisting unit, consisting of the two perforated drums and their suction inserts. Separated fibers flowing freely from drafting Unit II cover the core in the nip of the two twisting drums and produce the finished yarn by locking in the twist. This direct covering of the core in the twisting (spinning) unit results in a regular and positive combination of the two fibrous components.

In drafting Unit I, a sliver with a staple length of up to 60 mm (2.36 inches) and up to 3.5 kilotex (50 grains per yard) weight is drafted by a factor of 100 to 150. The structure may be further complemented by the inclusion of a filament core. The usual percentages of the yarn structure for each component are 20 to 50 percent sheath and 80 to 50 percent core, the latter being the major contributor to yarn strength.

As pointed out in Table 11.1, the delivery rate of yarn from this spinning unit is up to 300 meters per minute. Yarns produced on

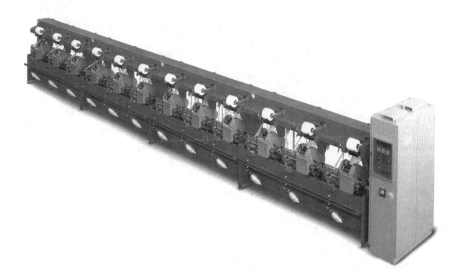

Figure 11.6. View of the DREF 3 friction spinning machine. (Courtesy Feher AG)

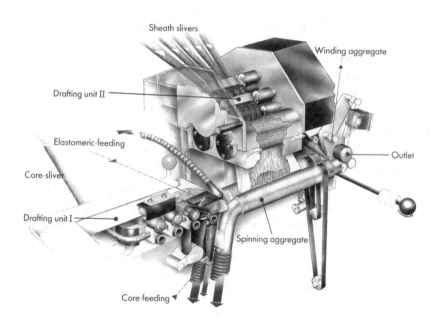

Figure 11.7. Illustration of the DREF 3 friction spinning unit. (Courtesy Feher AG)

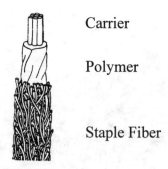

Carrier

Polymer

Staple Fiber

Figure 11.8. Model of the three components of the ICS yarn. (Courtesy Bobtex Corp., Ltd.)

the DREF 3 spinning machine are used in household textile products, leisure garments, textiles for outdoor use, and industrial fabrics.

The Bobtex Composite Spinning System

The Bobtex Mark I ICS (integrated composite spinning) system incorporates the means to combine continuous filament properties with polymer binder and stable fiber outer texture into a unique type of composite spun yarn, illustrated by the drawing in Figure 11.8. A diagram showing the main machine elements is depicted in Figure 11.9.

As illustrated in Figure 11.9, a continuous filament "carrier" is run through the die of an extruder which coats the filament with the adhesive polymer. While the polymer is still tacky, staple fibers from sliver are fed onto the strand as the covering sheath. The exiting yarn is wound onto a large cheese package. The covering texture fibers can be cotton, wool, man-mades, or other fibers in lengths up to 64 mm (2.5 inches). The bonding polymer is a thermoplastic resin such as an olefin, a polyamide, or a polyester.

Normal count range for DREF 3 is 30 to 300 tex (20's to 2's Ne). Output delivery rate is 600 meters per minute; production rate range is from 2.2 to 16.0 kilograms per hour (4.8 to 36.0 pounds per hour) depending on the yarn count. Applications for the composite yarns have been upholstery and drapery fabrics.

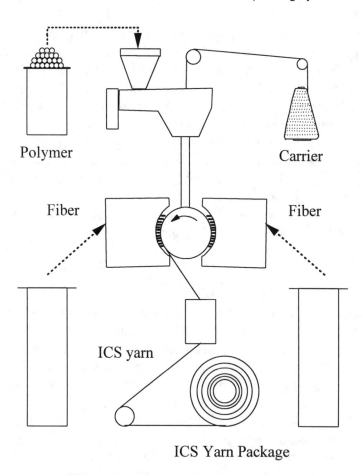

Polymer

Carrier

Fiber

Fiber

ICS yarn

ICS Yarn Package

Figure 11.9. Diagram of the ICS composite yarn production system. (Courtesy Bobtex Corp., Ltd.)

The flexibility of the ICS machine as to choice of the construction materials offers great opportunities for product applications.

Murata Twin Spinner (Ply Spinning)

The Murata 881 MTS spinning machine was introduced to the textile industry in 1989 by Murata Machinery, Ltd. of Kyoto, Japan. This design was a modification of Murata's earlier model MJS machines, which spin singles short staple yarns. The model

881 MTS is equipped to spin two separate yarns onto the same yarn package, ready to be ply twisted on a two-for-one twister; thus the name "Twin Spinner." This system of producing two-ply yarns offers substantial cost advantages compared to the traditional system of plied ring yarns. Another advantage is the lower twist required for plying as compared to ring yarns.

A diagram illustrating the 3-roll draft system, the twin nozzle, and the method of separating the two ends in the draft zone is shown in Figure 11.10. Delivery speed is normally 200 meters per minute. Suitable products for the Twin Spinner yarns are sewing thread, work clothes, outerwear, coat fabrics, shirting, knit fabrics, toweling, upholstery, and draperies.

Suessen Plyfil (Ply Spinning)

Another plied yarn spinning system that was developed to facilitate two-ply yarn manufacturing is the Suessen Plyfil system. The first time the Plyfil spinning machine was shown to the textile public was at the ITMA trade fair in Paris in October of 1987.

The Suessen Plyfil technology is suitable for both short and long staple fibers. The Plyfil 1000 machine, is designed for cotton, manmade fibers, and blends with fibers up to 60 mm (2.36 inches) in length with the nipping point method of drafting. Longer fibers are processed on the Plyfil Model 2000 machine.

The roving stage is eliminated in the Plyfil process because of the 5-roller drafting system with two pairs of control aprons; one pair in the pre-draft zone and one pair in the main draft zone. In the Plyfil Model 1000 machine this draft system is capable of drafts up to 350. The drafted fibers pass through air jets where they are consolidated and formed into a yarn by the wrapping of the fiber ends around the yarn core in one direction only. Two of these yarns are then brought together, and the combined two yarns are wound onto one cylindrical take-up package at delivery speeds up to 250 meters per minute. A simplified diagram showing the drafting and take-up mechanisms is provided in Figure 11.11.

The fibers obtain their coherence from the air jets with just enough stability as is necessary for the subsequent twisting operation. A pronounced wrapping is avoided deliberately. The wrapper

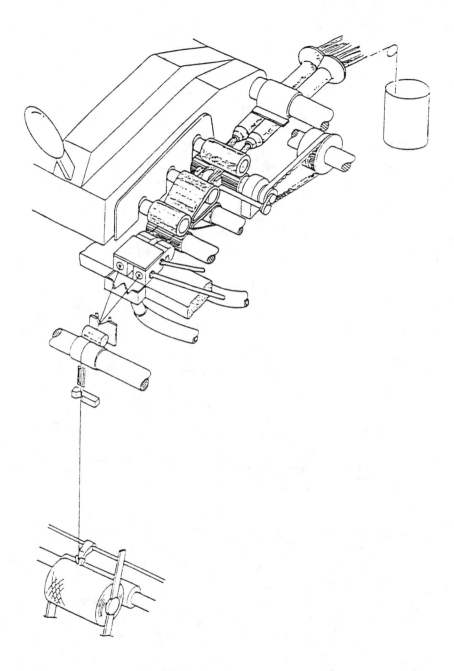

Figure 11.10. Diagram of the drafting and winding arrangement of the 881 MTS machine. (Courtesy Murata Machinery, Ltd.)

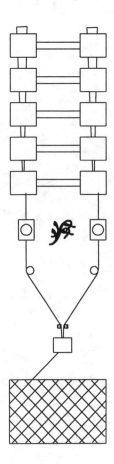

Figure 11.11. Simplified diagram of the Plyfil Model 1000 drafting and winding mechanisms. (Courtesy Spindelfabrik Suessen)

fibers release their grip on the yarn cores during the application of the ply twist which takes place on a two-for-one twister. This occurs because the twisting of the plies is in the direction opposite to the fiber wrapping. Thus, the resulting plied yarn is virtually free from constrictions caused by wrapper fibers, allowing the plied yarn to be soft and bulky.

The productivity of a Plyfil position is up to 25 times higher than that of a ring spinning delivery. Yarns from 100 percent combed cotton as well as blends of cotton and polyester are produced for end uses such as shirts, poplins, velours, and towels. Yarn count range is 24/2 to 72/2 Ne (25/2 to 8/2 tex).

Test Instruments and Quality Assurance Methods

Atmospheric Conditions in the Laboratory

Many important fiber and yarn properties are dependent upon prevalent atmospheric conditions. Since most textile fibers are hydroscopic, that is they absorb water from a moist atmosphere and lose it in a dry one, their physical properties can be greatly affected by the amount of water they contain. Because the moisture absorbed can influence the physical test results, it is necessary to record the atmospheric conditions prevailing during the testing of the material. In the USA the American Society for Testing and Materials (ASTM) states that the standard atmosphere for textile testing is 70 ± 2 degrees Fahrenheit (21 ± 1 degree Celsius) and 65 ± 2 percent relative humidity (RH).

It is recommended that samples be conditioned before performing any physical testing. Conditioning is the practice of placing relatively dry samples in a standard atmosphere until moisture equilibrium is reached so that there is no further gain or loss in weight because of moisture pickup. Some laboratories condition samples for as much as 24 hours prior to testing. Testing in standard conditions will result in comparable test data from one laboratory to another, and from day to day in the same laboratory. Some of the physical properties most affected by changes in atmospheric conditions are dimensions, tensile strength, elongation, electrical resistivity, stiffness, bending, and abrasion.

The most popular instrument for determining the atmospheric conditions in the laboratory is the wet and dry bulb hygrometer. If the bulb of a thermometer is covered by a wet fitted sock in an atmosphere that is not saturated with water vapor, water from the sock evaporates into the surrounding air at a rate proportional to the difference between the actual humidity and 100 percent humidity. Since the water being evaporated by moving air precipitates cooling, the temperature shown on the thermometer will be less than the temperature of the surrounding air. A hygrometer contains two identical thermometers, so both the wet and dry bulb temperatures can be read direct. The difference between the two readings is an indication of the relative humidity. Tables and psychometric charts aid in the derivation of the relative humidity level.

Sampling

A very important part of any quality assurance function is the method of obtaining the samples for testing. Basically, there are two kinds of samples, biased and random. A random sample is one where every individual in the population has an equal chance of being included in it. An example of a truly random sample follows. In the winding area there are 20 carriages of rotor spun packages of yarn representing ten frames. The yarn is all 20's Ne (20/1), 100 percent cotton. We want to sample this production to determine the breaking tenacity of the yarn. It would be easy to select 20 packages for this test from one or two of the carriages, but doing so would bias the results towards the production of one or two frames. A truly random sample would be to take one package from each of the 20 carriages, some from the top layer and others from down inside the other layers in the carriages. The 20 packages selected for testing would then truly represent the total population of the 20's, 100 percent cotton rotor spun yarn production. The number of samples to take is determined using acceptable statistical formulas for calculating sample size based on the confidence level and risk of error desired. Refer to formula 4.1 and associated comments in Chapter IV under the section titled "Laydown Size".

Fiber Testing Instrumentation and Methods

Fiber Fineness

The most commonly used procedure for determining cotton fiber fineness is the micronaire method. This very fast instrument oper-ates on the principle of resistance to air flow. Using 50 grains of cot-ton, which contains many thousands of fibers, the sample is com-pressed into a cylinder one inch long and one inch in diameter. Air is then forced through the cotton plug, and the resistance to air flow is measured to give an indication of the fineness value of the fibers.

The amount of air that can pass through a given weight of cot-ton fiber is dependent upon the surface resistance of the fibers. Fine fibers resist air flow greater than coarse fibers because of the increased ratio of surface area to weight. The typical range of mi-cronaire values for USA cottons is from 3.0 to 6.0. The larger the value, the coarser the cotton (less resistance to air flow).

It is important to know and control the cotton fineness value that is being used from day to day in order to produce quality yarn. Very fine cottons can be immature and thus lead to dye uptake problems, are prone to nepping, and are more easily broken during processing. On the other hand, cottons that are too coarse will produce yarns that have lower yarn strength and elongation levels, and higher un-evenness and thick and thin defect levels. The number of fibers in the cross section of a yarn is important with respect to all yarn qual-ity parameters, and is important to in-process performance.

For man-made fibers the fineness or denier is normally deter-mined in a textile plant using a Vibroscope. This method is based on the principle that a given fiber length, when held under known tension, has a natural frequency of transverse vibration. The vi-bration frequency is a function of the weight per unit length or de-nier. The basic formula[12.1] for linear density is as follows:

$$d = \frac{T}{4L^2f^2} \times 980 \quad (12.1)$$

Where: d = density in grams per centimeter (g/cm),
   T = tension in grams,
   L = length in centimeters, and
   f = the frequency in cycles per second.

Strength and Elongation

Fiber strength is an important property because of its effect upon yarn and fabric strength. It has a direct relationship to in-plant processing efficiencies, and to ends down and stop rates. Fiber strength can also influence the hand, drape, and similar characteristics of fabrics.

The most widely used instrument for measuring cotton and man-made fiber strength and elongation, information which can be used to calculate a fiber's "toughness", is the Zellweger Uster Stelometer. The constant-rate-of-load Stelometer normally uses a 3.2-mm gauge length enabling it to measure the quality referred to as elongation-at-break.

After removing the short fiber by physically combing a small amount of fiber, a fiber bundle or ribbon of parallel fibers 15 mm (0.591 inch) long and approximately 6 mm (0.236 inch) wide is prepared using a number of auxiliary sample preparation devices. This sample is then placed in a clamp whose jaws are separated by a 3.2-mm (0.126-inch) spacer. Referring to Figure 12.1, one of the jaws, J_1, is mounted in the adjustable jaw holder carried by the beam. The second jaw, J_2, is attached to the top end of the pendulum, which is pivoted at point 0. The beam's center of gravity is located to the right of its axis of rotation, A. When the arresting latch of the instrument is released, the beam rotates in a clockwise direction. The center of gravity of the pendulum arm coincides with the axis of rotation, A. Therefore, the heavy mass of the pendulum bob has minimum movement, eliminating inertia effects. Breaking load in kilograms and elongation in percent are indicated by the pointers P_1 and P_2, respectively.

After the fiber bundle has been broken, the two parts of the 15-mm x 6-mm fiber specimen are collected and weighed in milligrams (mg). The tensile strength of the material is then calculated by the following equation:[12.2]

$$\text{Tenacity, grams / tex} = \frac{\text{Breaking load in kg} \times 15}{\text{Sample weight in mg}} \quad (12.2)$$

Note: 15 in the formula is the total sample length in mm when using a 3.2-mm gauge spacer.

There are other bundle tests, as well as single fiber tests, for determining the strength and elongation of textile fibers, but their use can be declared as minor in comparison to the Stelometer (e.g., the WIRA Single-Fiber Strength Meter, the Pressley Dynamometer, and the AFIS-Manta Single-Fiber Tester from Zellweger Uster).

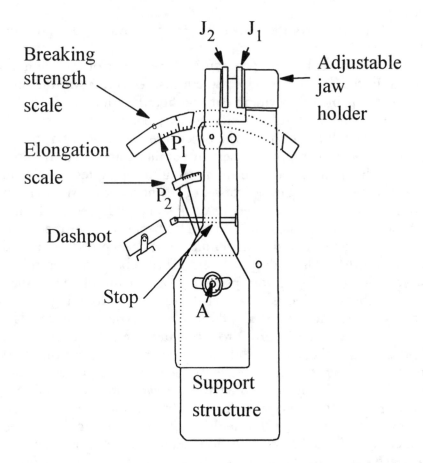

Figure 12.1. Schematic representation of the Stelometer instrument.

Fiber Length

Most of the equipment used in the manufacture of short staple spun yarns is dependent upon component settings based on the length of the fiber being processed. It is not surprising then that of the many fiber properties the industry should measure, length is perhaps the most important regardless of the spinning system used.

There are four stand-alone type instruments used in the USA for measuring fiber length properties, the Suter-Webb Array Method, the Zellweger Uster Fibrograph, the Zellweger Uster AL-101 with automatic fiber aligner, and the Zellweger Uster

AFIS L & D (Advanced Fiber Information System, Length and Diameter).

The Suter-Webb array procedure is the oldest of the methods listed. In the days before electronic measurement, it was the only method capable of providing data on fiber length and length distribution, and it is still considered as the most accurate of all methods developed to date. The procedure makes use of a sorting apparatus consisting of two sets of parallel combs that are used to straighten and parallelize the fibers in a 75-mg test specimen. After some manual parallelization of the fibers, the test specimen is placed into one of the sets of combs and is depressed slightly below the teeth of the combs. The front combs are dropped out from the specimen until a small number of fibers protrude beyond the comb that will be dropped next. These fibers are then transferred, using wide-grip forceps, to the other bank of combs such that a fiber baseline is established. Successive dropping of the combs allows access to all of the fibers in the specimen. The transfer is repeated in order to straighten the other ends of the fibers. As the fibers are withdrawn from the combs for the third time, they are arranged in sequential length order on velvet covered boards. The fibers extracted from each dropped comb are measured using a special ruler (also rule), and those that fall within the 3.2-mm (0.126- inch) length group for each comb are collected and weighed on a torsion balance having a sensitivity of at least ± 0.05 mg. From the weight-length data, the upper quartile length, mean length, coefficient of length variation, and percentage of fibers shorter than 12.7 mm (0.5 inch) (considered as non-spinnable) or any other specified length group can be calculated. (Reference: American Standard for Testing and Materials (ASTM) Method Designation D1440-90, "Standard Test Method for Length and Length Distribution of Cotton Fibers, Array Method," 1994 Annual Book of ASTM Standards, Section 7, Textiles.)

The array technique was the first in the USA to use the terms for fiber length and length distribution that are still in use today. From ASTM D1440 they are defined as:

Upper Quartile Length   That length which is exceeded by 25 percent of the fibers by weight in the test specimen.

Mean Length   The average length of all the fibers in the test specimen based on weight-length data.

Coefficient of Length Variation   A measure of the dispersion of the observed values equal to the standard deviation of the values divided by the average of the values, expressed as a percentage.

Percent Fiber Less than 12.7 Millimeters (0.5 Inch)   The percentage of fibers, by weight, in the specimen that are shorter than 12.7 mm.

This method is tedious and time-consuming and not usually used in a textile plant laboratory. However, it is highly accurate and is used for research purposes or whenever an accurate assessment of fiber length and length distribution is desired.

The Zellweger Uster Fibrograph instrument was developed in the 1950s by Dr. K.L. Hertel at the University of Tennessee. Originally marketed by Spinlab of Knoxville, Tennessee, the instrument gained in popularity globally. Although the results obtained from the Fibrograph are not as detailed or accurate as the results from the Suter-Webb Array method, the speed of testing is very high. The Fibrograph found its niche in the routine area of in-plant testing and quality control.

The instrument makes use of a tuft of "truncated" fibers prepared for the Fibrograph by a Fibrosampler device. The Fibrosampler specimen has the form of a "beard" of fibers held in a comb. The fibers are placed on the comb in such a manner that they are caught at random points along their lengths. The comb of fiber is then placed into the Fibrograph instrument where it is photoelectrically scanned from base to tip, the amount of light passing through the beard being used as an indicator of the number of fibers that extend various distances from the comb. The Fibrograph displays the amount, or optical density, and the length data from the fibrogram. The fibrogram is the curve representing the second cumulation of the length distribution of the fibers sensed by scanning the fiber beard. Three different test data are normally obtained, (i.e., the 50 percent and 2.5 percent span lengths and the 50/2.5 Uniformity Ratio). These measures are defined in ASTM Method Designation D1447 as follows:

50 Percent Span Length   The distance spanned by 50 percent of the fibers in the test beard.

2.5 Percent Span Length   The distance spanned by 97.5 percent of the fibers in a test beard.

50/2.5 Uniformity Ratio   The ratio between the 50 percent and 2.5 percent span lengths expressed as a percentage of the 2.5 percent span length.

It is widely understood in the short staple spinning sector that the upper quartile length as determined by the Suter-Webb Array method closely agrees with the 2.5 percent span length measurement from the Fibrograph. These two measures also agree somewhat with the upper half mean length from high volume instrument (HVI) test lines, and with the 25 percent point from the Zellweger Uster AL-101 device. The mean length from the array and the 50 percent length from the Fibrograph are quite different from each other because of the differences in how they are measured, weight of fibers versus number of fibers.

The Uniformity Ratio (50/2.5 span length x 100) is said to be a measure of the length uniformity in the specimen. The higher the Uniformity Ratio, the more the fibers are of the same length in the specimen. Although not direct, it is felt that this is an indicator of the short fiber content, or non-spinnable fiber, in the specimen tested.

The Zellweger Uster Fibroliner, FL-101, is used with the fiber length measuring unit, Almeter, AL-101, to determine the fiber length and length distribution of staple fibers 10 cm (4 inches) or less in length. It can be used on loose fiber samples as well as on slivers and rovings.

Approximately 100 mg of fiber from the sample are manually drawn into a hand sliver form. This bundle is then transferred directly into the needle field of the FL-101, making certain that approximately one half of the fibers face opposite from the other one half of the fibers. This is done to reduce or eliminate the effect of any fiber hooks that may have been introduced during processing. A specimen of approximately 100 mm (near 4 inches) in width is considered to be optimum. The FL-101 automatically prepares the sample for the AL-101 unit, similar to the method used for the Suter-Webb Array test. The specimen, whose fibers lie along a baseline such that end-to-end distance can be measured, is transferred to the bottom foil of the AL-101 with the uneven fiber ends facing the instrument sensor condenser. The top foil is lowered over the test specimen for protection and to hold it in place. The condenser scans the test specimen every 0.125 mm (0.005 inch). The usual length parameters calculated and reported by the instrument computer are by weight and number, mean length, upper quartile length, coefficient of length variation, and percent short fiber content less than 12.7 mm, and the one percent length which is the length exceeded by one percent of the number of fibers in the test specimen. The accuracy of this method in a laboratory is within approximately 0.5 mm (0.02 inch) for the mean and upper quartile lengths, and two percentage points for coefficient of variation and short fiber content.

One of the newest techniques for measuring fiber length and length distribution is the single fiber measuring system from Zellweger Uster called AFIS L & D, Advanced Fiber Information System, Length and Diameter. This system uses an electro-optical method of measurement. The base unit, shown in Figure 12.2, requires a fiber sample of approximately one half gram (g) in

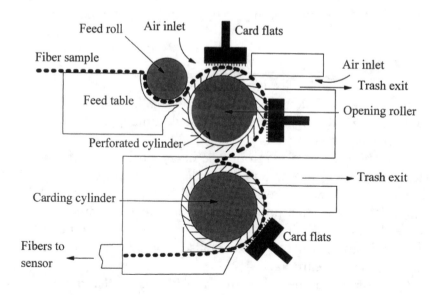

Figure 12.2. Schematic diagram of the principle of fiber sample opening with the AFIS base unit. (Courtesy Zellweger Uster AG)

weight. The sample specimen is manually fed between the feed roller and feed table of the aeromechanical opening unit. The opening roller and carding cylinder open the fiber specimen and separate it into fibers, neps, trash, and dust. The trash and dust are suctioned off to exhaustion. On their way through transportation and acceleration channels, the fibers and neps pass through an optical sensor which determines the fiber length and length distribution of the single fibers passing the sensor. The sensor can differentiate between single fibers and neps through impulse conversion. The statistical data are calculated and printed out on a printer. Data available include by weight and number, mean length, coefficient of variation of fiber length, and short fiber content less than 12.7 millimeters. The upper quartile length is available by weight only. Length qualities by number only are the 2.5 percent and 5 percent lengths. Frequency distributions and staple diagrams can be printed out by weight or by number. Also available from the D portion of the AFIS L & D are the mean fiber diameter in microns, by number, and the coefficient of variation of fiber diameter. The time required for each specimen is from 2 to 3 minutes. A picture of the AFIS Module, fitted with an Autojet 30-position rotating cassette that provides for automatic feeding of preformed slivers to the unit, is shown in Figure 12.3.

Maturity

Immature cottons are cottons that have not fully developed a secondary wall structure of cellulose. They have a propensity to rupture easily during processing through ginning and yarn manufacturing equipment, thus causing the short fiber content to increase. This plays a deleterious role in the production of a quality yarn. Immature cottons have a tendency to nep more than fully developed cottons. Because of their less-than-complete cellulose wall thickening, immature cottons do not absorb as much dyestuff which leads to objectionable white, undyed specks, especially in dark colors. Immature cottons are also more difficult to clean during opening and carding.

There are a number of methods available for determining cotton fiber maturity. The most widely used is a procedure for determining both fineness and maturity index by the IIC-Shirley Fine-

Figure 12.3. Photograph of the AFIS instrument fitted with the autojet feeding mechanism. (Courtesy Zellweger Uster AG)

ness/Maturity Tester. This method is reported in ASTM's Book of Standards as Method Designation D3818.

The procedure makes use of four grams of cleaned and well-opened cotton fiber. This specimen is placed into a holder and compressed by air to a fixed volume. Air is drawn through the compressed plug of fiber at a flow rate of 4.0 liters per minute, and the pressure drop is measured. The specimen is further compressed into half the initial volume, and the pressure drop is measured at 1.0 liter per minute. From the regression equations in the instrument computer, the two pressure readings are used to calculate the micronaire fineness, the linear density in millitex, and the maturity index in percent.

Although discontinued by ASTM in 1992, a method that remains popular also uses air flow for the determination of cotton fiber maturity, the Causticaire method. A sample of approximately 3.2 grams of cotton is tested for its micronaire value before and after it is subjected to swelling in an 18 percent solution of sodium hydroxide (NaOH). The percentage ratio of the untreated value divided by the treated value is taken as the index of maturity.

The Institute of Textile Technology used the Causticaire index of maturity to determine the levels of secondary wall thickness in a large number of cottons from around the world. These cottons were then used as a robust data set for calibrating a near-infrared (NIR) instrument. Once the calibration was placed into the NIR computer, studies were performed on other cottons of known maturity index level. The coefficient of determination (r^2) between NIR maturity index and Causticaire maturity index was 0.92, and with the IIC-Shirley Fineness/Maturity Index Tester it was 0.85. The relationship between NIR maturity index and dye uptake, k/s, was defined by an r^2 of 0.88. As mentioned earlier, the higher the secondary wall development, maturity, the more dye molecules can attach themselves to the cellulose.

Color

Color, along with leaf and preparation, is one of the primary factors of cotton grade. Color measurements on raw stock are useful in controlling the color of manufactured greige, bleached, or dyed yarns and fabrics. Normally, opened cotton is white in color. Continued exposure to weathering and micro-organisms can cause this white cotton to lose its brightness and become darker. Under extreme conditions of weathering, the color may become a very dark bluish gray. Cotton affected by frost or drought may have a yellow color that varies in depth. Spotted cottons could be from insects, fungi, or soil stains. There are 5 color groupings of Upland and American Pima cottons; white, light spotted, spotted, tinged, and yellow stained.

The only stand-alone instrument for measuring the color of raw cotton is the Nickerson-Hunter Cotton Colorimeter. This unit is currently marketed by Zellweger Uster, Inc., Knoxville, Tennessee. A smooth representative surface of a cotton sample is placed over a sample window, the cotton is then pressed down on the surface of the window to provide an unbroken surface of cotton, the instrument is started, and the two indicator arms move to a balance point under the lighted glass color diagram. The color is read at the cross-point of the arms in terms of Rd reflectance, or degree of grayness, and +b yellowness. The +b yellowness value represents the magnitude called "chroma" in the Munsell system. From the diagram of Rd and +b it is possible to determine the color of the USA standard (e.g., Strict Low Middling White, Low Middling Spotted, and so forth). The diagram for determining color grade from Rd and +b values is shown in Figure 12.4.

Non-Lint Content

In the manufacture of a short staple spun yarn, especially from natural fibers such as cotton and flax, non-lint content, which is defined as trash, dust, microdust, and respirable dust, plays a major role in process efficiencies, and in yarn and fabric quality. This is especially true when spinning on the rotor and air jet yarn systems where buildup in the rotors or nozzles prevents optimal performance. In all yarn types non-lint content can affect appearance, evenness, and defect levels.

According to the International Textile Manufacturers Federation (ITMF) in Zurich, Switzerland, trash is defined as having particle sizes greater than 500 microns, dust has particles between 50 and 500 microns in size, microdust has particles between 15 and

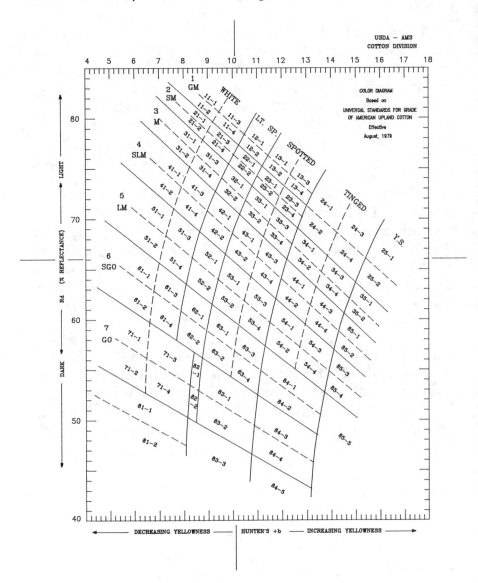

Figure 12.4. The Nickerson-Hunter Colorimetric Diagram.

50 microns, and respirable dust has particles smaller than 15 microns in size.

ASTM Method Designation D2812, Non-Lint Content in Cotton, mentions two instruments available for making this measurement (i.e., the Shirley Analyzer and the SRRL Non-Lint Tester). The SRRL Non-Lint Tester is not in general use, and very few units exist in industry. The Shirley Analyzer, on the other hand, has been in use for over 50 years and remains as the referee method to which all newly developed non-lint content instruments must compare.

The modern Shirley Analyzer principle is based on mechanical-pneumatic separation of non-lint and lint. An approximate 100-g specimen is opened between the action of the feed roll/streamer plate and an opening lickerin cylinder. Successive separation of the non-lint impurities from the fibers is accomplished by flotation in an air current due to their different specific weights. The specimen is separated into 5 fractions (i.e., cotton fiber, heavy impurities, finer impurities, very fine impurities larger than 150 microns, and very fine impurities between 50 and 150 microns). The latter two separations are accomplished through a system of filters. Gravimetric measurements must be made to obtain the percentages of the various 5 factions. The working elements of the Shirley Analyzer, Model MKII, are shown in Figure 12.5.

An instrument available for determining the number and size of particles of foreign matter, dust and trash, is the Zellweger Uster AFIS-T (Advanced Fiber Information System, Trash). This unit is capable of measuring samples containing 100 percent cotton, and man-made/cotton fiber blends. The principle of measurement is identical to the AFIS L and D instrument discussed in the length section of this chapter. The data available from the AFIS-T include

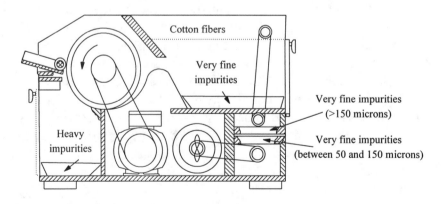

Figure 12.5. Schematic diagram of the working elements of the Shirley Analyzer, Model Mk II.

trash and dust particle count per gram, mean size of trash and dust particles in microns, dust particles per gram in count per gram, trash particles per gram in count per gram, and calculated gravimetric total visible foreign matter (VFM). The user can also obtain a graphic output of results including frequency distribution. The statistical values for the various measurements are mean, standard deviation, and coefficient of variation (%CV).

The Zellweger Uster MDTA3 Microdust and Trash Analyzer was developed by researchers at the Institut Fur Textil-Und Verfahrenstechnik Denkendorf in Denkendorf, Germany, and is marketed in the USA as the ITV Tester by J. D. Hollingsworth on Wheels, Greenville, South Carolina. It is also available from Zellweger Uster as part of their Quickspin System.

The basic ITV or MDTA3 is a very accurate instrument for measuring lint content, trash level, fiber fragments, and microdust less than 40 microns in size in a sample of cotton raw stock, or in samples of cotton or blended in-process sliver and roving. The precision of measurement is offset somewhat by the time required to perform a test at 20 minutes per specimen. Much of the time is in the weighing of filters before and after running the test and in the calculation of the percent of each fraction measured.

Today's cottons contain smaller trash particles because of the emphasis being placed on cleaning at the gin. The lint cleaners, following the gin stand where the lint and seed are separated, break many of the large particles during cleaning into smaller fragments which are more difficult to remove in the textile plant because of their lower relative weight. Also, because of certain seed breeding programs, where the search for more cottonseed oil reduced the thickness of the seed coat, the seeds in today's cottons tend to be weaker and thus break more easily during the actual ginning function. Seed coat fragments cause ends down in spinning if not removed, and cause defects in fabrics. These two happenings in today's ginning facilities make knowing more about the cottons being processed in the plant imperative, especially in plants with rotor or air jet spinning.

The use of non-lint testers in the textile plant laboratory allows for the calculation of cleaning efficiency, one of the more important measures that can be made in the opening room. Cleaning efficiency is generally defined by the equation[12.3]below. This calculation is explained in more detail in Chapter IV in the section titled "Cotton Cleaning Efficiency Calculation."

$$\text{Cleaning Efficiency (\%)} = \frac{\text{Trash In} - \text{Trash Out}}{\text{Trash In}} \times 100 \quad (12.3)$$

Nep Count

Neps are small knots of entangled fibers that usually will not straighten to a parallel position during carding or drafting. Neps are serious defects when processing cotton because they generally rise to the surface of the yarn during spinning and detract from the appearance of the yarn. Neps also limit the yarn count that can be spun, increase the number of yarn breakages, and increase correction costs during finishing. In addition, when a manufacturer attempts to remove neps, the amount of waste generated increases. Neps are generally formed during the stretching and sudden release of a fiber during processing. Once the fiber is released, it can move more freely and the free end may wrap around itself, around other fibers, or around other particles to form a loose nep. With continuing stretching and release, the loose nep may be transformed into a real nep. During each processing step, more neps can be formed.

In 1953 the American National Standards Institute introduced a test method for determining the number of neps in cotton samples. The method was revised in 1971 and basically involved visually counting the neps in a handmade or carded web. This method was time-consuming, very subjective, and its reproducibility was poor.

In 1978 a new test method was developed by the American Society for Testing and Materials for analyzing cotton samples for nep content. This method was very similar to the previously discussed test method, except in this method the web was compared to photographs of different cotton specimens to determine the level of neps in the sample. This method was also quite subjective.

A state-of-the-art means of nep measurement involves the use of an electro-optical instrument, the AFIS-N. The principle of measurement of this unit is identical to that mentioned in the length section of this chapter for the AFIS L and D. The data available from the AFIS-N instrument include nep count per gram, nep diameter in microns, coefficient of variation of nep count, and a histogram of nep diameter.

Yarn Quality Instrumentation

One of the indicators of how efficient a short staple manufacturing unit is in producing a yarn that meets the needs of the plant or its customer is its quality. Most in-plant and central corporate laboratories are well-equipped with modern instrumentation for characterizing the properties and variabilities of a yarn. Generally, a yarn having good properties and low variability of those properties will perform well in weaving and knitting. The instrumentation discussed in this section is basic for a short staple yarn quality assurance laboratory in the USA.

Linear Density - Sliver Weight, Roving Hank Size, and Yarn Count

There are three systems in general use in the USA for specifying a yarn's linear density, or weight per unit length. The most commonly used system for short staple yarns is the English count (Ne), also known as the cotton count system. This is an indirect system limited in application to roving for ring spinning, and to yarn size for all systems. It is based on the number of 840-yard hanks in one pound of material, in which the ratio of yards to hanks is 840. In this system, the larger the count or hank size value, the finer the material. To understand the relationship, a 30's (30/1) English cotton count yarn contains 25,200 yards of yarn (30/1 x 840) per pound; a 20's (20/1) English cotton count yarn contains 16,800 of yarn (20/1 x 840) per pound. As mentioned above, the English cotton count system can only be used for roving and yarn size expression. For card sliver, comber and drawing slivers, and comber laps, the size or weight per unit length is usually expressed in grains per yard (gr/yd) or grams per meter (g/m).

The second most widely used yarn numbering system is the Denier system. This is a direct numbering system based on weight per unit length. The definition of this system states that denier is the number of grams per 9,000 meters. One pound of 1 denier yarn would contain about 4,464,483 yards. Finer yarns or deniers have smaller numbers (e.g., a 70 denier yarn contains 63,778 yards per pound, a 150 denier yarn contains 29,763 yards per pound or 4,464,483 ÷ 150).

The Tex system is much like the Denier system. It is also based on weight per unit length, except that this system uses grams per 1,000 meters instead of grams per 9,000 meters. Since the Tex system is metric and decimal, it makes it a universal one that can be used to express material weights of fibers, laps, slivers, rovings, and yarns. The relationship to denier is 1 to 9; the yards per pound calculation for the material is 496,054 ÷ Tex.

In the textile plant it is common to physically weigh a certain yardage of roving, sliver, and comber laps to determine grain weight per yard. Reels and 1-yard cutting boards help facilitate this work. For rovings, laboratories in the USA normally weigh 12 yards from each of a number of bobbins to obtain a value for hank roving size. For yarn, using the English or cotton count system, it is common practice to reel a 120-yard skein from each of several packages and separately weigh them on a balance that reads directly in English count. The calculation of the mean, standard deviation, and coefficient of variation is usually performed automatically by the weighing balance computer. For those using either the Denier or Tex system, 100 meters of yarn are reeled and weighed on a balance that reads directly in Denier or Tex.

There are several commercial automatic balances that are currently being used in industry for yarn count determination (i.e., the Autosorter III by Zellweger Uster, the Automatic Yarn Count Tester L290 from Zweigle of Germany, the Electronic Yarn Count and Fabric Analysis System from Shirley Developments, Ltd. in England, and the Yarn Count Analyzer from Lawson-Hemphill).

It is mentioned later in the section on yarn evenness and strength that there is instrumentation that combines these measurements with a simultaneous, automatic determination of yarn linear density. Yarn number conversion factors in and between the three systems discussed are given in Table 12.1. Other discussion of yarn numbering systems is contained in the introductory chapter of this textbook, and a long list of yarn count conversion factors is available in Appendix II.

Table 12.1. Yarn Numbering Conversion Table

Using	Perform This	To Obtain
English Cotton Count	Divide 5314.84 by Cotton Count	Denier
Denier	Multiply by $1/9$	Tex
Tex	Divide 590.54 by Tex	Cotton Count

Yarn Tenacity and Elongation-at-Break

In most textile plant operations, strength has been considered to be one of the most important yarn properties. High strength yarns generally produce high strength fabrics. Also, yarns of high strength process more efficiently at spinning and at subsequent fabric manufacturing operations. Strength and elongation-at-break combine to equate to toughness, or the ability of a yarn to resist stresses and strains without rupturing.

There are two different approaches to testing yarns for strength. The oldest procedure, which continues to be popular today, is the multiple-end test, or skein break method. The other, and judged to be more valuable because of the individuality of the information and the fact that a distribution of the breaks and elongations at break is possible, is the single-end or single-strand test.

For the skein method, textile plants in the USA that use the English count system normally wind a 120-yard skein on a wrap reel. A 100-meter skein wound on a 1-meter circumference reel is used for those textile operations using the Denier on Tex yarn numbering system. Very coarse yarns may require a 60-yard or 30-yard skein because the tester that will eventually be used to determine the breaking load has limited full-scale load capability. If a 60-yard skein is used, it is common to multiply the skein strength value by 2 to obtain an equivalent 120-yard skein test result. If a 30-yard skein is used, multiply the strength value by 4. This adjusting of the skein strength has been found to be quite satisfactory for in-plant use.

The skein break procedure generally used is that skeins are wound on a wrap reel from a number of packages, are conditioned under slight tension in a standard atmosphere, and are then broken on a pendulum tester having a constant-rate-of-traverse loading mechanism. After the skein is broken and the result is recorded, the skein is usually weighed to obtain yarn count on a balance similar to those mentioned in the above linear density measurement section.

The two types of strength testers used in the skein method of strength analysis are constant-rate-of-traverse and constant-rate-of-extension. A constant-rate-of-traverse machine is one in which the pulling jaw moves at a uniform rate, and the force is applied through the other jaw. The rate of increase in force or elongation is usually not constant and is dependent upon the extension characteristics of the yarn sample. A constant-rate-of-extension machine is one where the moving jaw applies a uniform rate of extension on the yarn sample.

The constant-rate-of-traverse tester used in the USA is almost exclusively the Scott Pendulum Tester. Pendulum testers for skeins make use of two jaws (spools), the bottom one traversing at a constant rate of speed to exert the tensile force on the yarn specimen, and the top jaw (spool) holding the other end of the skein operates against an appropriately weighted pendulum arm. The skein is pulled at a constant rate of speed so that it is elongated at a constant rate, except for the effect of the movement of the top jaw (spool) which is attached to the pendulum and provides the load to move the load recording dial. When the yarns in the skein begin to fail, the pendulum arm slows and eventually stops. The lower jaw (spool) direction is reversed for return to its original position. The breaking force is then read from the dial and recorded. As a result of successive breaking of ends in the 160-end skein, the strength obtained by the skein test is always lower than the combined strengths of the individual yarns.

The constant-rate-of-extension test is exactly the same as the constant-rate-of-traverse test except that the top jaw (spool) does not move to actuate a recording dial. The top jaw (spool) is fixed to a stationary load cell that facilitates the recording of the breaking load either on a chart or on a computer printout. The most widely used tester with this loading principle is the table or floor model Instron Tester, fitted with spool-type jaws over which the skein is placed for testing.

The most common method of expressing skein strength is a term called "break factor". Break factor is the product of the breaking force in pounds times the actual English cotton count.

Break factor is intended to represent the length of yarn whose weight is equal to the breaking strength of the same yarn.

The more useful measurement of yarn strength is the testing of single ends, because single ends are run in the textile plant, not skeins. All modern single-end tensile testers can automatically evaluate yarns for strength and elongation-at-break using the constant-rate-of-extension, mainly, and the constant-rate-of-loading principles. A constant-rate-of-load machine is one in which the rate of increase in force is uniformly applied to the yarn sample.

The most popular constant-rate-of-load tester in use in the USA for determining yarn strength and elongation-at-break is the Zellweger Uster Dynamat II. This unit makes use of an automatic package changer and a means for winding up the broken length of yarn so that it will not interfere with the next yarn break. The Zellweger Uster Dynamat II uses a 500-mm (19.7-inch) gauge length and can test yarns up to a maximum 2000 grams strength and 40 percent elongation-at-break. The normal time-to-break for this instrument is 20 ± 3 seconds, which is in accordance with ASTM and ISO recommendations for constant-rate-of-load testers. Coupled to the Dynamat II tester is a calculator for automatically determining the mean values and the coefficients of variation for both force-to-rupture and elongation-at-break. Also displayed on the printout is the mean time-to-break and the number of tests performed.

Many of the world standards organizations make reference only to constant-rate-of-extension dynamic strength and elongation-at-break testers. The instruments generally used in the USA are the Zellweger Uster Tensorapid 3, the Textechno Statimat ME, the Instron Tester, and the Zellweger Uster Tensojet.

The Zellweger Uster Tensorapid 3 is an instrument that can measure breaking force up to 1000 Newtons, and corresponding elongation-at-break of up to 140 percent when using the standard 500-mm gauge length. The acceptable cotton count range is from 1 to 3000 Tex. Of particular interest concerning the Tensorapid 3 is its ability to test a yarn specimen quickly (i.e., up to 5000 mm/minute rate of elongation). The information possibilities from this instrument are single overall values for force, elongation-at-break, and work, and their standard deviations, coefficient of variations, 95 percent confidence levels, minimum and maximum values, and the weakest value for breaking force. Available also are force and elongation-at-break, and modulus and elongation-at-break characteristic curves. A photograph of the Uster Tensorapid 3 is shown in Figure 12.6.

One revolutionary instrument for determining a yarn's strength and elongation-at-break characteristics is Zellweger Uster's Tensojet. This constant-rate-of-extension unit was developed to provide extremely fast testing for short staple fiber yarns in the average count range of from 10 to 150 Tex. It has the capability of test strength up to 50 Newtons and elongation-at-break up to 70 percent. The rate of elongation is extremely fast compared to standard dynamometers at up to 400 meters per minute. It is possible to test up to 30,000 tests per hour with this tester, providing for statistically reliable results. The detection of seldom-occurring weak places because of its speed is a valuable contribution of the Tensojet. The high rate of elongation of 400 meters per minute

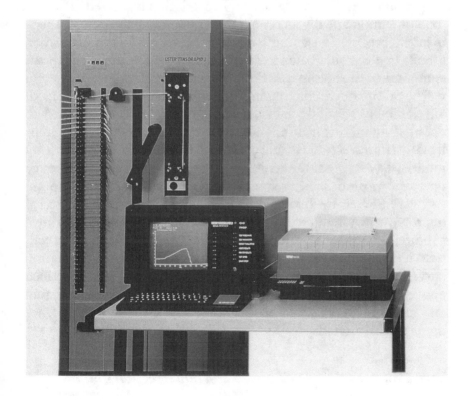

Figure 12.6. Photograph of the Tensorapid 3 instrument. (Courtesy Zellweger Uster AG)

corresponds approximately to the conditions to be expected on the weaving machine. Thus, some researchers say that there is a good correlation between Tensojet results and weaving stops.

The data available from a Uster Tensojet yarn analysis include mean, coefficient of variation, standard deviation, 95 percent confidence limit, and maximum and minimum values for breaking force, elongation-at-break, tenacity in centi-newtons/tex, and work-to-break expressed in one of five different ways. Also, the mean value for the weakest 0.01, 0.05, and 0.10 percent of the breaks can be included in the printout. Stroke and scatter diagrams and histograms of breaking force and elongation-at-break for visual analysis are added features of the Tensojet.

The Textechno Statimat ME constant-rate-of-extension dynamometer is unique in that it incorporates an Autocount measuring unit for determining the linear density of the tested yarn. The yarn count is required if a tenacity calculation is desired. The Autocount automatically determines the count of the tested specimens by precisely weighing a machine-measured length of yarn on a built-in electronic balance. The weight data are then sent to the computer of the Statimat ME for tenacity calculation.

The Statimat ME maximum load capability is 1000 Newtons. It has infinitely variable clamp speed, from 1 to 5000 mm (up to about 200 inches) per minute, and at infinitely variable gauge length, from 50 to 500 mm (about 2 to 20 inches). As with all modern-day strength testers it has automatic package changing for up to 20 positions in its standard creel. The standard computer software offered by Textechno allows the following data concerning the yarns tested: breaking force, breaking elongation, work-to-rupture, intermediate values for specified force or elongation values, modulus values at selected points on the force/elongation curve, and when fitted with the Autocount system described above, the yarn count. Relative to these data, the mean, maximum and minimum, coefficient of variation, and 95 percent confidence limits are calculated.

Yarn Evenness (Mass Variation) and Imperfection Count

The evenness, or unevenness, of the mass or linear density of a yarn has a major influence on its performance and properties.

Yarns that are uniform in mass along their length usually have higher strength, process better in spinning and in the manufacture of woven and knitted fabrics, and exhibit better appearance in fabric structures. As a matter of fact, the more uneven a yarn is in its linear density, the more variable its other yarn properties are likely to be.

The subject of evenness testing with respect to the electronic technical information and theoretical aspects that pertain to it are beyond the scope of this text. Zellweger Uster has done an outstanding job issuing bulletins that discuss the theory and electronics of evenness measurement and the many things that can influence the results. The particular two-volume book that is recommended as reading for evenness measurement scientific details is, Evenness Testing in Yarn Production, by Richard Furter of Zellweger Uster AG, Switzerland. These two volumes were edited by the Textile Institute of Manchester, England, in 1982 (ISBN 0900739 49 5), and are an excellent source of information on the measurement of variations in mass per unit length.

There are two instruments in general use for yarn mass variation measurement (i.e., the Zellweger Uster UT3 and the Keisokki KET-80II/B). They both use the electronic capacitance principle, which consists of two parallel plate air capacitors. When non-conducting materials such as staple fiber slivers, rovings, or yarns are drawn through the two plates of the measuring slot being used, changes in the capacity of the capacitors are proportional to the weight of the material present, thus a measure of the evenness of the material. The unit length of the measuring plates for yarn is the same for both the Zellweger Uster UT3 and Keisokki KET-80II/B at 8 mm. This means that the tester is electronically calculating the coefficient of variation (%CV or %CVm) of the "weights" of individual 8-mm lengths along the entire test length, which is normally 1000 meters for yarn, and at least 125 meters for sliver and roving. The mean deviation, or coefficient of variation, %CVm, is the common term used to express yarn evenness; the higher the %CVm value, the more uneven the yarn.

Both testers have the capability of simultaneously measuring yarn mass variation over cut lengths longer than 8 mm (0.315 inch). It is possible to also obtain the %CV of mass variations using electronic cut lengths of 1, 3, 5, and 10 meters. The Zellweger Uster UT3 can also measure evenness of 50-meter lengths. Use of longer electronic cut lengths enables the Quality Assurance function to evaluate such things as, for example, long-term mass variations which can reduce the yarn overall quality, the effect of the number of doublings at drawing, the effectiveness of card and draw frame autolevelers, and variations in weight per unit length from processes prior to spinning.

While the measurement of evenness %CV (%CVm) is being performed, both instruments can determine the relative count of the yarns being evaluated. This relative count measurement uses the value of 100 as its base. The average relative count of any group of yarns will always be 100. Each individual sample of yarn being tested will have a value either greater, equal, or less than 100, depending upon the yarn count of that sample. It is cautioned that this is not an actual count measurement. The relative count value can be used to show the variability that exists in yarn count between samples.

An important feature of both testers is a chart printout called the spectrogram. The spectrogram chart is a graphical wavelength representation of the material tested and has been developed to show cyclic or repeating variations usually caused by process mechanical problems. The spectrogram also has the capability to show drafting faults. Mechanical defects are noted as very distinct spikes on the chart; drafting faults are abnormal "humps" on the chart. The spectrogram chart can show the length and severity of periodic defects, multiple periodic defects, and drafting faults. Examples of spectrograms are presented in Chapter VI.

While the evenness testers are measuring mass deviations, it is possible to observe the changes in cross-sectional mass of a sample over the test length measured by using a graphical diagram chart on the instrument computer screen or a printout from the instrument printer. One can observe extreme thick and thin places in the yarn; random weight per unit length variation; immediate changes in the yarn mass; periodic or cyclic variations that are too long for the spectrogram to display; and short-term, random, and periodic variation.

The Zellweger Uster UT3 and Keisokki KET 80II/B both have included in their installations a unit that measures thin places, thick places, and nep-like imperfections. Imperfections measured on these instruments are considered to be frequently occurring faults and are normally reported as defects per 1000 meters. It has been found from research that thin place and thick place faults tend to be about the same length as the mean staple fiber length from which the yarns were made. Neps are defined as any thick place less than 4 mm in length. On the imperfection indicator of both instruments there are four sensitivity levels for each of the three classes of imperfections. They are -30, -40, -50, and -60 percent of the mean yarn diameter for thin places; +35, +50, +70, and +100 percent of the mean yarn diameter for thick places; +140, +200, +280, and +400 percent of the mean yarn diameter for neps. In the USA it is common to test ring and air jet spun yarns using sensitivity settings of -50 percent for thin places, +50 percent for thick places, and +200 percent for neps. For rotor spun yarns the values are -50, +50, and +280 percent for thin places, thick places, and neps, respectively.

It was mentioned earlier in this chapter in the discussion of linear density that it is possible to obtain the linear density of a tested yarn simultaneously while measuring evenness and imperfection count. The Zellweger Uster UT3 can be fitted with a Uster Autosorter III balance for measuring 100-meter or 120-yard lengths. The length chosen will be automatically obtained from the Zellweger Uster UT3, transferred to the balance, and the results sent to the UT3 to appear on the printout protocol. The form of the data are mean count, standard deviation of count between tests, and the 95 percent confidence level of the mean yarn count. A photograph of a complete Zellweger Uster UT3 installation is shown in Figure 12.7.

An evenness tester based on the optical measurement of the profile of a yarn is being marketed by Zweigle of Reutlingen, Germany. The present model is the G580 Yarn Structure Tester. The optical sensor determines the yarn diameter using infrared light according to the shadow casting principle. Deviations in the yarn diameter are used by the instrument computer to calculate optical %CVm, and the number of thin places, thick places, and neps. The sensing length for optical %CVm is 1 mm when using a test speed of 400 meters per minute, and 2 mm when using a test speed of 800 meters per minute. For the imperfection count, the diameter deviations and defect length category combinations total

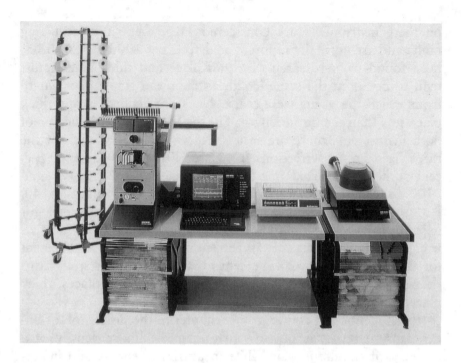

Figure 12.7. Photograph of the Evenness Tester 3-B. (Courtesy Zellweger Uster AG)

64. All are measured and reported simultaneously in the data printout. The defect lengths measured are 2, 5, 10, 20, 40, 80, 160, and 320 mm, and -40, -20, +25, +40, +80, +120, +160, and +200 percent diameter deviation. It is also possible to obtain spectrogram and optical variance length curve data from the G580. Because of its measuring principle, the G580 Yarn Structure Tester works independent of humidity, temperature, and color, and is capable of measuring the yarn profile deviations in diameter of metallic yarns without electronic interference.

Classifying Seldom Occurring Faults

More and more it has become important that yarns produced for the weaving and knitting trades contain fewer and fewer objectionable defects. Defects can cause stops in manufacturing which translate into lost efficiencies, higher costs, and off-quality fabrics.

It is imperative that these defects not be made in the first place, but because it is nearly impossible to produce a defect-free yarn, the classification of defects as to their length and mass aids in the correction of some of the reasons for their existence. The original intent of a defect classification system, however, was not to troubleshoot the processes producing the defects, but to obtain the defect mass and length distributions so that electronic yarn clearers could be scientifically set to sensitivity levels to optimally clear yarns at winding. Today's quality control programs use instruments for classifying seldom occurring faults for three purposes (i.e., troubleshooting processes, checking the quality of purchased incoming yarn, and establishing electronic clearing settings).

There are two widely accepted instruments on the market in the USA for analyzing and classifying seldom occurring thin and thick faults in staple spun yarns. They are the Zellweger Uster Classimat 3 and the Keisokki Classifault II. A photograph of the Zellweger Uster Classimat 3 is shown in Figure 12.8. Both units automatically classify seldom occurring thin and thick defects according to size classes. Because the faults detected by the classifying systems men-

Figure 12.8. Photograph of the Classimat 3 instrument without the measuring heads which attach to the winder. (Courtesy Zellweger Uster AG)

tioned above are infrequent and seldom occurring ones, compared to the very frequent ones measured using the imperfection counters of the three evenness testers discussed, this testing requires that a minimum of 100,000 meters be analyzed, instead of 1000 meters.

To facilitate the analysis of such a large quantity of yarn, 6 or 12 electronic measuring heads are mounted on a laboratory winder in a standard atmosphere. The winder speeds to accommodate defect measurement can range from 400 to 1250 meters per minute for the Zellweger Uster Classimat 3, and from 100 to 1500 meters per minute for the Keisokki Classifault II.

There are some small differences between the two available instruments, but basically they measure identical things. There are 23 fault classes on the Zellweger Uster Classimat 3, and 40 on the Keisokki Classifault II. The 17 extra classes on the Keisokki Classifault II are additional length groupings for slubs, thicks, and thins. Also, the Keisokki Classifault II allows the operator to select the counting sensitivity in ±5 percent increments up to ±95 percent. This feature is important when end product objectionable defect requirements are known.

Yarn Twist

The amount of twist placed in a staple spun yarn is important from a technical viewpoint because of its effect on physical properties and performance, and on finished product appearance. It is also important from a production standpoint because with every turn of twist there is an accompanying loss in productivity and an increase in cost. Twist also impacts fabric luster, fullness, hand, weight, and strength.

There are two possible twist directions, Z and S. Yarns twisted clockwise have Z twist, and yarn twisted counter-clockwise have S twist. In the USA most single yarns are Z twist.

Many of the instruments available to measure both Z and S twist are of the adjustable gauge length type, are non-automatic with respect to thread-up, and have motor drive capability on the rotating clamp. The most common twist measuring method used, which is applicable to ring, rotor, and air jet spun yarns, is the untwist-retwist method. This method is based on the premise that the contraction of a specified length of singles yarn is the same for any amount of twist, whether Z or S. A 50-cm (19.7-inch) length is untwisted under light tension, and then is retwisted in the opposite direction. The retwisting is continued until the contracted length is the same as the original specimen length. The total twist is the sum of the untwisted and retwisted turns. The actual turns of twist per 50 cm is the total twist divided by two.

A second method in popular use is the direct counting method. Here it is customary to use a 2.54-cm (1-inch) gauge length and merely untwist the specimen until no twist remains in the yarn structure. This is determined by piercing the fiber bundle with a needle-type pointer and running it through the bundle to ascertain that no twist remains. The number of turns shown on the twist counter is the actual number of turns per 2.54 cm in the yarn. The high variability of twist using this procedure requires that a large number of specimens be analyzed. It is for this reason that the industry prefers the untwist-retwist procedure.

Simple motor-driven twist testers are available from a number of suppliers (e.g., Lawson-Hemphill; Shirley Developments, Inc.; and Zweigle, represented in the USA by W. Fritz Mezger, Inc.).

A twist testing instrument that is fully automatic is the Zweigle D301 Automatic Twist Tester. This instrument is capable of measuring both Z and S twist using the two different popular methods mentioned above (i.e., untwist-twist and direct counting). The D301 printout of data contains values for the number of specimens tested, average twist, standard deviation, coefficient of variation, and the 95 percent confidence level of the twist average.

Yarn Hairiness

Hairiness is characterized as the quantity of freely moving fiber ends or loops projecting from the surface of the yarn. It has become increasingly important in the last few years because of air jet weaving where shed clinging can be devastating with respect to warp and filling stops. It is also known that hairiness can cause barré and shade variation in greige and dyed fabrics.

The Zweigle G565 Hairiness Tester uses the photometric principle as its measuring method. The yarn body and the fibers projecting from it interrupt a light beam which causes fluctuation in the measurable luminance of the beam. In order to allow the number of hairs in several length zones to be counted, the yarn is scanned by four photo-transistors. This allows the G565 to simultaneously count the hairs in 12 different length zones. The tester uses a speed of 50 meters per minute (m/min) and has the capability of adjustable test length of from 10 to 9,999 meters in one-meter increments. The hair lengths measured are 1, 2, 3, 4, 6, 8, 10, 12, 15, 18, 21, and 25 millimeters. The 80-mm x 100-mm (nominally 3 x 4 inches) graphic screen on the tester enables the operator to display a bar graph of the hair distribution by length group, as well as the number of hairs counted in each length group. For the packages of yarn tested in a test series, the data available, by hair length, are mean hairs per meter, standard deviation, coefficient of variation, 95 percent confidence level, and minimum and maximum package hairs per meter. These data can be transmitted to a computer for hard copy.

The Shirley Yarn Hairiness Tester is considered to be the established laboratory method for setting hairiness standards. It also uses optics to facilitate the counting of the number of hairs of a certain length. One major difference between this tester and the Zweigle G565 mentioned earlier is that it measures one pre-selected hair length per test. If the number of hairs of different lengths is desired, the yarns must be tested for each individual length. The length capability is up to a maximum of 10 mm (0.394 inch). In the USA it has been found that a 3-mm (0.118-inch) hair length is the critical length for estimating the pilling propensity of a yarn.

The optical head of the Shirley Yarn Hairiness Tester does not inspect the total circumference of the yarn. Those hairs within an angle of approximately 35 degrees on each side of a vertical plane along the yarn axis are counted. Appropriate statistical information is available from software in the computer/printer.

A portable Shirley Yarn Hairiness Monitor is also marketed by Shirley Developments, Ltd. This unit was developed for use on the production floor for observation of wear on guides and travelers, doffing tube condition, effect of rotor dust buildup, and so forth. The hair length setting for measurement is fixed at 3 mm, but other hair length possibilities exist if requested.

The most recent offering with respect to measuring yarn hairiness comes from Zellweger Uster. Used is an infrared laser optical measuring system that can be fitted to the Uster Tester 3. This allows a measure of hairiness simultaneously while measuring mass variation and thin, thick, and nep-like defects.

The hairiness measurement zone is formed by a field of homogeneous, parallel light beams. When the yarn passes through this field, only scattered light finds its way to the receiver. The fibers that protrude outside of the yarn body are responsible for scattering the light, which results from refraction, diffraction, and reflection, making them appear to be luminous. This scattered light is taken as a measure of the hairiness of the yarn. The amount of scattered light reaching the receiver is connected into an electrical signal and after evaluation, the hairiness of the yarn is specified as total hair length, in centimeters (cm), per cm of yarn. The yarn hairiness data available from the Uster Tester 3 are hairiness index average and standard deviation, diagram charts, and periodic hairiness variation spectrograms. Frequency distribution diagrams and hairiness variation using different cut lengths can also be obtained.

Because some dyes can absorb infrared light, dyed yarn hairiness values can be somewhat lower than those of white yarns having the same hairiness.

Yarn Appearance Grade

Analyzing yarns for visual appearance grade is based on a combination of factors, for example, evenness, hairiness, neppiness, slubs, and foreign matter. Yarns from several packages are normally wound on rectangles of heavy cardboard, wallboard, or light plywood, at least 140 mm x 250 mm (5.5 x 9.8 inches) in size, and finished in dull black. The winder used has the capability of traversing the yarn along the board so that the number of wraps per cm is such that the yarns have the required spacing between them that is based upon the count being studied. Once the boards are wound, they are visually compared to the appropriate yarn size series of ASTM Spun Yarn Appearance Standards. One-half grades are possible from A+ to below D. Index values are assigned to each letter grade to facilitate statistical analyses. It is recommended in ASTM Method Designation D2255 that three independent graders grade the same set of yarns and assign a letter and index value for appearance. This highly subjective test is in widespread use in the USA, but recently electronic yarn appearance grading has been developed and is creating much interest in staple spun yarn plants.

At the 12th International Exhibition of Textile Machinery held in Milan, Italy, in the fall of 1995, three companies showed electronic systems for determining what a yarn might "look" like when wound onto a yarn board and/or when processed into a knitted or woven fabric. These companies were Lawson-Hemphill, Zweigle, and Zellweger Uster.

The Electronic Inspection Board for Spun Yarns (EIB-S) system of Lawson-Hemphill uses an optical yarn diameter sensing device called a Profiler to grade yarn on appearance, based on ASTM Standard D-2255. Two modes of operation, yarn profile and EIB software, are present within one unit. The information gained using the two modes is outlined in Table 12.2.

Besides being able to grade yarn with no subjectivity, it is also possible to use the EIB to aid in yarn clearer settings and to help in predicting fabric appearance. The operator can select yarn board widths onto which the yarn is electronically "wound." If, for example, moiré effect is present at a yarn board width of 30 cm (11.8 inches), it can be hypothesized that problems may occur if the yarn is woven on a loom whose width is 30 cm.

The Zweigle unit for simulating a yarn board for appearance grading and for fabric appearance is called Cotton Yarn Rating On-Line Simulation (CYROS). The operator of this instrument is able to

Table 12.2 Two Modes of EIB Operation (Lawson-Hemphill)

Yarn Profiler Mode
- Shows image of yarn
- Counts thick and thin places
- Gives average yarn diameter
- Neps and hairiness can be measured

EIB Software Mode
- Histogram showing distribution of defects in yarn
- Letter grade of appearance based on ASTM standards
- Printout simulating yarn wound on an appearance grading board

simulate a yarn appearance board similar to that specified in ASTM Standard D-2255 for single-jersey knit fabric and plain-weave fabric appearance. The operator can also choose the knit fabric stitch density, woven fabric picks per inch, and fabric width. Another important feature is that the average optical evenness and defect data for a group of yarns (10 packages, for example) can be stored and used in the appearance simulations. The Zweigle G580 optical evenness discussed in the yarn evenness testing section is used in conjunction with special software to accomplish the CYROS.

Zellweger Uster's Expert software for yarn testing is used in conjunction with the Uster Tester 3 (UT3) as a source for Uster world-wide yarn quality statistics and for locating possible causes of periodic defects detected by the UT3. In addition to these functions, the software can be used for yarn appearance board and for woven and knit fabric appearance. The operator is able to see on-screen what a package of yarn with certain tested characteristics might look like in a single-jersey knit construction, a plain-weave woven construction, or on a yarn appearance board. The present software does not allow the technician to change any elements of the simulated fabric construction.

High Volume Instrumentation for Fiber Property Measurement and Establishing Cotton Mix Laydowns

The first test instrument that most cottons grown in the USA come in contact with is a high volume test instrument (HVI) for measuring fiber fineness, strength, length and length uniformity, color Rd reflectance, color +b yellowness, and trash grade. Some HVI units also have the capability of measuring elongation, maturity, and percent fibers less than 12.7 mm (0.5 inch)(non-spinnable fiber). Cottons being marketed that are eligible for government loan must be HVI tested by United States Department of Agriculture classing offices located throughout the Cotton Belt. In 1995 approximately 16 million bales of the 17.9 million bales harvested were HVI tested.

The single builder of HVI lines is Zellweger Uster, Knoxville, Tennessee. Their first unit was sold in 1985. In late 1995 there were 940 systems in place in 62 countries. The HVI system is a modular arrangement of four different instruments. Thus, the 940 systems in existence in 1995 were each configured to the needs of the user. The module for length and length uniformity operates on the principle mentioned earlier for Zellweger Uster's Digital Fibrograph, and the strength and elongation test, included in the length module, uses the same principle as the Stelometer discussed earlier. The micronaire fineness module's principle of measurement was mentioned in the fineness section, and the color and reflectance module's principle was discussed in the Nickerson-Hunter Colorimeter method. The trash grade module uses an optical sensor to determine the number and proportion of the sample surface (percent area) covered by trash particles. From these values a trash grade is derived which is similar to the USDA standards.

The HVI system from Zellweger Uster uses the bundle test rather than single fiber test to establish property and variability values. If two laboratory technicians operate an HVI 900 testing system, one sample passes through the complete test cycle in approximately 20 seconds. Zellweger Uster's HVI 900A version with automatic sample preparation for the measurement of fiber length, length uniformity, strength, elongation, color, and trash grade can be operated by one technician at comparable test speed. A photograph of Zellweger Uster's HVI 900A is provided in Figure 12.9.

The cost share for raw materials is between 40 and 70 percent of the total yarn cost. Wise selection and use of cottons can improve the competitiveness of a spinning plant. It is in this niche that HVI plays an extremely important role in purchasing the exact qualities of cotton for the end product being manufactured, and for establishing cotton mix laydowns that have essentially the same average fiber properties and variabilities from mix to mix, day after day. Consistent mix laydowns equate to even and efficient running performance in a spinning plant.

Quality Assurance Practices and Procedures in Short Staple Spinning

The measurement and control of the various products in a textile plant are of utmost importance if the final product is to meet acceptable quality levels and performance expectations. In a cotton

Figure 12.9. Photograph of the HVI 900A instrument. (Courtesy Zellweger Uster AG)

mill the very first quality consideration is the establishment of the bale mix laydown to ensure that the yarns made from it are satisfactory and satisfy given final product specifications. Additionally, there are quality measurements that should be performed on a routine basis to ensure proper processing machine function as well as the quality properties and variabilities of the stock in process.

In every short staple spinning plant there exists some type of program for testing materials to determine if they comply with the quality demands of the end product being produced. Obviously, a testing program for a plant producing yarn for pocketing material or denim will not be as all encompassing as a testing program for a plant producing yarn for expensive upholstery fabric or high fashion ladies knitwear.

A typical quality assurance or control program for a combed cotton and polyester draw frame ring spun blend yarn, 65 percent polyester and 35 percent cotton, going into an end use such as low-end knitwear is given in Table 12.3. This example illustrates how some of the testing equipment discussed in this chapter can be used in a quality testing program.

Table 12.3. Typical Quality Assurance Testing Program for a Combed Cotton and Polyester Blend Ring Spun Yarn for Low-End Knitwear

Process	Test Description	Frequency of Test	Sample Size	Method of Test
Incoming Cotton	Physical Properties for Mix Laydown	Two Determinations/Bale	Classer's Sample	Uster HVI
Incoming Polyester	Finish Level	10% of Shipment	Ten grams/bale	Tube Elution
	Crimps Per Inch	10% of Shipment	Ten fiber chips per bale	Direct Counting
Opening & Cleaning (Cotton)	Non-lint Content of Waste under Machines	Quarterly	Seven 0.5-gram samples per waste type	AFIS-T
	Spinnable Fiber in Waste	Quarterly	Seven 0.5-gram samples per waste type	AFIS - L&D
Opening (Polyester)	Spinnable Fiber in Waste	Quarterly	Two 100-gram samples per waste type	AL-101
Carding (Cotton and Polyester)	Sliver Weight Per Meter	All cards daily	One 100-meter length per card	Reel, cut, weigh
	Sliver Uster Evenness, %CV (includes Spectrogram)	20% of cards daily	50 m/min, 2.5 minutes	Uster Tester 3
	Sliver Nep Count	All cards weekly	Seven 0.5-gram samples/per card	AFIS-N
Pre-Draw (Cotton)	Sliver Weight per Meter	20% of deliveries per day	One 5-meter length per delivery	Reel, cut, weigh
	Uster Evenness, %CV (includes Spectrogram)	20% of deliveries per day	50 m/min, 2.5 minutes	Uster Tester 3

Table 12.3 continued.

Process	Test Description	Frequency of Test	Sample Size	Method of Test
Lap Winding (Cotton)	Lap Weight, grains/meter	50% of deliveries per day	One 5-meter length per delivery	Unroll, measure, cut, weigh
	Total Lap Weight, kg.	50% of deliveries per day	Two full laps per delivery	Plant scale
Combing (Cotton)	Sliver Weight per Meter	20% of deliveries per day	One 5-meter length per card	Reel, cut, weigh
	Uster Evenness, %CV (includes Spectrogram)	20% of deliveries per day	50 m/min, 2.5 minutes	Uster Tester 3
	Percent Noil Removal	20% of frames per day	Plant established total nips	Clean, run, weigh sliver and noils, calculate percentage
Intermediate Drawing (Cotton/polyester blend)	Sliver Weight per Meter	20% of deliveries per day	One 5-meter length per delivery	Reel, cut, weigh
	Uster Evenness, %CV (includes Spectrogram)	20% of deliveries per day	50 m/min, 2.5 minutes	Uster Tester 3
Finisher Drawing (Cotton/polyester blend)	Sliver Weight per Meter	All deliveries daily	One 5-meter length per delivery	Reel, cut, weigh
	Uster Evenness, %CV (includes Spectrogram)	All deliveries daily	50 m/min, 2.5 minutes	Uster Tester 3
	Blend Analysis	Twice per week	Correct weight from composite of sliver from all frames	Chemical Analysis

Table 12.3 continued.

Process	Test Description	Frequency of Test	Sample Size	Method of Test
Roving (Cotton/polyester blend)	Roving Size	40% of frames per day	Two front and two back line packages per frame	Reel, cut, weigh
	Uster Evenness, %CV (includes Spectrogram)	20% of frames per day	One 12-meter length per package	Uster Tester 3
Ring Spinning (Cotton/polyester blend)	Yarn Number	25% of frames per week	8 bobbins per count, 100 meters per bobbin	Reel, weigh
	Uster Evenness, %CV with Spectrogram and Imperfection Count	All counts per week and all frames per month	4 bobbins per count, 400 m/min, 2.5 minutes	Uster Tester 3
	Single-End Strength and Elongation	All counts per week	5 bobbins per count, 20 breaks per bobbin	Uster Tensorapid
	Defect Classification	All counts per week	100,000 meters minimum per count	Uster Classimat
	Yarn Appearance Grade	All counts per week	4 bobbins per count	Compare to ASTM standards

Short Staple Spun Yarn Manufacturing Costs

The cost of producing short staple spun yarns is largely dependent upon raw material prices, capital depreciation, local power costs, the capability of process machines, and the manufacturing skills of the people managing and operating the processes. There is in practice a range of costs among spinners for each yarn type produced. Data in this chapter are intended to provide a general understanding of the relative contributions of certain cost factors to the total cost for producing yarns on each main spinning system. The following definitions of cost categories help to explain the tables of data that are presented:

1. Labor costs—for hourly paid production workers including machine operators, machine technicians and overhaulers, and support labor such as cleaners, transporters, and traveler changers;
2. Supply costs—for manufacturing parts and repairs, plant facility, auxiliary equipment, and office supplies;
3. Utility costs—for electrical power, other energy fuels, and for water;
4. Benefit costs—for personnel costs such as payroll taxes, pensions, vacations, holidays, and insurance;
5. Management & Administrative costs—for manufacturing site management, supervision, technical and administrative personnel;
6. Depreciation costs—for writing off the cost of capital equipment and facilities; and
7. Corporate & Other Overhead costs—for Corporate office

functions, general corporation and administrative sales functions and property taxes.

The percent contributions to total cost of various cost categories are presented for each spinning system in Table 13.1. These data are averages based on an industry survey that included over 50 short staple spinning systems.

Relative operating costs can be compared among the spinning systems across a range of counts in the form of cost indices. Such a comparison is presented in Table 13.2 and assumes these costs for producing a 20's Ne carded ring spun yarn to be an index of 1.00. The analysis indicates that with the machines in place in 1996, rotor and air jet yarns are on average less costly to manufacture than are ring yarns, and rotor and air jet yarns have similar cost in the 20's and 30's count range.

Table 13.1. 1996 Average Percent Contribution to Total Costs for the Production of Various Carded Spun Yarns

Cost Category	% Contribution to Cost by System and Count (Ne)								
	Ring				Rotor			Air Jet	
	10's	20's	30's	40's	10's	20's	30's	20's	40's
Total Hourly Labor	17.9	16.5	15.0	14.6	7.8	8.0	8.3	13.3	11.9
Benefits	3.6	3.3	2.9	2.8	2.1	1.9	1.7	2.0	2.4
Supplies	4.5	5.2	5.4	6.0	5.1	4.4	3.5	3.0	4.7
Utilities	3.6	5.0	8.1	8.3	3.6	4.2	4.9	4.1	5.5
Management/Admin.	5.7	5.7	5.1	5.0	1.8	1.6	1.4	2.5	2.1
Depreciation	3.6	7.4	11.4	13.7	4.5	8.0	10.5	3.5	11.5
Corp./Other Overhead	6.4	8.3	8.7	10.1	7.1	6.6	6.4	3.6	3.4
Raw Materials (Fiber)	54.7	48.6	43.4	39.5	68.0	65.3	63.3	68.0	58.5

Table 13.2. Indices of 1996 Relative Average Operating Costs (Excluding Fiber) for Surveyed Ring, Rotor, and Air Jet Carded Spun Yarns

Operating Cost Category	Cost Indices (1.00 = Base for Comparison)									
	10's Ne		20's Ne			30's Ne			40's Ne	
	Ring	Rotor	Ring	Rotor	Air Jet	Ring	Rotor	Air Jet	Ring	Air Jet
Labor										
	0.30	0.08	0.32	0.10	0.16	0.34	0.12	0.18	0.36	0.20
Benefits										
	0.06	0.02	0.06	0.02	0.03	0.06	0.02	0.03	0.06	0.04
Supplies										
	0.09	0.05	0.10	0.05	0.04	0.12	0.05	0.06	0.14	0.08
Utilities										
	0.06	0.05	0.10	0.06	0.06	0.17	0.07	0.07	0.20	0.09
Mgmt./Admin.										
	0.09	0.02	0.11	0.02	0.03	0.11	0.02	0.03	0.12	0.03
Depreciation										
	0.06	0.06	0.15	0.11	0.05	0.24	0.15	0.11	0.33	0.19
Corp./Other OHD										
	0.11	0.09	0.16	0.08	0.05	0.18	0.09	0.05	0.23	0.06
Totals	0.77	0.37	1.00	0.44	0.42	1.22	0.52	0.53	1.44	0.69

Chapter 14

Strategies for Competitiveness in the Manufacturing of Short Staple Yarns

Yarn manufacturing staff members at the Institute of Textile Technology were recently asked to develop a strategy for spun yarn companies operating in a tough and globally competitive marketplace. The goal of the strategy, to be developed with emphasis on the technical perspective, would be to assist the short staple yarn manufacturer in becoming *the* preferred supplier of their products. It is assumed that state-of-the-art technology is in place as needed for competitiveness. The following other key areas comprise the major strategy emphasis:

1. Reduce variabilities in the yarn.
2. Monitor important aspects to ensure compliance.
3. Think and analyze costs systemically.
4. View process engineering as a dynamic and never-ending activity.
5. Develop partnerships in order to acquire new knowledge.
6. Confirm or improve on machinery setup recommendations.
7. Have on-going and well-designed experiments.
8. Involve everyone!

Reducing variabilities deserves an in-depth discussion because of the numerous influences on product and process variability in a spun yarn manufacturing operation. This subject is explored in great depth after brief comments about the other seven strategy tenets.

Measure Everything

An important philosophy is that, "if you do not measure it, you can not control or improve it." Some key manufacturing items to monitor are exposed in the discussion on reducing variabilities. In some cases, new and practical measurement techniques must be created. First, there has to be an understanding of what and why to measure. This should be one expectation of any technical education offering considered.

Analyze Costs Systemically

Thinking about and analyzing total cost, from planning to product use, is becoming more and more necessary. Focusing on an individual cost element or on a single process or department can lead to the failure to make the right decision. Examples would be decisions regarding (a) waste removal in opening and carding, (b) labor content for cleaning machines, (c) machine speeds, (d) soft cots at ring spinning, and (e) choice of wire for the rotor spinning combing roll.

Another very important cost analysis component should be input from the customer. How else will the sales yarn producer confirm that the knitter's number one priority is perhaps package quality rather than a specific yarn property? This piece of information would make for better decisions regarding investments in package quality monitoring instrumentation.

Dynamic Process Engineering

View the engineering of the process as a never-ending activity. It would surprise many to learn that there are cards and draw frames that have had the same settings for years. This can not be ideal, since from year to year cotton frictional and other properties change, and occasionally there are merge or other changes in synthetic fibers that affect the way the fibers draft. Measuring the right things at every process will focus timely attention on the need for process specification modification.

Develop Partnerships

No one group of persons has all the knowledge necessary for manufacturing short staple yarns in the best way possible. Reach out to the supplier community, to customers, to universities, and to industry consortia with a true partnership mentality. This means giving to the partnership as well as getting from the partnership.

Test Machinery Setup Recommendations

Test all recommendations regarding machinery setup and operation. Use recommendations as starting points, and evaluate options for fine-tuning each process.

Continuously Improve

Once processes are set up, do not become complacent. Textile processes are dynamic and are sometimes affected by even slight changes in raw material, ambient conditions, substitute machine parts, and especially by changes in prior processes. There are always opportunities for honing a process and thereby improving productivity and quality.

Involve Everyone

Familiarize people with the appropriate diagnostic tools that are available. Educate them as to why every detail of manufacturing is important and how each can be measured so that control and improvement are made possible. The newest approach for engaging more persons in ownership of processes and products involves the creation and support of empowered teams. Whatever mechanism is used, broad participation of persons associated with the manufacturing system is the right thing. Give each person the necessary training and education and a stake in the business of continuous improvement toward being the preferred supplier.

Reduce Variabilities

Reducing variabilities in the yarn needs to become the focal point of those people responsible for the processes of yarn manufacturing. In order to be thorough and yet simple in listing the ways to reduce variability, the questions that follow are provided and should be asked and answered by those persons responsible for spun yarn manufacturing excellence. Reasons for and answers to these questions are intended to be exposed in the primary chapters of this textbook. Review the appropriate sections as necessary.

Opening Room

1. Are color, short fiber, maturity, and trash factored into laydown management?
2. Are the bales placed optimally in the laydown?
3. Is blending adequate as indicated by the machinery system blend factor number?
4. Is reworkable waste consistent, well opened, blended, homogeneous, and if fed back into the system, fed at a constant rate?
5. Are bales bloomed at least 24 hours?
6. Are ambient conditions controlled throughout the plant?
7. Is the opening line throughput rate low enough for quality?
8. Are there data to prove that the cotton cleaning machines (if applicable) are performing at their potential?

Sliver Manufacturing

9. Are all the card mats the same weight?
10. How is the card mat inch-to-inch consistency?
11. Are the card autolevelers calibrated correctly?
12. Is the sliver from each card slub-free, highly aligned, and free of impurities?
13. Is a system in place to ensure randomization in drawing creels?
14. Do all slivers have consistent bulk?
15. Do data exist to prove that roll spacings and draft distributions are optimum for the fibers being processed?
16. How is the quality of sliver documented? What must be known beyond Uster %CVm?
17. Is the right trumpet size being used at sliver processes?
18. Is the draw frame producing slubs?
19. Are can builds designed for quality and production?
20. Are draw frame autolevelers calibrated properly?

Roving

21. What is the best way to handle leftover piecings in the roving creel?
22. Is the mid-term weight variation of all fiber assemblies under control?
23. Is the tension consistent throughout the build of a roving package (if applicable)?
24. Are flyer speeds at the roving process established based on quality or production?
25. Is the roving twist appropriate for quality spinning?
26. Are position-to-position and machine-to-machine settings consistent?

Ring Spinning

27. What is the critical draft at ring spinning for each roving type?
28. How many factors at ring spinning affect yarn count variability? How many are just in the creel?
29. Is the traveler design suited to the spinning speed?
30. Is the front roll overhang *consistent* and set for best quality?
31. What can be done to prevent traveler loading?
32. How much do the yarns shed and what can be done to reduce shedding?
33. Is the relationship of wind-to-bind and total spinning coils appropriate for high speed unwinding and spinning?

Yarn Preparation

34. Are all wax discs turning?
35. Is winding tension consistent?
36. Are procedures in place to minimize backwinding? What percent of winding capacity is used for backwinding?
37. Is yarn conditioning consistent day to day? How is it checked?

Barré Prevention

38. Are barré prevention measures set up throughout the plant?
    - Micronaire checks at carding.
    - Black light color checks at packing.
    - No mixtures of backwound and straight-wound packages.
39. What procedure is in place to keep the customer from mixing old and new yarn?
40. Does package labeling ensure separation of yarns spun under different conditions?

Rotor Spinning

41. Is the evenness of the sliver fed to the rotor spinning machine equal to or less than 3.5% CV?
42. Has the non-lint content in the feed sliver used in rotor spinning been reduced to less than 0.10% as tested on the ITV Dust and Trash Tester (MDTA3)?
43. Has the draft setting on the rotor spinning machine been optimized to yield the best attainable yarn quality?
44. Does the yarn tension in rotor spinning fall below the maximum allowable for efficiency and yarn elongation?
45. Is the navel being used optimally for the yarn count and fiber system being spun?
46. Have the rotor spinning room ambient conditions been set and maintained consistently?

Air Jet Spinning

47. Have total draft and draft distribution been optimized on the air jet spinning machine?
48. Are all draft zone settings on air jet machines correct?
49. Are nozzle settings and pressures optimized?
50. Has feed ratio on air jet spinning machines been optimized?
51. Is winding tension uniform for all deliveries on the air jet machine?
52. Have maintenance procedures and frequencies been established and documented for all machines in the yarn plant?

Determining Fiber Alignment in Sliver Using the ITT Bulk Tester

Description

The ITT Bulk Tester (schematic shown in Figure A.1) is a trough 8 inches long, 1 inch wide, and 5 inches deep. It has a base block that is 4 inches long; thus the height of slivers under compression are measured over a 4-inch length. A weight holder,

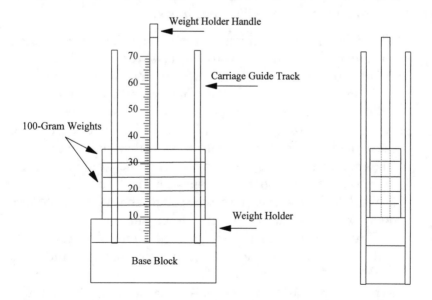

Figure A.1. Drawing of the ITT Bulk Density Tester, used for sliver fiber alignment measurement.

weighing 100 grams, is used to apply pressure to the top of the sliver bundle for compression. There are five additional 100-gram weights that can be sequentially and cumulatively applied to the top of the sliver bundle after it has been loaded into the trough.

The fixed scale on one side of the trough allows a quick and easy means to read the compressed height of the sliver after each increment of weight has been applied.

Principle

The basis for this test is that compressed volume of textile fibers in strand (sliver) form is related to fiber straightness/alignment. When the fibers are straight and aligned, the compressed volume is less. When the fibers are poorly aligned and hooked, such as in card sliver, the compressed volume is greater. With this test apparatus, sliver with straight and well aligned fibers will be compressed to a lower scale reading (height in trough) than slivers with hooked and unaligned fibers:

Suggested Procedure

The following procedure explains the methods and calculations for determining the Fiber Alignment Index:

1. Weigh five 10-yard lengths of the sliver to be tested and determine the grain weight/yard for the sliver.
2. Divide 500 by the sliver grain weight and round to the nearest whole number. This is the number of 12-inch pieces that will be needed for testing.
3. Cut the required number of 12-inch lengths and lay them in the trough. Care is needed to ensure that the sliver is not stretched or twisted when it is placed in the trough.
4. Place the weight holder with one additional 100-gram weight (a total of 200 grams) into the trough and allow it to come down slowly until the sliver bundle supports the weights. Record the height of the compressed bundle by reading the scale at the bottom edge of the weight holder. Remove the weight holder from the trough.
5. Repeat Step 4 using a total of 400 grams and 600 grams on the weight holder.

6. After the last reading, remove the sliver samples and weigh to within one grain.

7. Repeat steps 1-6 for at least four additional samples (5 samples total).

8. Calculate the average height readings for each weight category of each of the 5 samples. Calculate the average grain weight for the 5 samples.

9. Calculate the Equivalent Grain Weight/Yard: Equivalent Grain Weight = Measured Grain Weight of 12-Inch Lengths x 3.

10. A Correction Factor is necessary to adjust for specimens where the Equivalent Grain Weight is not 500 grains. The Correction Factor is calculated as follows: Correction Factor = 500/Equivalent Grain Weight.

11. Determine the Fiber Alignment Index (AI): AI=(Height Reading @ 200 grams + Height Reading @ 400 grams + Height Reading @ 600 grams) x Correction Factor.

Table A.1 is the form for use with the ITT Bulk Tester. An example set of completed test data is presented in Table A.2.

Table A.1. Form for Use with the ITT Bulk Test

BULK TEST FOR FIBER STRAIGHTNESS AND ALIGNMENT IN SLIVERS

Operator: _____   Date: _____

Sample Description: _____   Sliver Weight: _____

Number of Pieces: _____

Load on Sample (Grams)	Specimen Number										Average Value	Corrected to 500 Grains
	1	2	3	4	5	6	7	8	9	10		
200												
400												
600												
Specimen Wt. (Grains)											Alignment Index (Total of Corrected Values)	

Average Equivalent Weight: _____

Correction Factor: _____

Table A.2. Sample Results from the ITT Bulk Test

BULK TEST FOR FIBER STRAIGHTNESS AND ALIGNMENT IN SLIVERS

Operator: ___RT___

Date: ___3/21/97___

Sample Description: ___Finisher Sliver Draw Frame # 2___

Sliver Weight: ___335 grains/yd___

Number of Pieces: ___14___

Load on Sample (Grams)	Specimen Number										Average Value	Corrected to 500 Grains
	1	2	3	4	5	6	7	8	9	10		
200	12.5	12.5	12.5	12.0	12.5						12.4 x 1.01	12.5
400	10.0	10.0	9.5	9.5	10.0						9.8 x 1.01	9.9
600	8.0	8.5	9.0	8.5	9.0						8.6 x 1.01	8.7
Specimen Wt. (Grains)	164.8	165.3	166.2	165.1	165.6						165.4	
Alignment Index (Total of Corrected Values)												31.1

Average Equivalent Weight: ___165.4 x 3 = 496.2___

Correction Factor: ___500/496.2 = 1.01___

Converting Between Units Commonly Used in Staple Yarn Manufacturing

A. Starting With This	Multiply By This	To Convert To This
Bars	1.019716	Kilograms/Square Centimeter (kg/cm²)
Bars	14.5038	Pounds/Square Inch (PSI) or lb/in.²
Centimeters (cm)	0.39370079	Inches (in.)
Centimeters (cm)	0.01	Meters (m)
Centimeters (cm)	10	Millimeters (mm)
Centimeters (cm)	10000	Microns (&)
Cubic Centimeters (cm³)	0.061023744	Cubic Inches (in.³)
Cubic Feet (ft³)	28316.847	Cubic Centimeters (cm³)
Cubic Feet (ft³)	0.028316847	Cubic Meters (m³)
Cubic Feet (ft³)	28.316847	Liters (l)
Cubic Inches (in.³)	16.387064	Cubic Centimeters (cm³)
Cubic Inches (in.³)	0.016387064	Liters (l)
Cubic Inches (in.³)	16.387064	Milliliters (ml)
Feet (ft)	30.48	Centimeters (cm)
Feet (ft)	0.3048	Meters (m)
Feet/Minute (FPM) or ft/min	0.3048	Meters/Minute (m/min)
Feet/Minute (FPM) or ft/min	0.00508	Meters/Second (m/sec)
Feet/Second (ft/sec)	18.288	Meters/Minute (m/min)
Grains (gr)	0.06479891	Grams (g)
Grains (gr)	0.0022857143	Ounces (oz) (avdp.)
Grains (gr)	0.00014285714	Pounds (lb) (avdp.)
Grams (g)	15.432358	Grains (gr)
Grams (g)	0.001	Kilograms (kg)
Grams (g)	1×10^6	Micrograms (μg)
Grams (g)	0.035273962	Ounces (oz) (avdp.)
Grams (g)	0.0022046226	Pounds (lb) (avdp.)
Inches (in.)	2.54	Centimeters (cm)
Inches (in.)	0.0254	Meters (m)
Inches of H_2O (4 °C)	0.03612628	Pounds/Square Inch (PSI) or (lb/in²)

Inches/Minute (in./min)	152.4	Centimeters/Hour (cm/hr)
Kilograms (kg)	15432.358	Grains (gr)
Kilograms (kg)	2.2046226	Pounds (lb) (avdp.)
Kilograms (kg)	0.001	Metric Tons
Liters (l)	1000	Cubic Centimeters (cm^3)
Liters (l)	61.02545	Cubic Inches (in.^{3})
Liters (l)	0.2641794	Gallons (U.S. Liquid)
Meters (m)	100	Centimeters (cm)
Meters (m)	3.2808399	Feet (ft)
Meters (m)	39.370079	Inches (in.)
Meters (m)	1000	Millimeters (mm)
Meters (m)	1.0936133	Yards (yd)
Meters/Minute (m/min)	3.2808399	Feet/Minute (FPM) or (ft/min)
Meters/Minute (m/min)	0.054680665	Feet/Second (ft/sec)
Meters/Second (m/sec)	196.85039	Feet/Minute (FPM) or (ft/min)
Meters/Second (m/sec)	3.2808399	Feet/Second (ft/sec)
Micrograms (µg)	1×10^{-6}	Grams (g)
Micrograms (µg)	0.001	Milligrams (mg)
Microns (µ)	0.0001	Centimeters (cm)
Microns (µ)	0.001	Millimeters
Millibars	0.001	Bars
Millibars	1.019716	Grams/Square Centimeter (g/cm^2)
Millibars	2.088543	Pounds/Square Foot (lb/ft^2)
Millibars	0.0145038	Pounds/Square Inch (PSI) or lb/in.^{2})
Newtons	0.22480894	Pounds (lb)
Ounces (oz) (avdp.)	437.5	Grains (gr)
Ounces (oz) (avdp.)	28.349523	Grams (g)
Pounds (lb) (avdp.)	7000	Grains (gr)
Pounds (lb) (avdp.)	453.59237	Grams (g)
Pounds (lb) (avdp.)	0.45359237	Kilograms (kg)
Pounds (lb) (avdp.)	0.00045359237	Metric Tons
Pounds/Square Foot (lb/ft^2)	0.000478803	Bars
Pounds/Square Foot (lb/ft^2)	0.48824276	Grams/Square Centimeter (g/cm^2)
Pounds/Square Foot (lb/ft^2)	0.192227	Inches of H$_2$O (39.2 °F)
Pounds/Square Foot (lb/ft^2)	4.8824276	Kilograms/Square Meter (kg/m^2)
Pounds/Square Inch (PSI) or (lb/in.^{2})	0.0689476	Bars
Pounds/Square Inch (PSI) or (lb/in.^{2})	70.306958	Grams/Square Centimeter (g/cm^2)
Pounds/Square Inch (PSI) or (lb/in.^{2})	27.6807	Inches of H$_2$O (39.2 °F)
Pounds/Square Inch (PSI) or (lb/in.^{2})	0.070306958	Kilograms/Square Centimeter (kg/cm^2)
Square Centimeters (cm^2)	0.001076391	Square feet (ft^2)
Square Centimeters (cm^2)	0.15500031	Square inches (in.^{2})
Square Feet (ft^2)	929.0304	Square Centimeters (cm^2)
Square Feet (ft^2)	0.09290304	Square Meters (m^2)
Square Inches (in.^{2})	6.4516	Square Centimeters (cm^2)

Square Inches (in.²)	0.00064516	Square Meters (m²)
Square Inches (in.²)	645.16	Square Millimeters (mm²)
Square Meters (m²)	10.76391	Square Feet (ft²)
Square Meters (m²)	1550.0031	Square Inches (in.²)
Square Meters (m²)	1.19599	Square Yards (yd²)
Square Millimeters (mm²)	0.0015500031	Square Inches (in.²)
Square Yards (yd²)	8361.2736	Square Centimeters (cm²)
Square Yards (yd²)	0.83612736	Square Meters (m²)
Tons (metric)	1.1023113	Tons (avdp.)
Tons (avdp.)	0.90718474	Tons (metric)
Yards (yd)	91.44	Centimeters (cm)
Yards (yd)	0.9144	Meters (m)

B. Starting With This	Perform This	To Convert To This
Degrees Celsius	9/5 (°C) +32	Degrees Fahrenheit
Degrees Fahrenheit	5/9 (°F - 32)	Degrees Celsius
English Count (Ne)	590.54/(Ne Count)	Tex
English Count (Ne)	5314.84/(Ne Count)	Denier
English Count (Ne)	(Ne Count) x 1.5	Worsted Count (Nw)
English Count (Ne)	(Ne Count) x 1.69336	Metric Count (Nm)
English Count (Ne)	(Ne Count) x 0.525	Woolen Run
English Count (Ne)	(Ne Count) x 2.8	Woolen Cut
Tex	590.54/Tex	English Count (Ne)
Tex	Tex x 9	Denier
Tex	885.81/Tex	Worsted Count (Nw)
Tex	1000/Tex	Metric Count (Nm)
Tex	310.03/Tex	Woolen Run
Tex	1653.51/Tex	Woolen Cut
Denier	5314.84/Denier	English Count (Ne)
Denier	Denier/9	Tex
Denier	7972.26/Denier	Worsted Count (Nw)
Denier	9000/Denier	Metric Count (Nm)
Denier	2790.29/Denier	Woolen Run
Denier	14881.55/Denier	Woolen Cut
Worsted Count (Nw)	(Nw Count)/1.5	English Count (Ne)
Worsted Count (Nw)	885.81/(Nw Count)	Tex
Worsted Count (Nw)	7972.26/(Nw Count)	Denier
Worsted Count (Nw)	(Nw Count) x 1.1289	Metric Count (Nm)
Worsted Count (Nw)	(Nw Count) x 0.35	Woolen Run
Worsted Count (Nw)	(Nw Count) x 1.866667	Woolen Cut
Metric Count (Nm)	(Nm Count)/1.69336	English Count (Ne)
Metric Count (Nm)	1000/(Nm Count)	Tex
Metric Count (Nm)	9000/(Nm Count)	Denier
Metric Count (Nm)	(Nm Count) x 0.88581	Worsted Count (Nw)
Metric Count (Nm)	(Nm Count) x 0.31003	Woolen Run
Metric Count (Nm)	(Nm Count) x 1.65351	Woolen Cut
Woolen Run	(Woolen Run)/0.525	English Count (Ne)

Woolen Run	310.03/(Woolen Run)	Tex
Woolen Run	2790.29/(Woolen Run)	Denier
Woolen Run	(Woolen Run)/0.35)	Worsted Count (Nw)
Woolen Run	(Woolen Run)/0.31003	Metric Count (Nm)
Woolen Run	(Woolen Run) x 5.333333	Woolen Cut
Woolen Cut	(Woolen Cut)/2.8	English Count (Ne)
Woolen Cut	1653.51/(Woolen Cut)	Tex
Woolen Cut	14881.55/(Woolen Cut)	Denier
Woolen Cut	(Woolen Cut)/1.866667	Worsted Count (Nw)
Woolen Cut	(Woolen Cut)/1.65351	Metric Count (Nm)
Woolen Cut	(Woolen Cut)/5.333333	Woolen Run

Glossary of Terms

accelerator A device or an arrangement for controlling the yarn envelope (balloon) as it is being removed from a ring spinning bobbin on a high speed winder.

acetate A generic fiber category that has been defined by the Federal Trade Commission as "a manufactured fiber in which the fiber-forming substance is cellulose acetate. Where not less than 92% of the hydroxyl groups are acetylated, the term triacetate may be used as a generic description of the fiber."

acrylic A generic fiber category whose properties are: soft, wool-like hand; brilliant colors obtainable on certain types of fibers in this category; excellent sunlight resistance; good wash-and-wear performance; good wrinkle resistance. Fiber usually is used in staple form.

AFIS Acronym for Advanced Fiber Information System. High speed equipment developed by Zellweger Uster Corporation for automated determination of cotton characteristics including maturity and nep count.

air jet spinning A type of open-end spinning that utilizes a stationary tube in which jets of air are directed to cause fibers to twist, thereby forming a yarn.

apron A small conveyor belt driven by rolls in a drafting zone. Aprons support and help control the movement of fibers passing through the zone.

array In cotton fiber testing, a display that shows the weight percentages of fibers in a sample having different fiber lengths.

ASTM Acronym for American Society for Testing and Materials. Organization that sets standards for materials, products, systems, and services, including those for textiles through the ASTM Committee D13. Headquarters: Philadelphia, PA.

autodoffer Apparatus used mainly on yarn spinning machines and on winders to remove full bobbins and replace them with empty ones. An autodoffer can be attached to a machine or can be transported to various machines as needed.

autoleveler A device for improving the uniformity of sliver, roving, or yarn during manufacture by detecting size variations and feeding back control signals.

autoleveling The use of an autoleveler.

backwinding Winding yarn from one package to another, more commonly to a larger package. May be done to improve smoothness of unwinding.

bale A bag, sack, square, or oblong package into which staple fiber is compressed. The sizes and weights of bales vary; however, most American cotton bales weigh an average of 480 pounds.

balloon In spinning, twisting, winding, or unwinding of yarn, the curved path traversed by the running yarn between eyelet and package or between package and roller nip.

balloon control ring A metal ring positioned between the balloon eye and ring on a spinning or twisting machine.

batt A random arrangement of fibers, usually cotton fibers, creating a blanket in which the fibers hold together by entanglement and friction; synonyms: mat, fleece, lap.

benchmarking (1) Definitely stated stretch levels of quality and process parameters which can be obtained in producing competitive products for marketing. (2) The best achievable quality or process parameters observed.

bi-component yarn (1) Plied yarn where the singles components are different. (2) Corespun or wrapped yarn.

blend A textile containing two or more different generic fibers, variants of the same generic fiber, or different colors or grades of the same fiber. When different fibers are combined by weight in the Opening Room of the yarn spinning plant, the blend is an intimate blend. When different fibers are combined by weight at the drawing process, it is a draw frame blend.

blend factor (1) The number assigned to a mixing machine which indicates its fiber tuft mechanical mixing or folding capacity. (2) Also the number assigned to the cumulative mixing power of a group of mixing machines that are used in a common staple fiber processing line.

blending (1) The process of preparing homogeneous mixtures, usually with specific proportions, of two or more generic fibers or two or more variants of the same fiber. (2) An all encompassing term describing all actions necessary for creating such a homogeneous mixture of fibers including fiber sampling, testing, warehousing, computer laydown management, mixing, doubling, and cross-blending of slivers and laps.

blowroom British term for the opening, cleaning, and blending areas of a yarn mill.

bobbin In textile production, a cylindrical or slightly tapered wood, cardboard, or plastic core on which yarns are wound for various operations. It has a hole in the center so that it will fit on a spindle, skewer, shaft, or other holding device.

break draft The small amount of draft between the rear and next to last rolls in a drafting zone.

break factor In yarn testing, the strength of a standard skein in pounds, multiplied by the cotton count. Also known as count-strength product (CSP).

bulk density The apparent weight or mass per unit volume of a material such as a textile.

Bulk Density Tester A particular procedure for determining relative fiber straightness and alignment in a continuous strand, usually sliver, as developed by the Institute of Textile Technology.

burrs Rough protrusions or edges left on metal or other solid materials caused by scoring, scratching, cutting, or drilling.

calender roll Fluted roll that is used to compress fiber lap or sliver during processing.

card clothing A thick foundation material usually made with textile fabric through which many fine, closely spaced, specially bent wires project. Various surfaces of the card are covered with this material and it is these that perform the cleaning and equalizing functions of the card.

card web The thin sheet of fibers delivered by a card before condensation to sliver. Card webs may be combined to make dryland nonwovens. In woolen spinning system yarn manufacturing, the card web is spit into narrow strips by the tape condenser and then spun.

carding Preliminary process in spun yarn manufacture. The fibers are separated, distributed, and equalized, and formed into a thin web and condensed into a continuous, untwisted strand of fibers called a sliver. This process removes most of the impurities and a certain amount of short, broken, or immature fibers. The operation is performed on a card.

centrifugal force The outward force acting upon a mass as it is being rotated about a center.

cheese A cylindrical package of yarn firmly cross-wound on a paper or wooden tube or a spring. A cheese usually has a diameter far greater than the width of the traverse, as distinguished from the yarn package known as a "tube". Like the latter, the package has no flanges. Cheeses wound on springs can be package dyed.

Classimat defects A measurable disruption of a uniformily continuous strand as determined by the testing procedure and equipment as marketed by Zellweger Uster; defects (faults) are separated into 23 classifications for the Classimat II tester depending on mass change and length of defect.

cleaner Processing machine for cotton or wool, the most important function of which is to remove nonfibrous trash and, in the case of cotton, motes.

clearers (1) A term applied to particular devices on textile processing machinery that prevent damaging buildup of extraneous fibers on critical elements of the machines. (2) Sensors and associated components on winders that cut out defects of preset magnitude.

clearing Removing such blemishes and imperfections as unwanted slubs from a yarn; the imperfections are replaced by splices or knots.

clump (1) An irregular, tight grouping of fibers. (2) A defect resulting from insufficient fiber separation during web formation.

cobwebbing Bunching of yarn at the end of a wound package because of slippage; can produce yarn breaks in unwinding.

coefficient of determination (r²) The square of the statistically calculated correlation coefficient (r). A term applied to the relationship between two variables, which reveals the percentage magnitude of the effect of one of the variables on the other variable.

cohesiveness (cohesion) In textiles, a term that describes the ability of fibers to cling together or resist being pulled apart in processing. Fiber length, crimp, and presence of lubricants affect cohesiveness.

coiler Mechanical device that positions sliver in a cylindrical sliver can in helical coils one half the diameter of the can. Coiled sliver can be removed easily with a minimum of fiber disturbance.

colorfastness The resistance of a material to change in any of its color characteristics, to the transfer of its colorant(s) to adjacent materials, or both, as a result of the exposure of the material to any environment that might be encountered during the processing, testing, storage, or use of the material.

combing A step that is subsequent to the carding process in both cotton and worsted spinning system yarn manufacture. In the cotton spinning system, card slivers are combined into a sliver lap before combing. The combing process separates the long, choice, desirable fibers from the neps and shorter stock (noil). The comber straightens and arranges them in parallel order in the form of combed sliver. Practically all remaining foreign matter is removed from the fiber stock. Only the best grades of cotton and wool may be combed. Combed yarns are finer, cleaner, more lustrous, and stronger than carded yarns. Combing is necessary for the production of fine yarns and also is applied to coarser yarns when high quality is desired.

composite yarn A yarn comprised of two or more staple fiber and/or filament components that are combined in the spinning process.

condenser (1) The "trumpet" used to convert a web of fibers to a roving or sliver. The web enters the "bell" and exits in rope form. (2) Perforated cylinder used to collect airborne fibers as a means of fiber transport between processes.

cone (1) A hollow yarn holder or bobbin of conical shape used as a core on which to build a yarn package. Synonym: cone core. Cones may be made of cardboard or plastic; the latter type may be perforated for dyeing. (2) Yarn when wound on a package of conical form. Because of their large size and ease of overend unwinding, cones are popular as the yarn supply for weft knitting and for filling in shuttleless weaving.

Confidence Interval The magnitude of difference between comparative measurements that is necessary for establishing definite differences at a reliable expected percentage of assurance, usually 95 percent. Synonyms: Confidence Limit, d-min.

Confidence Limit See Confidence Interval, also know as d-min; the minimum amount of difference for stating that two numbers are statistically different at an expected percentage of assurance.

correlation coefficient (r) The relationship between two variables expressed in decimal proportions from zero to one, one being perfect correlation.

cot Covering material used on various fiber processing rolls, especially drafting rolls.

cotton A cellulosic vegetable seed fiber consisting of unicellular hairs attached to the seed of several species of the genus Gossypium, of the family Malvaceae. The length of the fiber is roughly 1 inch (2.5 cm) but varies from less than 1/2 inch (1.2 cm) to over 2 inches (5 cm), and the diameter is quite fine. The normal color of cotton is a light to dark cream, though it may be brown or green depending on the variety, weather, and soil conditions.

cotton dust In raw cotton, particles of trash that have diameters less than 0.5 mm. Consists largely of leaf debris and dirt. Respirable dust, on the other hand, consists of particles below diameter 0.015 mm.

cotton grading A system of evaluating cotton with respect to its cleanliness, color, and ginning preparation.

cotton spinning system A process of manufacturing staple fiber up to 64 mm or 2.5 inches in length into yarn. This includes the general operations of opening, cleaning, carding, drawing, roving, and spinning in the production of so-called carded yarns. For combed yarns, three steps culminating in combing are included after the carding operation.

cotton staple length An average length of a sample or bale of cotton fiber, determined by custom; a selected portion of the fibers is measured, and the length is assigned to the sample or bale as a whole. Not a mathematical average of the contents of the bale, but based on the longer fibers in the sample or samples selected.

count (1) The count or number of the yarn is the numerical designation given to indicate yarn size and is the relationship or ratio of length and weight. See Yarn Number. (2) Woven fabric count is indicated by enumerating first the number of warp ends per inch then the number of filling picks per inch. For instance, 68 x 72 means there are 68 ends per inch and 72 picks per inch.

cover (1) Degree to which surface fibers obscure the underlying structure of a fabric. (2) An appearance of fullness and density in fabric, especially napped fabric, with very little open space left between the yarns. A "well-covered" cloth has such an effect.

cradle The mechanism, usually molded plastic, for positioning and holding in place the aprons used for controlling fibers in the zones of higher drafts in processing machines that attenuate the fibrous strands.

creel (1) A spool rack on which to wind warp yarns. (2) A framework supporting packages of sliver, roving, or yarn so they can be drawn off smoothly without tangling.

creeling Process of loading sliver, roving, or yarn packages on a creel.

crimp Waviness in fibers, especially wool and manufactured fiber staples. This characteristic may be measured by the difference in distance between two points on the fiber as it lies in an unstretched condition and the same two points when the fiber is straightened under specified tension, expressed as a percentage of the unstretched length.

critical draft The term applied to the magnitude of draft at which the drafting force fluctuates rapidly and unevenly (slip-stick effect). It is measured on a device called a Draftometer developed by the Institute of Textile Technology.

crown The layers of sliver above the top rim of a sliver can.

crush rolls Pair of smooth pressure rolls positioned after the doffer in a cotton card to disintegrate remaining vegetable trash in the card web.

cut system Unit of yarn measurement used in the woolen yarn numbering system, especially in the Philadelphia area. It is the number of 300-yard (274-m) lengths per pound.

cylinder In a carding machine, a large iron shell completely covered with card wire clothing. Fiber stock is fed onto the cylinder by the lickerin, brushed against flats or workers, and is removed by the doffer as a card web.

defect Deviation from an intended requirement or reasonable expectation of use for a product or service.

denier An important international direct numbering system for describing linear densities of silk and manufactured filament yarns and fibers other than glass. Denier is equivalent numerically to the number of grams per 9,000 meters length of the material.

direction of twist Yarn or cord is held in a vertical position to determine the direction of twist; in S-twist, the spirals conform in slope to the central portion of the letter S; in Z-twist, the spirals conform in slope to the central portion of the letter Z. The Z-twist is known as regular twist; S-twist is known as reverse twist.

doff To remove the bobbin, beam, or package from a textile machine (a full bobbin usually is replaced immediately with an empty one).

doffer The final carding machine cylinder from which the web of carded fiber is removed.

donning The act of installing of empty bobbins, cones, or tubes onto a machine that winds yarn onto such bobbins, cones, or tubes. Donning is done manually or automatically.

doubling The operation of combining two or more strands of sliver, roving, or yarn in the yarn manufacturing process.

doublings This term is used to describe the number of fibrous units, such as laps, slivers, or rovings, that are layered in sandwich form or positioned side by side to be drafted out to a weight per unit length less than the total combined mass of the doubled arrangement. It refers to the combining of laps or strands which allows blending and drafting for improving quality and reducing variability.

draft (1) Reduction in linear density of a strand of material by attenuation. (2) The ratio between weight or length of stock fed into and the weight or length delivered from various machines in yarn manufacturing.

drafting The process of attenuating a strand of material by pulling or attenuating it between pairs of rollers. The material may also be supported between roller pairs by aprons or sets of pins.

drafting force The force required to pull fibers apart in drafting. It varies according to frictional characteristics of the fibers and their average orientation. Drafting force is high in card sliver and much lower in combed sliver. See Cohesiveness.

drafting wave Cyclic variation in the linear density of strands of textile fibers after drafting. It arises because shorter fibers tend to be pulled through drafting zones in clumps. Use of aprons or pins to support the material dampens drafting waves.

drafting zone The space between two pairs of drafting rolls. Between the back rolls and middle rolls of a 3-roll draft system is the breaker drafting zone; between the middle and front rolls is the finisher or main drafting zone.

Draftometer An instrument developed at the Institute of Textile Technology for measuring the force developed in drafting slivers or roving. This drafting force can be viewed as the resistance of the fiber bundle to the extension of the strand caused by drafting.

drawing A process in yarn manufacturing in which a group of slivers is elongated by passing them through a series of drafting rolls, each pair moving faster than the previous one.

elongation The difference between the length of a stretched textile specimen and its initial length, expressed as a percentage of the initial length. It is measured at any specified load or at the breaking point.

end down End broken during spinning or weaving. In weaving, an end down causes the weaving machine to stop when the drop wire held by that end falls and actuates a stop motion. In spinning, the operator or automatic device pieces up the end without stopping the machine.

evenness testing A measure of the variation in weight per length of slivers, rovings, or yarns.

fabric A flexible sheet material that is assembled of textile fibers and/or yarns that are woven, knitted, braided, netted, felted, plaited, or otherwise bonded together to give the material mechanical strength.

fabric cover The extent to which a fabric restricts the transmission of light; a tightly constructed fabric has greater cover than a loosely constructed fabric.

fiber alignment The degree to which the fibers are oriented in an attitude parallel to the longitudinal axis of the continuous strand generated by the fibers. Fiber alignment in slivers is measured by the ITT Bulk Tester, and the result is affected by the degree to which the fibers are straight along their entire length.

fiber fineness The linear density of a fiber expressed in such units as milligrams per centimeter (principally used in Great Britain), decitex, millitex, and denier. Cotton fiber fineness is also indirectly expressed by micronaire.

fiber hooks Fibers which are curved into small, medium, or large crooks which resemble a shepherd's staff; fibers are many times hooked on both ends.

fiber The fundamental component that is used in the assembly of textile yarns and fabrics. Fibers can be spun into a yarn or made into a fabric by interlacing in a variety of methods, including weaving, knitting, braiding, felting, and twisting.

fiber tuft A group of fibers which are entangled or clumped together causing tufts or flocks of various sizes.

fillet Narrow strip of wire clothing used mainly on the flats for short staple carding.

flat One of the bars covered with card clothing used on an endless chain surrounding the cylinder of a revolving flat card for processing cotton.

flat strips Fiber-containing waste removed from the flats of a revolving flat card.

flocks Small tufts of fibers.

fluted A surface which is covered with rounded grooves that are somewhat like gear teeth. Bottom rolls of a drafting zone are often fluted or knurled.

fly Cotton fibers shorter than 5 millimeters that may break loose during such standard processes as carding, combing, and spinning. Being so small and light, they may float or fly in the air before settling.

flyer Used in yarn making operations, it is an inverted V-shaped or U-shaped device that supports the strand of fiber above a spindle. It is used on cotton and worsted spinning system roving frames. Some twisting of yarn also is done by this method.

foreign matter See Trash.

frame A general term applied to many of the machines used in yarn manufacturing.

gin A machine employed to separate cottonseed from cotton fiber. There are two types of cotton gins: (1) saw gin, used for upland and other short staple cottons; and (2) roller gin, used for Pima and Egyptian long staple cottons.

ginning A process of separating cotton fiber from the seed.

grain One unit for expressing the weight of slivers. A grain is 1/7000th of a pound.

gram Metric unit of weight: (Rounded) 1 gram = 15.43 grains; 28.35 grams = 1 ounce; 453.6 grams = 1 pound.

grams per denier A measure of fiber tenacity that is expressed as the number of grams of force required to break a fiber divided by its denier. Denier is a measure of fiber or yarn size or linear density.

grams per tex A measure of fiber tenacity that is expressed as the number of grams of force required to break a fiber divided by its tex. Tex is a measure of fiber or yarn size or linear density.

grid bars Groups of metal rods, of angular cross section, set at a slight distance away from the path of beaters in fiber opening, cleaning, or carding machines. Trash flung from the fiber stock being processed passes through the grid bars for later collection as waste.

grist A small group of fibers in a continuous flow as they have been drafted to a relative light weight per unit length in a roller drafting system before twist is inserted to form a yarn.

hand A characteristic of fabrics that is perceived by touching, squeezing, or rubbing them. Describes tactile qualities such as softness, firmness, drapeability, fineness, and resilience.

hank (1) Textile material in coiled form. See Skein. (2) Cotton system roving count designation. (3) An 840-yard length of a textile strand of any weight per unit length formerly used for reporting production amounts.

hard end The term usually applied to the end emerging from a roller drafting system of the ring spinning frame in which the fiber bundle has not been drafted properly.

high volume instrumentation (HVI) An integrated set of electronic instruments that work semiautomatically to measure the following characteristics of raw cotton: length, length uniformity, micronaire fineness, strength, color, and trash (leaf). Near infrared measurement modules are also available for cotton maturity and/or sugar content determination. Virtually all bale samples of American Upland cotton and Pima cotton are classed by HVI so they will be eligible for government price supports.

histogram A representation of a frequency distribution by means of rectangles whose widths represent class intervals and whose areas are proportional to the corresponding frequencies.

hollow core spindle Spinning device for the creation of wrapped or core-spun yarns. The yarn to be wrapped passes through a hollow tube in the center of the spindle while a tube containing the wrapping yarn is placed over the hollow spindle. Both yarns are threaded through the hollow part of the spindle and a twisting mechanism wraps the wrapper yarn around the core or central yarn.

Institute of Textile Technology (ITT) An organization, totally supported by the American textile industry, which integrates graduate education in the textile sciences and practical research relevant to the primary textile industry. Also maintains a complete textile library and computerized information retrieval service. Located in Charlottesville, VA.

IPI defects Refers to the number of three specific yarn defects as determined by the Zellweger Uster Imperfection Indicator (IPI) device attached to the Zellweger Uster Evenness Tester; the defects are Thick Places, Thin Places, and Neps per 1000 meters of yarn. The degree of departure from the average yarn mass can be preset.

ITT openness index The relative openness of fibrous stock as determined by the Openness Tester developed at the Institute of Textile Technology. Normal indices range from 40 in bale stock to 165 for very open stock.

Kawabata Evaluation System (KES) Set of standardized machines and test procedures for measuring mechanical and physical properties of textile fabrics and converting the results into correlates with subjective judgement of fabric hand.

lap A continuous batt (blanket) of fibers that has been calendered to produce a stable sheet of fibers that can be formed into a roll for transport

lap-up Wrapping of fibers or fiber assemblies such as sliver, rovings, or yarns around a roll.

leveling This term is applied to a system that reduces the variation in mass of continuous mats or strands of fibers in yarn making processes prior to yarn spinning.

lint (1) A cotton staple; raw cotton that has been ginned. (2) The fiberous component of cotton raw material, which also contains trash and dust. (3) Loose, short fibers, fine ravelings, or fluff from yarn or fabric.

linters The short cotton fibers that adhere to the seed after the first ginning process. They are usually less than 1/8 inch (0.32 cm) in length, and are removed from the seed by a delinting process.

lumen The canal or cavity that runs longitudinally in each cell of a vegetable or animal fiber. In some instances it is large and continuous; in others, it is small and discontinuous.

luster The amount of light reflected from the surface of a fiber, yarn, or fabric. High luster generally is desirable.

main draft In a drafting system that encompasses more than one draft zone, the main draft is the zone with the greatest draft ratio, usually the final zone.

man-made fiber Fibers created through technology. The term "manufactured fiber" has replaced this term.

manufactured fiber Fiber defined in the Textile Fiber Products Identification Act as "any fiber derived by a process of manufacture from any substance which, at any point in the manufacturing process, is not a fiber."

mat An assembly of disoriented fibers being moved along a tray into a fiber processing machine. Synonyms: batt, fleece.

merge Used to classify lots of manufactured fiber based upon physical properties and dyeability. All shipments of fiber having the same merge number can be mixed or used interchangeably.

meter A standard unit of length. One meter equals 39.37 inches.

metric system A system of measurement in which the unit of length is the meter, the unit of weight is the gram, and the unit of time is the second. This system has superseded the English system (foot, pound, second) over most of the world.

metric ton A measure of weight equal to 1000 kilograms, or 2204.6 pounds.

micron (μ) In the SI system, this unit is a micrometer (μ), a unit of length equal to one millionth of a meter, 10^{-6} meter, 0.000039 inch. This unit of measure is employed to designate thickness of such natural fibers as wool and hair.

micronaire method A procedure for measuring the fineness of fibers, especially cotton, by determining the resistance of a plug of fiber to flow of air forced through it. Low micronaire values indicate fine and/or immature cotton; high values indicate coarse and/or mature cotton.

mill The building in which textile yarns are spun, woven, or knit.

mill trial A processing experiment done using full-scale machines in a textile mill.

mixing The mechanical stirring of fiber tufts to accomplish the intermingling of tufts from different sources with the objective of producing consistent fiber arrangements over extended time periods.

moisture content Percentage of water present in textile fiber, yarn, or fabric based on its weight including the water.

moisture regain Moisture present in textile material expressed as a percentage of the moisture-free (bone-dry) weight of the material.

monitoring Usually a mechanical or electrical or pneumatic device, or a suitable combination of systems, that senses the variations in mass or quality of machine production for recording or controlling.

natural fiber Fiber obtained from animal, vegetable, or mineral sources, as opposed to those that are synthesized from low-molecular-weight chemicals or regenerated (i.e., manufactured from some natural polymer). Major natural fibers are cotton, flax, silk, and wool.

navel The device in a rotor yarn spinning machine through which the spun yarn is withdrawn. It is positioned at the axis of the rotor.

nep A small knot of tangled fibers. In cotton, neps can consist of dead or immature fibers. Natural neps are short, undeveloped fibers that become embedded in yarn.

newton A force equal to 101.97 grams, or 0.2248 pounds.

nip (1) A flaw in a yarn consisting of thin places. (2) The line of contact between two contiguous rollers. (3) The line of contact between the jaws of a clamp, as in a tensile testing machine.

nips per minute For the cotton combing machine, a nip refers to one complete cycle of separating the fibrous lap, combing through the resulting beard, reassembling the combed beard to the previously combed stock, and exiting the assembled (pieced) section. Thus, nips per minute is the production speed of the comber.

noil Short fibers removed during the combing operation of yarn making. This occurs in production of such yarns as worsted, combed cotton, and spun silk. The fibers sometimes are mixed with other fibers to make low-quality yarns or are used for purposes other than yarn making, such as padding, stuffing.

non-lint content The portion of a sample of cotton that can be mechanically separated from the fibers. Consists mainly of leaf and other trash.

non-reworkable waste Fibers that have separated from the processed fibrous stock and are so damaged or contaminated that they are considered to be non-reusable for the product being produced. Floor sweeps and filter waste are usually in this classification.

nylon A generic fiber category defined by the Federal Trade Commission as "a manufactured fiber in which the fiber-forming substance is a long-chain synthetic polyamide in which less than 85 percent of the amide linkages are attached directly to two aromatic rings." There are various types of nylon made from different carboxylic acid and amine monomers.

on-line testing Automated testing procedure for carrying out quality control measurements on materials as they move through production processes. Has become important in textile manufacture.

open-end spinning (rotor) The creation of yarn by transferring twist from the end of a previously formed yarn to fibers or clumps of fibers continuously fed from sliver to the spinning area where they are incorporated into the yarn end.

opening A preliminary step in processing of staple fiber. This operation separates the compressed staple fibers into loose tufts and removes heavier impurities.

parallel yarn A synonym for wrapped yarn or fasciated yarn.

parameter In statistics, a variable that describes a characteristic of a population or a mathematical model (ASTM). For example, twists per inch (or cm) in samples of yarn; weights per unit area of samples of fabric.

periodicity Regularly recurring defect or change in size in a yarn or fabric.

piecing A tick place in a spun yarn caused by poor piecing up. Considered a defect.

piecing up Joining an end of spun yarn from a bobbin to fibers issuing from the drafting zone of a spinning frame, so that yarn formation and twisting can proceed.

pill Bunch of tangled fibers that is held to the surface of a fabric by one or more fibers. See Pilling.

pilling A process of forming small tangles of fibers. These tangles are defects, which are produced when the surface of a material is rubbed either against itself or another substance or material. Short fibers pull out of the fabric yarns during wear and entangle themselves with the ends of one or more fibers that are still held in the yarn. The pill is held on the surface and can be removed only by breaking the fibers that hold it. Because manufactured fibers are stronger, pills do not break away from the fabric as readily as with cotton or wool.

pirn A bobbin usually with a conical base and a cylindrical or slightly tapered tube; can be made in various lengths.

piston top The cap, usually plastic, which fits over the top end of the helical spring mounted in a sliver transport can. The sliver is coiled onto the piston as it enters the can, and as the sliver builds up in the can, the piston moves down into the can as the sliver coils increase in number and as the added weight causes the helical spring to be compressed.

ply The number of single yarns twisted together to form a ply yarn; also the number of ply yarns twisted together to form a cord (ASTM).

ply yarn A yarn formed by twisting together two or more single yarns or strands in one operation. May be two-ply, three-ply, four-ply, or more.

polyester fiber A generic fiber category defined by the Federal Trade Commission as "a manufactured fiber in which the fiber-forming substance is any long-chain synthetic polymer composed of at least 85% by weight of an ester of a substituted aromatic carboxylic acid.

polymer A substance created by the reaction of monomers (simple molecular compounds) that have reactive groups that allow them to join to form long chain-like molecules. When all monomers in the polymer are of the same compound, a homopolymer is formed. When two or more different monomers join, a copolymer is formed.

presser foot On a roving machine, device that controls the roving as it is wound onto a bobbin.

quality Characteristics of a product or service that are related to its ability to meet the needs of a user.

quality assurance Systematic activities to assure that a product or service will fulfill requirements. See Quality Control.

quality control Continuous testing and inspecting of mill operations to ascertain that all products (e.g., yarns and fabrics) meet established quality standards. See Quality Assurance.

quill A light, tapered, wooden, cardboard, fiber, metal or combination tube upon which filling yarn is wound before shuttle weaving.

rail Metal bar on spinning or twisting machine that moves up and down to wind the yarns onto spindles. A rail may carry spindles, as in a downtwister, or spinning rings.

ratch Distance between pairs of rolls in a roller drafting system. The term used to indicate the nip-to-nip distance between two successive drafting rollers. Another term used is "roll spacing".

raw material The fiber, or fibers, used in the manufacture of staple yarn; the fiber is normally received at the yarn plant in bale form.

rayon A generic fiber category defined by the Federal Trade Commission as "a manufactured fiber composed of regenerated cellulose as well as manufactured fibers composed of regenerated cellulose in which substituents have replaced not more than 15 percent of the hydrogens of the hydroxl groups. Characterized by high absorbency, bright or dull luster, pleasant hand or feel, good draping qualities, and the ability to be dyed in bright colors.

reclaimable waste Material consisting of fiber and non-fibrous contamination such as pieces of plant stem and leaf, which material has been extracted from the usable fibrous material being processed into yarn. This reclaimable waste can be processed in special machines to extract the fiber from the waste, so that these extracted fibers can be used in certain product manufacturing.

relative humidity (RH) The proportion of the amount of water vapor present in air to the maximum amount possible at the same temperature; usually expressed as a percentage (technically, the proportion is between actual and maximum water vapor pressures). RH greatly affects the mechanical, physical, and processing characteristics of hydrophilic fibers.

resilient The term used to express the characteristic of a material that springs back to its original shape after being deformed to a degree.

resultant yarn number Actual yarn number of a plied, twisted, or textured yarn based on measuring weight/mass per unit length. Also is expressed as Equivalent Yarn Number.

reworkable waste Waste from textile processing that contains enough long fiber to be worth blending into fiber stock that will be used for the same or lower quality yarn.

ring spinning Twist is inserted into a yarn after it emerges the front drafting rolls by passing it through a yarn guide, to a traveler, and to a rapidly rotating bobbin, simultaneously and continuously. Normally, roving is used as the feed stock, but sliver-to-yarn spinning is possible.

rotor In open-end spinning machines, a device resembling a centrifuge, in which the fibers are assembled and in which, by virtue of its rotation, real twist is inserted in the forming yarn (ASTM). See Open-End Spinning.

rotor spinning A type of open-end spinning in which fibers are delivered into a rotor where they are collected by and twisted into the end of the yarn being formed.

roving (1) A loose assemblage of fibers drawn or rubbed into a single strand, with very little twist. In the modern cotton spinning system, it is an intermediate state between sliver and yarn. (2) The process for manufacturing roving from sliver.

run system One of the numbering systems for woolen yarn in which the unit of measure, or yarn number, is the number of 1,600-yard hanks of yarn in one pound.

sales yarn A trade term for yarn produced by a spinning company whose business exclusively is to spin yarn for weavers, knitters, and other manufacturers.

sample A portion of a lot of material which is taken for testing or for record purposes.

saw-ginned cotton Cotton that has been ginned by a saw gin as opposed to a roller gin; saw-ginning is used on all but the longest staple cottons.

seed coat fragment In cotton, a portion of a cotton seed, usually black or dark brown in color, broken from a mature or immature seed, and to which fibers and linters may or may not be attached (ASTM). Synonym: bearded mote.

Shore Hardness A specific number indicating the relative hardness of a deformable material, such as leather, hard rubber, or certain synthetic materials, that is tested with a special instrument called the Shore Hardness Tester.

short staple A classification for any fiber cut into lengths to process on the cotton spinning system. Fiber lengths up to 64 mm or 2.5 inches can be spun on the cotton system.

short staple spinning See Cotton Spinning System.

singles yarn The simplest strand of textile material suitable for operations such as weaving, knitting, etc. (ASTM). Includes yarn of a great variety of fibers from the spinning machine, the silk reel, or the spinneret that has not been subjected to twisting (plying) with other yarns.

skein A continuous strand of yarn in the form of a flexible coil having a large circumference in proportion to its thickness (ASTM). The circumference of the reel on which yarn is wound is generally 45 to 60 inches (114 to 152 cm), although in most silk reeling the diameter is less.

SKU Acronym for Stock Keeping Unit.

sliver A continuous, ropelike strand of loosely assembled fibers without twist that is approximately uniform in cross-sectional area. This is the usual state of fibers after they have been carded and passed through a condenser trumpet, after drawing, and after combing.

sliver can Cylindrical container, often on wheels, that holds coiled sliver.

Sliver Analyzer A testing device developed at the Institute of Textile Technology for drafting several slivers, arranged side by side, into a thin web which flows over a light table so that slubs and other imperfections can be identified and counted.

slough-off (sluff-off) Coils of yarn that, because of poor yarn winding on filling bobbins, come off a bobbin all at once and are woven into fabric as a group.

slub A thick, unevenly twisted place in yarn; may be deliberately inserted in a fancy yarn or a flaw in yarn that is supposed to be of uniform diameter. Synonym: Slug.

spanker plate The part of a chute feeding mechanism which oscillates slowly to slightly compress opened fibrous material so that it will flow downward to the exiting rolls of the chute.

specific gravity Expresses the density of a textile fiber in relation to the density of an equal volume of water at a temperature of 4 °C. Most fibers have specific gravity of more than one, that is, they are heavier than water.

specimen A specific portion of a material or a laboratory sample upon which a test is performed or which is selected for that purpose.

spectrogram An incremental mass chart generated on instruments such as the Zellweger Uster Evenness Tester that indicates the mass distribution of finite lengths of material, either 8 mm or 12 mm, and the frequency of such mass accumulations.

spider web See Cobwebbing.

spinnability A fiber characteristic evaluated by counting the number of yarn breakages during spinning of a certain total amount of yarn from the fiber material. See Spinning Limit.

spinning cradle The cradle as used on roller drafting spinning machines such as ring and air jet spinning machines. See Cradle.

spinning limit The finest yarn number that can be spun satisfactorily from a specified lot of fiber under specified conditions (ASTM).

staple Synonym for fiber. Term used to indicate lengths of fiber that require spinning and twisting in the manufacture of yarn.

stock A general mill term, which refers to raw materials or fibers that are involved in processing. Stock includes raw fiber, lap, sliver, roving, waste materials, blends, and staple.

stop motion Electrical or mechanical device on a textile machine that automatically arrests its operation whenever yarn breakage or some other defect is detected. Essential component of power looms and knitting machines.

strand A general term for one component of a rope, thread, or ply yarn, or any of the fibers that are twisted or plaited together to form the aforementioned. Sometimes the term also is applied to the entire rope, cable, thread, or ply yarn.

strength Resistance to deformation or breakage caused by application of a force.

synthetic fiber Term applied to any manufactured fiber other than rayon, acetate, or regenerated protein (azlon) fiber. The latter are classed as regenerated fibers.

take-up rollers Pairs of short cylinders that turn together to withdraw open-end spun yarn from a rotor. See Open-End Spinning.

tenacity The tensile stress of a material based on the linear density of the unstrained material (ASTM). The breaking tenacity of a yarn or fabric is the tensile force at rupture per unit cross-sectional area or per unit linear density.

tensile strength Ability of fiber, yarn, or fabric to resist breaking under tension as opposed to torsion, compression, or shear. Measured in total force or force per unit of the cross-sectional area or linear density of the original specimen (e.g., lb/in.^{2} or g/denier).

tension A force in one direction that tends to cause the extension of a body or the force within the body that resists the extension.

tensor bar A smooth rounded bar which guides long bottom drafting aprons near front drafting rolls.

tensor pin A smooth metal bar formed in a U-shape (hair pin) to guide both the top and bottom short aprons near the front drafting rolls.

tex system A method of numbering yarns, fibers, and all types of textile strands. It is part of the SI system and is intended to replace gradually the diverse numbering systems now used for cotton, woolen, and worsted yarns as well as the denier system. The tex number of a yarn, fiber, or other strand is the weight in grams of one kilometer length.

textile Derived from the Latin term "textilis", which is based on the verb "texere", to weave. A broad classification of materials that can be utilized in constructing fabrics, including textile fibers and yarns.

textile process Any mechanical process used to transform a textile fiber or yarn to a fabric (cloth) or any other textile material.

thick place A segment of yarn at least 1/4 inch (0.6 cm) long that is noticeably thicker than adjacent portions of yarn.

thin place A segment of yarn at least 25% smaller in diameter than adjacent portions of yarn.

throughput Total weight or length of material processed through a machine in unit time.

tow Large bundle of manufactured fiber filaments without definite twist collected in loose, ropelike form. Tow is the form that most manufactured filaments are prepared in before cutting into staple. Some tow is processed on tow conversion machinery into sliver and yarn. Some rayon tow is used to make rayon flock.

transfer tail Length of yarn place on a package holder, at the start of winding, beyond the main part of the wound package. This is tied to yarn from the next package when the package is double creeled. See Creel.

trash Finely divided foreign matter in raw cotton. Includes soil particles, bits of leaf, motes, and seed coat fragments. Trash content of cotton can be evaluated visually with a trash meter or quantitatively by weight as with the Shirley Analyzer.

traveler A C-shaped metal or plastic piece that rides on the spinning ring in a ring spinning or twisting machine. During machine operation, the yarn being twisted is guided onto the bobbin by the traveler. Sometimes spelled traveller.

traverse length Distance parallel to package axis between reversals of traverse direction.

traverse Movement of yarn during winding across the width of the package.

trumpet Flared metal tube used to bring textile material together such as card web to sliver.

tube A holder or bobbin of cylindrical shape used as a core for a yarn package of cylindrical form.

tuft See Fiber Tuft.

twist The number of turns about its axis per unit of length observed in a yarn or other textile strand (ASTM). Generally, this is indicated as turns per inch, or tpi. It is expressed also as turns per centimeter or meter, or by helix angle in a structure of known diameter.

two-ply Composed of two singles yarns, as two-ply yarn.

uniformity index In fiber length testing of cotton, this term denotes the ratio between the mean length and the upper-half-mean length expressed as a percentage of the upper-half-mean length (ASTM).

upper quartile length That length which is exceeded by 25% of the fibers, by weight, in the test specimen (ASTM). Applicable to cotton fiber length testing.

upper-half-mean length In fiber length testing of cotton, the mean length by number, of the longer one half of the fibers by weight (ASTM).

virgin stock Refers to the fibrous raw material that is for the first time being processed progressively into a yarn product.

visible foreign matter (VFM) The sum of all the non-fibrous material that can be separated from the fibers on a test instrument similar to the Dust and Trash Tester developed at the Denkendorf Institute for Textile Research and Technology.

waste Byproducts created in the processing of fibers, yarns, and fabrics. Waste is classified broadly as soft waste and hard waste and specifically by the stage of manufacturing in which it occurs.

web Fibers of cotton, wool, or manufactured staple as they are doffed from the card cylinder and before they are condensed or processed further.

wild yarn Short pieces of yarn that occur on several processing machines, especially on high speed winders, that get snatched back into wound yarn packages and get carried along into section and warp beams. These pieces of wild yarn cause defects and stoppages in the processes used in fabric formation and finishings.

winding Process of transferring yarn or thread from one type of package to another (e.g., from cakes or bobbins to cones or tubes).

worsted count The number of 560-yard lengths of worsted yarn that weigh a pound. See Yarn Number.

wrap spinning See Hollow Core Spindle.

yarn A continuous strand of textile fibers that may be composed of endless filaments or shorter fibers twisted or otherwise held together.

yarn count (size) See Yarn Number.

yarn number A measure of the fineness or linear density of a yarn. May be expressed in indirect units (length per unit of weight or mass) or direct units (weight per unit of length).

Indirect yarn numbers have been used for most spun yarns (e.g., cotton, wool, linen). Lower numbers designate heavier or thicker yarns while higher numbers refer to finer yarns. An incomplete list of indirect yarn number systems includes American asbestos and glass (number of 100-yd lengths/lb), cotton count and spun silk (840-yd lengths/lb), linen lea and woolen cut (300-yd lengths/lb), metric (kilometer lengths/kilogram), woolen run (1600-yd lengths/lb), and worsted (560-yd lengths/lb).

Direct yarn numbers are used for filament yarns, manufactured fibers, and jute. Lower numbers indicate finer sizes of yarn and higher numbers heavier yarns. Among the direct yarn numbering systems are tex (number of g/1000 m), decitex (g/10,000 m), denier (g/9000 m), linen and jute spyndle (lb/14,400 yd), and grain system (grains/120 yd). The international standard for yarn numbering is now the Tex System.

yarn twist, z and s See Direction of Twist.

Literature Cited

Anderson, P.N. (1995). *The effects of nozzle components on the cover and hand of fabrics produced with Murata Jet Spun yarns.* Unpublished master's thesis, Institute of Textile Technology, Charlottesville, VA.

Arzt P., Azarschab M., & Maidel, H. (1990). Trash content of card slivers related to the frequency of broken ends in rotor spinning. *Textil-praxis International, 45*(11), 1146.

Barnes, D.K. (1996). *The influence of modern carding parameters on fiber alignment in slivers and the resulting impact on open-end yarn quality.* Unpublished master's thesis, Institute of Textile Technology, Charlottesville, VA.

Butenhoff, A.N. (1995). *The effect of polyester fiber variants, spinning parameters, and rotor speed on lint shedding of 50/50 polyester/cotton open-end yarns.* Unpublished master's thesis, Institute of Textile Technology, Charlottesville, VA.

Copeland, A. (1996). *The effect of navels, washers, and rotor speeds on open-end yarn characteristics.* Unpublished master's thesis, North Carolina State University, Raleigh, NC.

Dailey, T.A. (1996). *The effects of combing roll wire, surface, and speed on quality properties of 100% cotton and 50/50 polyester/cotton open-end yarns.* Unpublished master's thesis, Institute of Textile Technology, Charlottesville, VA.

Duessen, H. (1993). *Rotor spinning technology.* Monchengladback, Germany: W. Schlafhorst and Company.

Fite, D.A. (1994). *The effect of laundering, spinning system, and blending method on the physical, asesthetics, and wash-down characteristics of workwear fabrics.* Unpublished master's thesis, Institute of Textile Technology, Charlottesville, VA.

Frye, J.S. (1994). *The effects of card mat weight, openness, and tension draft on silver and yarn quality.* Unpublished master's thesis, Institute of Textile Technology, Charlottesville, VA.

Leifeld, F. (1990). Modern, sophisticated preparation plants for ring spinning. *Textil-praxis International 45*(2), 107-111.

Mackey, B.E. (1995). *An investigation of the influence of the properties of 50/50 polyester/cotton yarns on filling insertion performance in high speed air jet weaving.* Unpublished master's thesis, Institute of Textile Technology, Charlottesville, VA.

McInnish, R.J. (1963). *The cause of hooked fibers and means for minimizing their formation in carding.* Unpublished master's thesis, Institute of Textile Technology, Charlottesville, VA.

Research on tandem carding: Part IV. (1989). *Textile Topics, 17*(10).

Stone, F.M. (1993). *The effect of total draft, draft distribution, and twist on the quality of 100% combed cotton yarns and 50/50 polyester/cotton yarns spun on the Suessen RingCan silver-to-yarn spinning system.* Unpublished master's thesis, Institute of Textile Technology, Charlottesville, VA.

Tutterow, M.N. (1992). *Determining the practical spinning limits of open-end carded and combed yarns produced from Pima, Upland, and California cottons at various rotor speeds.* Unpublished master's thesis, Institute of Textile Technology, Charlottesville, VA.

Index